Human Factors in Vehicle Design: Lighting, Seating, and Advanced Electronics

SP-1088

All SAE papers, standards, and selected books are abstracted and indexed in the Global Mobility Database.

Published by:
Society of Automotive Engineers, Inc.
400 Commonwealth Drive
Warrendale, PA 15096-0001
USA
Phone: (412) 776-4841
Fax: (412) 776-5760
February 1995

ISBN 1-56091-638-9
SAE/SP-95/1088
Library of Congress Catalog Card Number: 94-74741

Printed in USA

PREFACE

The automotive customer's increasing demand for vehicles designed with good human factors continues to grow. Customers consistently want better and better products that are easier and safer to use.

A comfortable, well designed seat using state of the art design and manufacturing techniques is a high customer want. Comfort evaluations such as those presented in this volume are an absolute requirement if seat suppliers intend to satisfy customer expectations in the area of seat comfort. Issues such as occupant posture, lumbar support, and the distribution of pressures on the body induced by the seat must be considered.

The customer perceived quality of the seat comes not only from his or her comfort evaluation of the seat, but also from the quality of the design and construction. Proper application of the seat trim material as well as seat frame design and testing, issues which are addressed in this volume, are only a part of what needs to be done in the design process.

In the area of lighting and visibility, today's customer wants to not only see down the road safely, but he also wants a beam pattern that is pleasing to the eye when he uses it. The complex lens and reflector designs that are appearing in today's headlamps are making these customer demands easier to meet. As the technology for designing these reflectors becomes more feasible, headlamps, as we used to know them, will disappear. Customized lamps designed specifically for the vehicle and the customers who buy them will predominate. New, state of the art headlamp reflector design and evaluation techniques described in this volume will take us a long way toward realizing this goal.

Visual performance with mirrors must also be addressed. Blind spots exist, and strategies to overcome them should be developed and evaluated. This publication addresses these issues and presents potential solutions.

Direct field of view is also becoming more of a challenge to engineers as car and truck designs become more stylized and futuristic. Evaluation of the field of view provided to a driver, along with all the targets he needs to see is discussed with the use of a polar plotting technique.

The development of Intelligent Vehicle Highway Systems (IVHS) introduces additional human factors challenges to vehicle design. Such systems will offer considerably more information than drivers have in their current vehicles. If these systems are to be effectively used by drivers, careful consideration must be given to the way the information is managed and presented to drivers. A well designed driver interface will be essential in order to provide a system that can be easily used and understood while driving.

The safety implications of these advanced technologies must be carefully assessed before they can be included on production vehicles. Methods for identifying safety related issues are described in the volume.

Two road studies of drivers' acceptance and confidence in vehicle control systems are also discussed in this volume. Driver acceptance of automatic control and lead vehicle target acquisition is addressed in an evaluation of an Autonomous Intelligent Cruise Control (AICC) system. A comparison of driver confidence when driving with four wheel drive, traction control, and standard front wheel drive is also included. These studies begin to assess the human factors issues of automating portions of the vehicle control task.

This SAE special publication, Human Factors in Vehicle Design: Lighting, Seating, and Advanced Electronics (SP-1088) addresses these customer issues. The papers deal with the design, development, and evaluation of a variety of products and user requirements that must be addressed in today's vehicles.

David H. Hoffmeister
Ford Motor Co.

Session Organizer for Lighting and Seating

Gretchen Paelke-Zobel
Ford Motor Co.

Session Organizer for Human Factors of Advanced Electronics

TABLE OF CONTENTS

950140 **Development of the New Generation Ergonomic Seat Based on Occupant Posture Analysis** 1
Michihiro Katsuraki, Toshimichi Hanai, Kouichi Takatsuji, and Atsuhiko Suwa
Nissan Motor Co., Ltd.
Hideyuki Nagashima
Ikeda Bussan Co., Ltd.

950141 **Some Effects of Lumbar Support Contour on Driver Seated Posture** 9
Matthew P. Reed, Lawrence W. Schneider, and Bethany A. H. Eby
University of Michigan Transportation Research Institute

950142 **Evaluation of Objective Measurement Techniques for Automotive Seat Comfort** 21
Janilla Lee and Ted Grohs
Lear Seating Co.
Mari Milosic
Wayne State University

950143 **The Use of Electromyography for Seat Assessment and Comfort Evaluation** 27
Tamara Reid Bush and Frank T. Mills
Michigan State Univ.
Kuntal Thakurta
Johnson Controls, Inc.
Robert P. Hubbard and Joseph Vorro
Michigan State Univ.

950144 **Evaluating Short and Long Term Seating Comfort** 33
Kuntal Thakurta, Daniel Koester, Neil Bush, and Susan Bachle
Johnson Controls, Inc.

950146 **Seat System Fatigue Test** 39
Russ Davidson, Janilla Lee, and Ed Pan
Lear Seating Corp.

950147 **The Effects of Fabric Backcoatings on Automotive Interior Manufacturing Processes** 45
John Olari
Lear Seating Corp.

950148 **Magnetic Induction Heating for Automotive Seat Trim Bonding** 49
Clarice Fasano
Lear Seating Corp.

950590 **Computer Assisted Headlight Design and Research** 55
Han-Wen Tsai and Chien-Ping Kung
OES, ITRI

950591 **High Intensity Discharge Headlamps (HID) - Experience for More Than 3-1/2 Years of Commercial Application of Litronic Headlamps** 63
Wolfgang Huhn
BMW AG
Guenter Hege
Robert Bosch GmbH

950592 **Headlight Beam Pattern Evaluation Customer to Engineer to Customer - A Continuation** 69
Daniel D. Jack, Stephen M. O'Day, and Vivek D. Bhise
Ford Motor Co.

950593 **Application of Free Form Reflectors in Modern Headlamp Systems** 81
Henning Hogrefe and Rainer Neumann
Robert Bosch Corp.

950594 **New Optical Simulation Systems Revolutionise Headlamp Development** 87
John F. Monk
Magneti Marelli

950595 **Development of High-Speed Photometric Measurement System** 93
M. Sasaki
Koito Manufacturing Co., Ltd.

950597 **Evaluation of the SAE J1735 Draft Proposal for a Harmonized Low-Beam Headlighting Pattern** 105
Michael Sivak and Michael J. Flannagan
The University of Michigan Transportation Research Institute

950599 **Dual-Filament Replaceable Bulb Headlamp Using a Multi-Reflector Optimized with a Neural Network** 119
Yutaka Nakata
Ichikoh Industries, Ltd.

950600 **On-the-Road Visual Performance with Electrochromic Rearview Mirrors** 133
Michael J. Flannagan, Michael Sivak, Masami Aoki, and Eric C. Traube
University of Michigan Transportation Research Institute

950601 **The Geometry of Automotive Rearview Mirrors - Why Blind Zones Exist and Strategies to Overcome Them** 143
George Platzer
Consultant

950602 **Automotive Field of View Analysis Using Polar Plots** 157
E. J. McIsaac and V. D. Bhise
Ford Motor Co.

950967 **Systematic Development of a Complex MMI-Interface Shown by the Example of an Integrated Display and Information System** 167
Ebner Roland and Spreitzer Wilhelm
Siemens AG

950968 **Using the Safety State Model to Measure Driver Performance 175**
Edward J. Lanzilotta
Massachusetts Institute of Technology

950970 **Automatic Target Acquisition Autonomous Intelligent Cruise Control (AICC): Driver Comfort, Acceptance, and Performance in Highway Traffic 185**
James R. Sayer, Paul S. Fancher, Zevi Bareket, and Greg E. Johnson
The University of Michigan Transportation Research Institute

950971 **SUSI Methodology: Evaluating Driver Error and System Hazard 191**
Richard Stobart and Jeremy Clare
Cambridge Consultants Ltd.

950972 **Correlation of Driver Confidence and Dynamic Measurements and the Effect of 4WD 199**
Midori Kubota and Takayuki Ushijima
Fuji Heavy Industries
Jac Brown
Subaru Research & Design

950140

Development of the New Generation Ergonomic Seat Based on Occupant Posture Analysis

Michihiro Katsuraki, Toshimichi Hanai, Kouichi Takatsuji, and Atsuhiko Suwa
Nissan Motor Co., Ltd.

Hideyuki Nagashima
Ikeda Bussan Co., Ltd.

ABSTRACT

In this study, the functions required of automotive seats were analyzed from the standpoint of occupant posture. The results have been incorporated in the development of the New Generation Ergonomic Seat, which better fits the contours of the human body and prevents a stooped posture that places a greater load on the lumbar region, thereby reducing fatigue during long hours of driving. The new seat adopts the concept of "combined pelvic and lumbar support," based on an analysis of the muscular and skeletal structure of the human body, sitting posture and body pressure distribution.

INTRODUCTION

Automotive seats, which are in constant contact with vehicle occupants, play an important role in improving the safety and comfort of the car interior. In the research and development work described here, the functions required of automotive seats from the standpoint of occupant posture were pursued in an improved seat that has now been adopted in production vehicles. Those functions are ample space for occupants, (seating arrangement), support for proper sitting posture and suitable body pressure distribution.

The new seat, called the New Generation Ergonomic Seat, was developed specifically to provide improved sitting posture support, which influences the load placed on the human body and the driver's position for operating vehicle control systems. This new seat adopts the concept of "combined pelvic and lumbar support," based on an analysis of the muscular and skeletal structure of the human body, sitting posture and body pressure distribution. With this concept, continuous support is provided from the pelvis to the lower part of the thorax, which works to avoid a stooped posture without any feeling of discomfort. In addition, the upper portion of the seat back has also been designed with sufficient softness to accommodate individual differences in build and posture, when occupants are firmly supported in their pelvic and lumbar regions and along a broad area of their back by this concept of combined support. This results in an improved fit with the contours of the body for a better distribution of body pressure along the entire back.

The new seat was assessed in tests involving three hours of continuous driving. With the new seat, lumbar discomfort, which correlates closely with perceived fatigue, was reported less often and not until longer time had elapsed, compared with an existing seat.

CAR INTERIOR COMFORT AND FUNCTIONS REQUIRED OF SEATS

The functions indicated in Fig. 1 are required of automotive seats with respect to interior comfort.

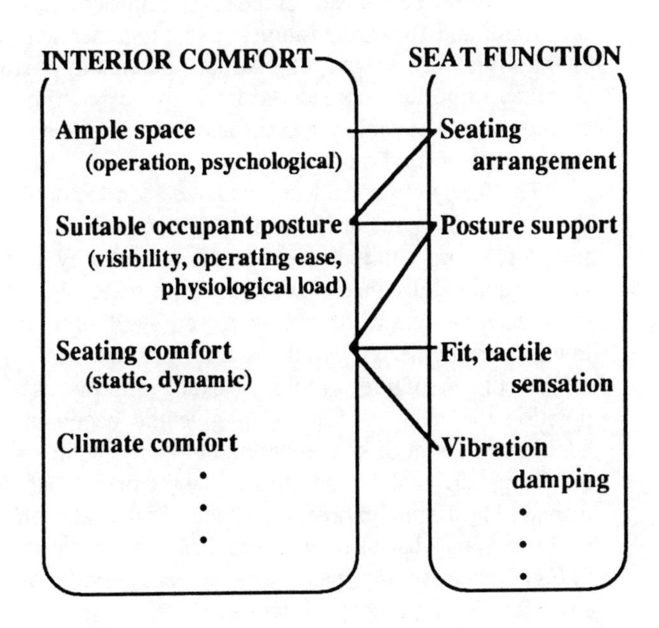

Figure 1 : Seat functions required for interior comfort

The following discussion will focus on the importance of a seating arrangement that assures ample space for the occupants, the posture support provided by the seat itself and body pressure distribution. These functions are especially important from the standpoint of reducing the load on the occupant's body.

The design of the seating arrangement should be focused on the occupants instead of merely trying to cram the seats into the passenger compartment after the interior dimensions have been determined. Important design considerations here include the provision of sufficient space around the occupants, the relationship between a proper posture and operation of the control systems and assurance of adequate space needed to improve overall seat performance.

With front seating systems having an independent driver's seat and passenger's seat, slide rails are of course used to allow forward/rearward adjustments to suit individual preferences. Beyond this, it is also necessary to provide ample and well-balanced space between the two occupants and between them and the doors, as this has a direct effect on the sensation of being confined. An important design factor for bench-type rear seats is to give the occupants ample knee room. Allowing a generous distance between the rear-seat occupants and the back of the front seats will give them an unrestrained feeling of being able to stretch their legs. It will also enable them to assume a sitting posture with a large open angle between their lower limbs and torso, which is effective in preventing a stooped posture that oppresses the abdomen and places a load on the lumbar area. In addition, providing ample longitudinal space is also helpful in improving seat comfort because it allows greater freedom for increasing the height and thickness of the seat back.

The vertical dimension of the seats influences available headroom and forward visibility, and, together with the longitudinal dimension, it affects occupant posture. Another important consideration is to design the seat cushion with a suitable thickness and construction so as to improve the absorption of vibration.

The foregoing considerations have been factored into the design of a recently developed compact car. For instance, in order to increase the interior width by 30 mm and expand the distance between the front seats by 20 mm, the overall vehicle width was increased by 20 mm and an inward-curving design was devised for the door trim. The major portion of the wheelbase extension was used to position the rear seats so as to give the occupants an additional 57 mm of knee room and to increase the size of the front-seat back. The overall height of the car was increased by 10 mm in order to expand front-seat headroom by 11 mm and that of the rear seats by 9 mm. As shown in Fig. 2, the resulting seating arrangement provides ample space for enhanced occupant comfort.

The posture support provided by a seat is a key factor in preventing the occupant from slipping into a slouched or stooped position. This function is important in assuring good visibility for the driver and a suitable position for operating the control systems during extended periods of continuous driving. In addition, it is also essential for a seat to support a sitting posture that reduces the load on the lumbar region, which tends to receive the largest load over long periods of time.

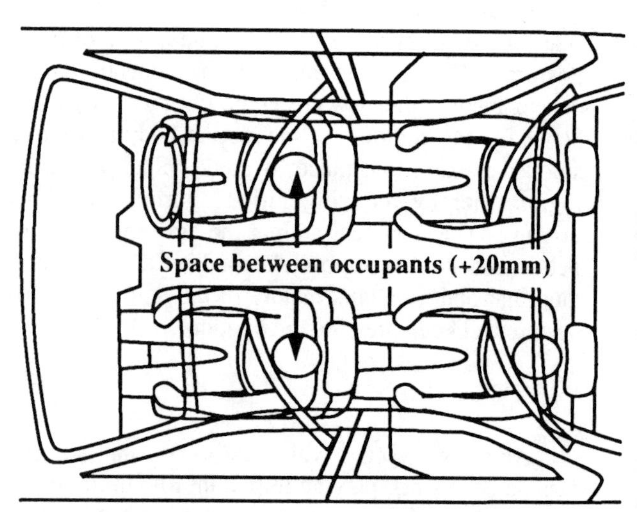

Dimensions are compared with those of previous model (manufacturer's data)

Figure 2 : Seating arrangement

The body pressure distribution characteristic indicates the condition of contact between the occupant and the seat. Not only does it affect the perceived fit with the body and natural feel, a poor pressure distribution can obstruct blood circulation, thereby promoting localized numbness or a feeling of fatigue. Therefore, a suitable level of pressure should be applied to the occupant's body at the right places for supporting the desired posture.

The concept of combined lumbar and pelvic support, which provides superior support for maintaining a proper posture with respect to the human skeletal construction, was adopted to improve the posture-support function and body pressure distribution characteristic. A target sitting posture was defined on the basis of measurements made of the shape of the human back under a posture that minimized the load on the lumbar region. That target posture and considerations for individual posture differences were taken into account in developing the New Generation Ergonomic Seat. A detailed description of the development process is given in the following section.

NEW GENERATION ERGONOMIC SEAT

CONCEPT OF COMBINED SUPPORT - As seen in the previous section, the posture-support function of automotive seats is a key performance requirement with respect to safety, ease of operation and comfort.

Human Skeletal Structure - Conventional thinking on posture support has long held that support should be provided to maintain the natural S-curve of the spinal column and for the lumbar area [1]. While firm localized support has been used as a way of maintaining the S-shaped curvature, it has been found that this approach is apt to cause a feeling of discomfort. Thus, seat designers have been faced with the problem of how to optimize the firmness of seat support. In this work, a thorough review was made of the human skeletal structure in order to devise a seat design that would improve the posture-support function.

Dividing the human body into the head, torso and limbs, it is the torso that primarily receives the load when a person is sitting. The skeletal structure of the body has the following characteristics (Fig. 3). In the natural posture of the human body, the spinal column traces an S-shaped curve, with the lumbar vertebrae thrust forward to form a curved shape [2]. In contrast, a stooped posture is characterized as an unnatural state where the lumbar vertebrae are straight vertically or curved in the rearward direction. As a result, a large load is placed on the lumbar region.

Looking at the skeletal structure near the lumbar region, we see that the pelvis serves as the base of the spinal column and that the thorax is formed by the ribs which restrict the movement of the chest. The pelvis and the thorax are connected by very flexible lumbar vertebrae.

Since the thoracic vertebrae, lumbar vertebrae and pelvis form a continuous structure, their respective motions and curvatures naturally affect one another. As seen in Fig. 4, the skeletal motions that result in an unnatural stooped posture include (1) the rearward inclination of the pelvis, (2) the rearward inclination of the lumbar vertebrae and, (3) the forward inclination of the thorax.

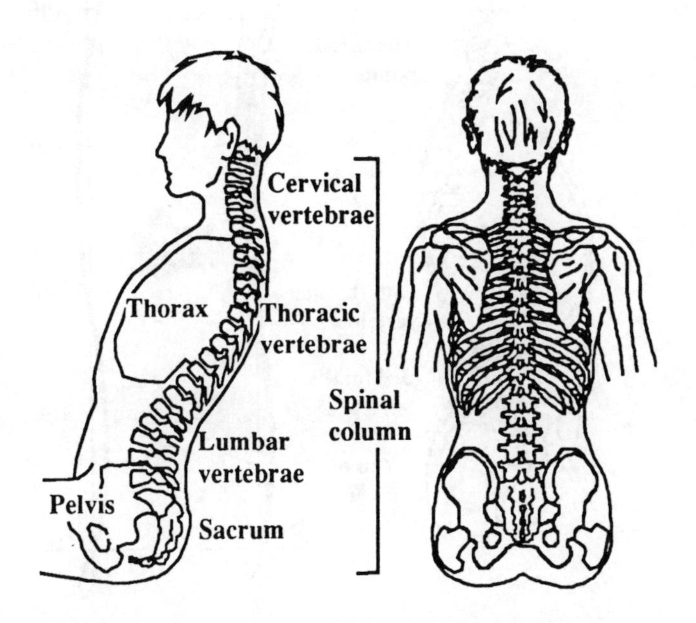

Figure 3 : Skeletal structure of the human body(2)

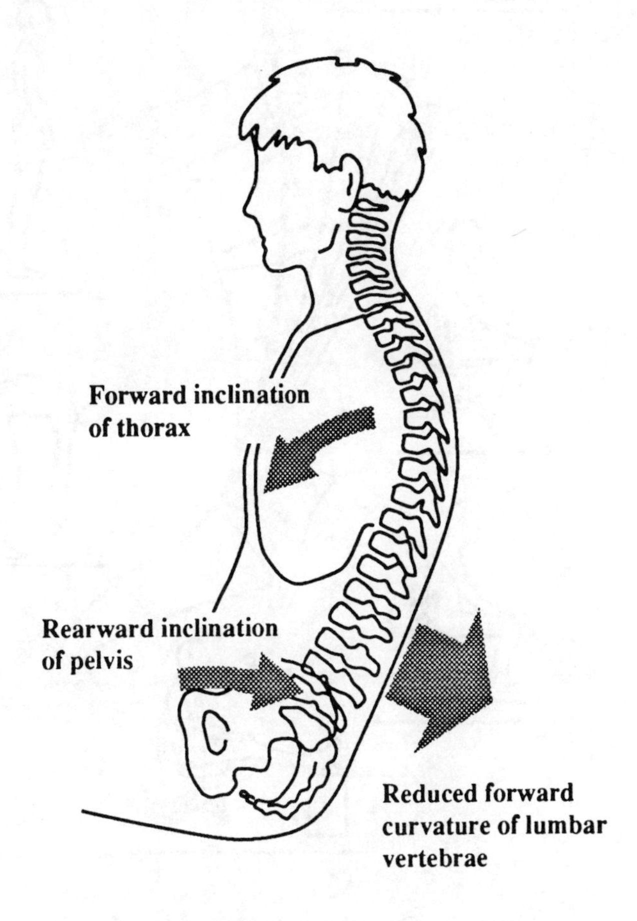

Figure 4 : Factors causing a stooped posture

Measurement of Less Load Posture and Definition of Target Sitting Posture -- The natural shape of the spinal column is an S-like curvature, as noted above. In that shape, the intervertebral disks that function as the joints of the spinal column are in an intermediate position characterized by a balance of pressure and an absence of strain. This is thought to be an ideal posture that placed less load on the lumbar region than other postures (denoted as B in Fig. 5) [3].

Measurements were made of the shape of the human back under a condition that reproduced the natural shape of the spinal column with the intervertebral disks in an intermediate position. Based on the measured results, a target sitting posture was defined as the desired shape to be reproduced when an occupant is sitting in the seat.

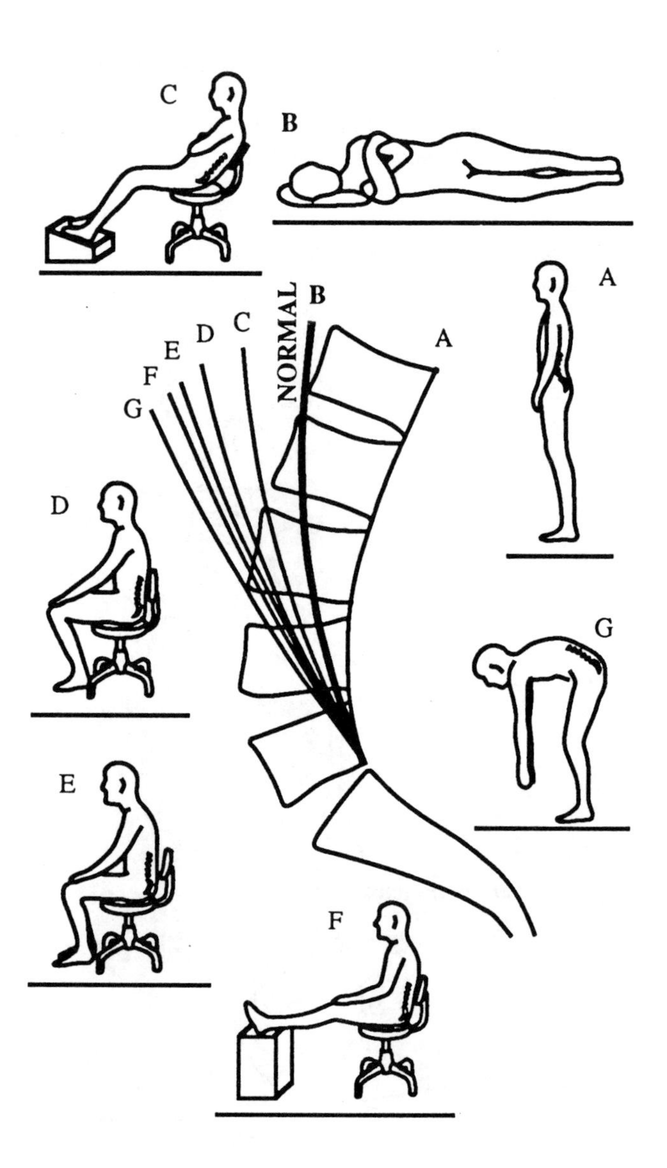

Figure 5 : Condition of lumbar vertebrae in various postures(3)

Seven subjects participated in these measurements. They were asked to lie down on their side and to bend their hip joints and knee joints slightly at approximately a 135° angle (B in Fig. 5). In that posture, the shape of the back was measured lengthwise along the center where the shape of the spinal column would be readily reflected. Measurements were made at the top of the sacrum, at the third lumbar vertebra and at the twelfth thoracic vertebra, points of projection that represent distinct locations along the spinal column.

The measured results showed virtually no variation in the shapes obtained from the same individual, but the subjects showed different degrees of forward curvature in the lumbar region. A target sitting posture was then defined as the desired shape to be reproduced when an occupant is sitting in the seat (Fig. 6). This posture represents a composite of the measured shapes obtained from four subjects who showed the natural S-shaped lumbar curvature.

It was concluded from the measured results that the natural shape of the spinal column was reproduced in this test. That conclusion was based on the stability seen for the measured shapes obtained from the same subject. The differing degrees of forward curvature seen for the lumbar region were attributed to individual differences among the subjects with respect to the shape of their spinal column.

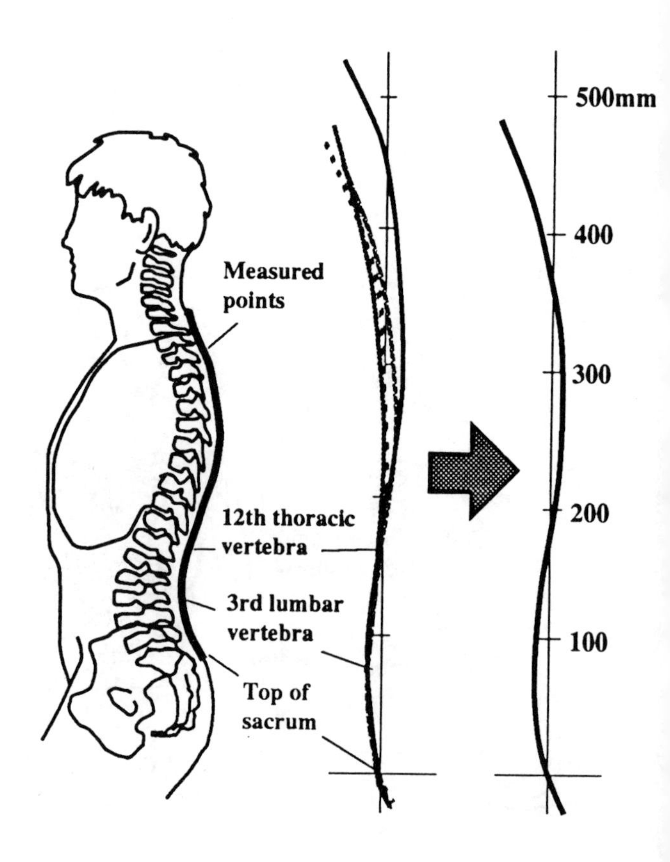

Figure 6 : Definition of target sitting posture

Combined Pelvic and Lumbar Support -- It is believed that continuous support, taking into account the movements of each part of the skeleton, is more effective than localized support in holding the spinal column whose elements are interrelated as described above. With the concept of combined pelvic and lumbar support, continuous support is provided from the pelvis to the lower part of the thorax. Therefore, in addition to supporting the pliant lumbar vertebrae, this design also works to prevent the rearward inclination of the pelvis and the forward inclination of the thorax. As a result, it prevents a stooped posture and stably maintains the desired sitting posture for improved comfort (Fig. 7).

In addition, because support is provided over a wide area from the pelvis to below the thorax, this support concept disperses the load to dispel any sensation of discomfort. Further, even if occupants change their sitting position somewhat, the locations that should be supported will still receive effective support.

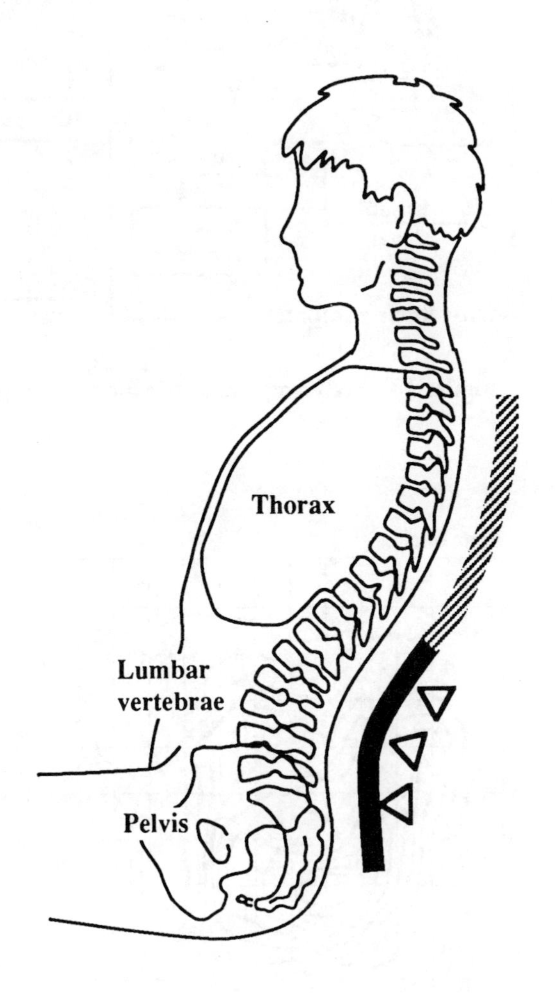

Figure 7 : Concept of combined pelvic and lumbar support

DEVELOPMENT OF NEW SEAT -- The concept of combined pelvic and lumbar support, derived from the re-examination of the human skeletal structure, was then incorporated in the New Generation Ergonomic Seat to achieve the target sitting posture that had been defined on the basis of the measured shapes of the spinal column in a natural position. Figure 8 shows the system used to measure the shape of the human back in a sitting position. This system was employed to confirm whether the target sitting posture was achieved with the new seat.

Curvature-sensitive tape was placed on the seat to serve as strain gages. When a subject sat in the seat, the output of the strain gages made it possible to determine the curvature at various locations of interest. Three-dimensional curved surfaces were then obtained by a process of interpolation [4].

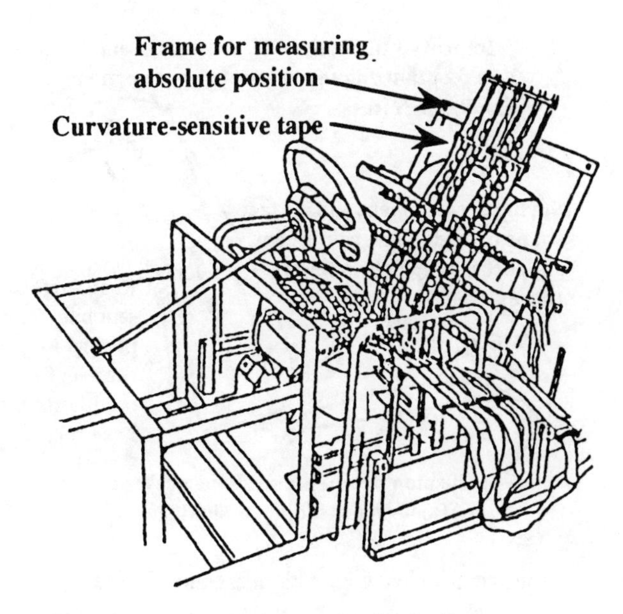

Figure 8 : System for measuring back shape when sitting (4)

The results obtained with this measuring system for the new seat indicated that a spinal column shape close to that of the target sitting posture was achieved in the case of the subjects having an S-shaped lumber region. However, for the subjects characterized as having a straight type lumber region, the measured curvature did not reach the target shape. In their case, the results did not go beyond the straight shape observed in the measurements obtained when each subject's spinal column assumed its natural shape.

The seat was then improved so that it would accommodate slight individual differences in sitting posture and thereby provide a better fit with the contours of the body. As seen in Fig. 6 where the measured spinal shapes are superimposed in the lumbar area, the improved seat design allows sufficient curvature in the middle back area (near the shoulder blades) in relation to the firm support provided for the pelvic and lumbar regions.

In terms of construction, one feature of the new seat is the use of a zigzag spring arrangement in the seat back to provide firm support throughout the pelvic and lumbar areas. To improve the fit with various body sizes, the seat back has been enlarged and the thickness of the upper portion has been optimized to achieve suitable softness. Meanwhile, the rear surface of the seat back curves inward to allow additional knee room for the rear-seat occupants. The shape and firmness of the seat cushion are designed to help maintain the occupant's posture by preventing the buttocks from sliding or sinking (Fig. 9). In addition, the so-called "alpha knit" has been adopted as the seat upholstery offered in some model grades. This fabric consists of a mixture of soft and firm fibers to provide a firm yet gentle feel for an improved tactile sensation.

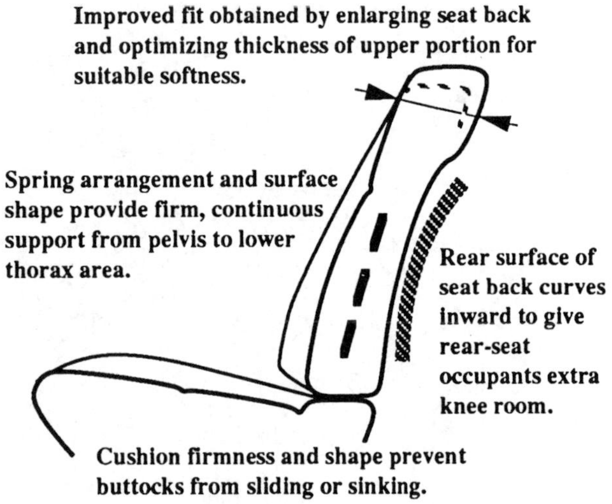

CONFIRMATION OF SUPPORT AND FIT BASED ON BODY PRESSURE DISTRIBUTION -- A system for measuring the body pressure distribution was used to verify the posture-support function of the new seat and the fit with the contours of the human body (Fig. 10). A mat containing several rows of load cells arranged in a grid fashion was placed on the seat to detect the load at numerous locations. The body pressure distribution was evaluated on the basis of the detected loads and pressure values and the resulting isobar patterns.

Measurements were made of the body pressure distribution on the seat back of the new seat and on that of an existing seat. Typical examples of each are shown in Fig. 11. The isobars in the left figure for the new seat show that the main pressure distribution extends upward continuously from a position corresponding to the pelvic and lumbar regions to the lower portion of the thorax.

Improved fit obtained by enlarging seat back and optimizing thickness of upper portion for suitable softness.

Spring arrangement and surface shape provide firm, continuous support from pelvis to lower thorax area.

Rear surface of seat back curves inward to give rear-seat occupants extra knee room.

Cushion firmness and shape prevent buttocks from sliding or sinking.

Figure 9 : New Generation Ergonomic Seat

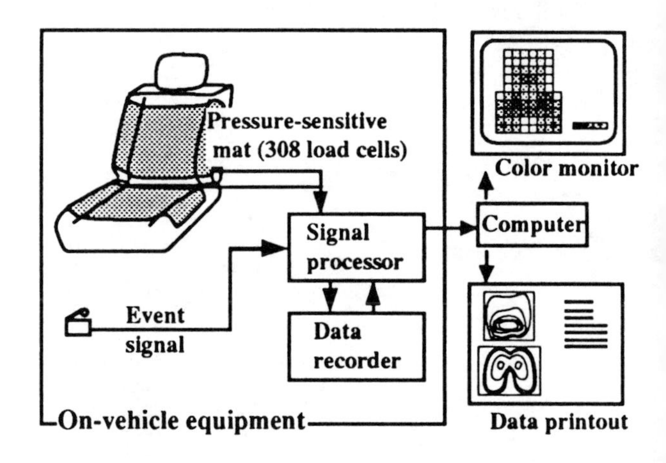

Figure 10 : Body pressure distribution measurement system

Seat back

New Generation Ergonomic Seat

Figure 11 : Comparison of Body pressure distribution

This indicates that the concept of combined pelvic and lumbar support achieves the intended support condition. The outermost isobar indicates the area of contract between the subject's back and the seat back. The broad expansion of this isobar shows that there is a good fit between the seat design and the shape of the human back.

EVALUATION OF OCCUPANT POSTURE AND FATIGUE DURING LONG-TERM DRIVING -- Subjective evaluations were conducted to assess the degree of fatigue experienced with the new seat and the existing seat during long-term driving. Five subjects participated in driving tests that were conducted over a distance of approximately 70 km, involving two to three hours of continuous driving on ordinary roads, including expressways. The aspects evaluated included support firmness, feeling of a stooped posture and feelings of discomfort in various parts of the body [5]. The subjects were ask to make the evaluations at specified time intervals.

The evaluation results for the stooped posture feeling are shown in Fig. 12 for the new seat.

The horizontal axis shows the elapsed time over a three hour period of sitting in the seat. The vertical axis indicates the subjects' evaluation of their sitting posture. The middle point of the vertical axis represents an assessment that the posture was just right or ideal, while descent along the scale indicates an increasing sensation of a stooped posture. The results show that an ideally perceived posture was maintained without hardly any feeling of a stooped posture even after three hours of driving.

The evaluation results for the degree of back and lumbar region fatigue perceived by the subjects are shown in Fig. 13 for both seats.

Attention was focused on the back and lumbar region because, among the feelings of discomfort generally mentioned for various parts of the body, these locations are thought to correlate closely with fatigue. The horizontal axis shows the elapsed time, while the vertical axis indicates the degree of fatigue based on an analysis of the level of discomfort reported by the subjects for their back and lumbar region. The degree of fatigue increases as one moves up the vertical scale.

A comparison of the data shows that feelings of back and lumbar region discomfort were expressed at a much later time and with less frequency for the new seat, thereby indicating a lower level of fatigue compared with the existing seat.

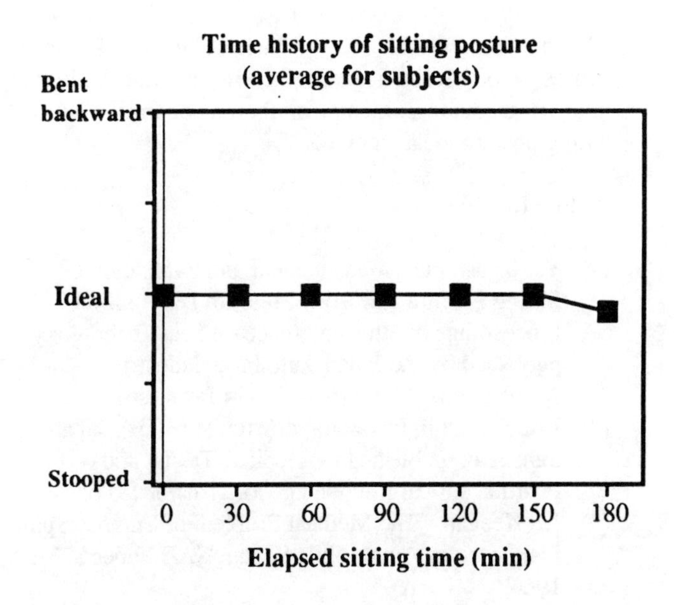

Figure 12 : Evaluation results for stooped posture feeling with new seat

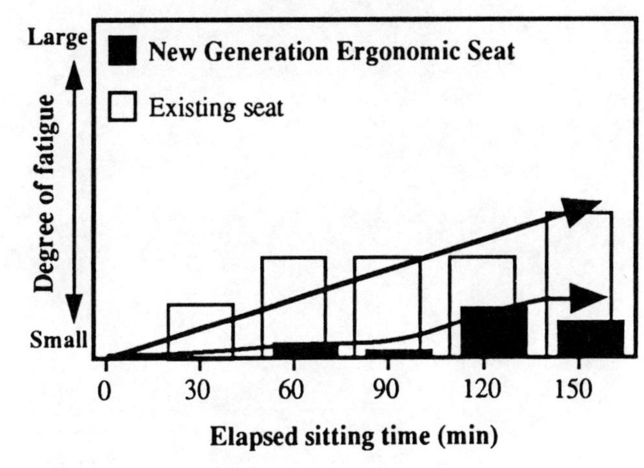

Figure 13 : Time history of lumbar region fatigue

CONCLUSION

This paper has described the development of the New Generation Ergonomic Seat, which is designed to provide improved occupant comfort. This new seat incorporates the concept of combined pelvic and lumbar support, based on the muscular and skeletal structure of the human body and an analysis of sitting posture and body pressure distribution. It better fits the contours of the human body and is effective in maintaining a proper occupant posture so that a stooped condition is not likely to occur even after long hours of sitting. Further, it is also effective in mitigating lumbar region discomfort, which is largely responsible for fatigue.

ACKNOWLEDGMENTS

The authors would like to thank Dr. N. Yamazaki, Professor of Mechanical Engineering at Keio University, and his graduate students for their cooperation with the sitting posture measurements.

REFERENCES

(1) Y. Nagashima, "Measurement and Evaluation of Sitting Posture and Body Pressure Distribution," Proceedings of 5th Conference on Seat Technology, published by the Japan Auto Parts Industry Association, 1989, pp. 34-36 (in Japanese).
(2) I.A. Kapandji, Physiologie Articulaire (III), Japanese translation published by Medical, Dental and Pharmaceutical Publishing Co., Tokyo, 1986.
(3) J.J. Keegan, "The Medical Problem of Lumbar Spine Flattening in Automobile Seats," SAE Paper 838A, 1964.
(4) T. Terauchi, et al., "Measurement of Contact Surface Between Sitting Man and Driver's Seat," Human Engineering, Vol. 25, April 1989 (in Japanese).
(5) Thomas B. Sheridan, et al., "Physiological and Psychological Evaluations of Driver Fatigue During Long Term Driving," SAE Paper 910116, 1991.

950141

Some Effects of Lumbar Support Contour on Driver Seated Posture

Matthew P. Reed, Lawrence W. Schneider, and Bethany A. H. Eby
University of Michigan Transportation Research Institute

ABSTRACT

An appropriately contoured lumbar support is widely regarded as an essential component of a comfortable auto seat. A frequently stated objective for a lumbar support is to maintain the sitter's lumbar spine in a slightly extended, or lordotic, posture. Although sitters have been observed to sit with substantial lordosis in some short-duration testing, long-term postural interaction with a lumbar support has not been documented quantitatively in the automotive environment. A laboratory study was conducted to investigate driver posture with three seatback contours. Subjects† from four anthropometric groups operated an interactive laboratory driving simulator for one-hour trials. Posture data were collected by means of a sonic digitizing system. The data identify driver-selected postures over time for three lumbar support contours. An increase of 25 mm in the lumbar support prominence from a flat contour did not substantially change lumbar spine posture.

INTRODUCTION

The appropriate design of lumbar support is the most widely discussed issue in seating ergonomics. The concept of support for the lower back in sitting is certainly not new. Åkerblom is widely credited with beginning the modern study of seating with his 1948 monograph (1*), but he cited more than 70 previous works related to the subject. Åkerblom formulated chair design recommendations after extensive investigations of spine anatomy, muscle activity, and force balance in sitting. Since Åkerblom's work, there have been hundreds of papers published on seating ergonomics, many of which include recommendations for lumbar support configuration that do not differ substantially from earlier recommendations (2, 3, 4). In view of this body of work, one might question the need for further research on lumbar support. However, some research suggests that current lumbar support recommendations based on physiological considerations do not adequately take into account the behavior of the sitter in the driving environment (5). This paper describes preliminary results from a series of experiments that are now under way to provide a new research basis for lumbar supports in auto seats.

DEFINITION OF LUMBAR SUPPORT – An important preliminary issue is the definition of the term "lumbar support." For the purposes of this paper, lumbar support will be defined geometrically, using a method similar to that employed by Andersson and others (6, 7, 8, 9, 10). Figure 1 shows the sagittal contour (profile) of a seated person. The lumbar support reference line is tangent to the posterior curves of the buttocks and thorax. The lumbar support prominence is defined as the maximum deviation of the profile curve from the reference line. If the resulting depressed seat contour is convex, as shown in the Figure 1, the lumbar support prominence is positive. The construction is slightly more complicated for negative

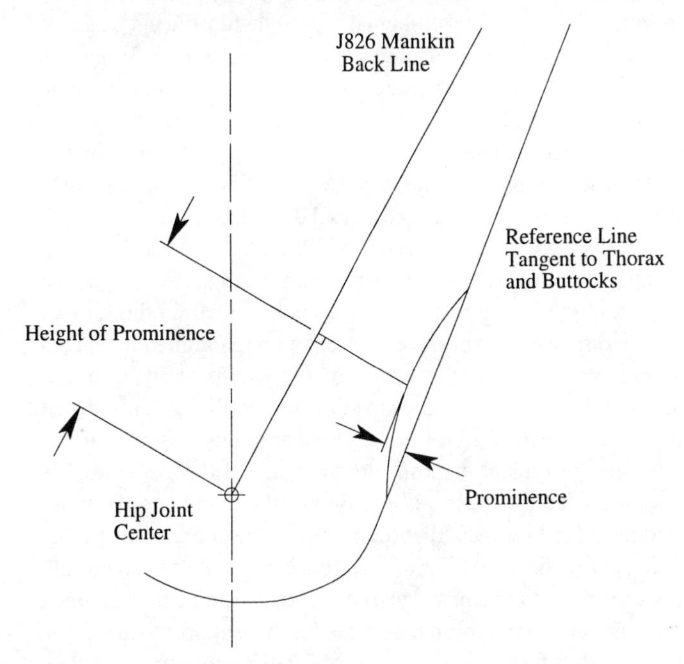

Figure 1. Geometric definition of lumbar support, adapted from Andersson (6).

† The rights, welfare, and informed consent of the volunteer subjects who participated in this study were observed under guidelines established by the U.S. Department of Health, Education, and Welfare (now Health and Human Services) on Protection of Human Subjects and accomplished under medical research design protocol standards approved by the Committee to Review Grants for Clinical Research and Investigation Involving Human Beings, Medical School, The University of Michigan.

* Numbers in parentheses signify references at the end of the paper.

prominences. If the lumbar spine is kyphotic, then the lumbar reference line is constructed in the position it would occupy if the sitter's back were straight, and the (negative) lumbar prominence is the maximum deviation from the reference line in the low-back region. The height of the lumbar support is defined as the height along the SAE J826 manikin back line (11) of the point of maximum prominence above the sitter's hip joint centers. A laterally symmetric posture is assumed.

This method of defining lumbar support supposes a particular seated posture and a particular sitter. A different posture, or a different sitter, could produce a different measure of lumbar support. We would like lumbar support to be a quantitatively measurable property of the seat, yet we also desire a measure that reflects the manner in which the seatback design affects the experience of the sitter. A depressed contour must be used, but a depressed contour requires a sitter, whether a human or a weighted surrogate like the SAE J826 H-point manikin (11). The J826 manikin, which is the only available anthropomorphic tool for measuring auto seats, does not have an articulated spine and hence cannot respond to different seat designs with different torso curvatures. Consequently, in the absence of a standard sitter, the lumbar support characteristics of a seat must be described statistically, with reference to the sitting behavior of a particular population.

For some laboratory seats, the definition given above is readily applied. Andersson *et al.* (6) used a wooden laboratory chair with a flat seatpan and seatback. Each subject was instructed to sit with the hips as far to the rear on the seat as possible so that the buttocks firmly contacted the seatback, and with the back of the thorax touching the upper part of the seatback. The plane of the seatback thereby represented the lumbar reference line depicted in Figure 1. When the position of the lumbar support was changed relative to the reference plane, the change in lumbar prominence was directly measurable, since the relative positions of the upper and lower tangent points, and the apex of the lumbar curve, were determined by the apparatus. In later studies on a car seat, Andersson *et al.* (9) used an identical definition for lumbar prominence, although they did not describe the method used for determining the reference plane. When the reference plane cannot be accurately determined, the absolute magnitude of the lumbar prominence cannot be determined, but relative measurements can be made if the reference plane is assumed to remain constant. In view of the findings of this study, this is probably not a generally valid assumption.

PURPOSE OF LUMBAR SUPPORT – Åkerblom (1), Keegan (12, 13, 14), and others recommended that a firm pad be located in the lower part of the seatback to restrain the lumbar spine from flexing excessively. Åkerblom recommended a firm support beginning at the height of the fourth or fifth lumbar vertebra, *i.e.*, at or below the top of the pelvis. Keegan suggested that seats be designed to produce a lumbar lordosis about midway between the typical standing lordosis and a flat contour. He recommended this posture because he found that people under treatment for low-back disorders were often more comfortable sitting in a reclined posture with lumbar lordosis than in an upright posture with a flat spine curvature. Both recommended an open space about 115 mm high below the lumbar support to allow the pelvis to shift forward and backward for different spine postures.

By the mid-1970s, most lumbar support recommendations were strongly influenced by physiological studies of the load on the lumbar spine. Andersson *et al.* (6, 7, 8, 9) used quantitative measurements of back extensor muscle activity and internal lumbar-disc pressure to assess spine loads for a range of postures. Andersson *et al.* found that disc pressure was lower in standing than in a wide range of seated postures, both unsupported and supported. Back extensor muscle activity was also low both in standing and supported sitting with reclined back angles. The experiments of Andersson and his coworkers suggest that lumbar intradiscal pressure is primarily affected by three factors: (a) quantity of body weight supported by the lumbar spine, (b) the tension exerted by the paraspinal musculature, and (c) the curvature of the spine. In both standing and sitting with a vertical torso angle (*i.e.*, upright), the lumbar spine sustains an axial load that supports most of the weight of the upper body, contributing to the lumbar disc pressure. The back extensor muscles, notably the erector spinae, have lines of action largely parallel to the spine. Consequently, tension developed in these muscles adds to the axial load on the lumbar discs. As a person reclines, some of the upper body weight is supported by the seatback, reducing the axial load on the lumbar spine slightly. Reclining also moves the upper body masses rearward relative to the lumbar spine, reducing the extensor moment supplied by back muscles and the muscle-tension contribution to axial spine load. Lumbar muscle activity is typically near zero when the sitter is reclined more than 20 degrees from the vertical for relaxed upper-body postures.

The curvature of the lumbar spine is the third important contributor to intradiscal pressure. Åkerblom, in his own work and in citations from previous researchers, identified a "natural form" for the spine. When the spine is excised with its ligaments intact, the unloaded lumbar spine assumes a posture Åkerblom describes as similar to the standing lordotic curvature. Keegan identified a similar spine posture, obtained by a recumbent subject with a torso-thigh angle of about 135 degrees, which he called the neutral spine posture. Andersson and others have noted that this "natural" spine curvature is produced by the wedge shape of the lumbar discs, which are taller anteriorly than posteriorly. The paraspinal ligaments hold the discs in compression. Åkerblom reported that removing the ligaments, leaving only the discs between vertebrae, caused an increase in spine length of 37 mm in one preparation. The "natural" spine posture, therefore, represents a spine posture in which the forces and moments on the vertebral bodies due to tension in the ligaments and compression of the discs are in equilibrium. Keegan's studies show that a similar spine posture results from passive equilibrium when the musculature is included. Deviations from this posture (*i.e.*, flexion or extension of the spine) result in increased stress in the spine and paraspinal tissue.

The research of Andersson and his coworkers shows that the disc pressure changes from standing to supported sitting result from alterations of spine posture as well as from changes in the amount of body weight supported by the spine and tension in the paraspinal musculature. When the seatback is reclined 20 degrees from the vertical, the back extensor muscles are virtually inactive, and therefore do not contribute significantly to the intradiscal pressure. However, at all seatback angles, including 20 degrees, changes in the lumbar spine curvature affect the intradiscal pressure. Since the amount of upper body weight borne by the lumbar spine probably does not change substantially when the lumbar curvature is varied, the reduction of disc pressure with increased lumbar support prominence is due primarily to the change in lumbar posture. In general, Andersson and his coworkers found that, for reclined postures, increasing the lumbar lordosis toward the standing posture decreases lumbar intradiscal pressure. In subsequent experiments with a car seat, Andersson *et al.* (9) found the lowest levels of back extensor muscle activity and intradiscal pressure with a seatback angle of 30 degrees and a lumbar support prominence of 50 mm. "Based on the assumption that low myoelectric activity and disc pressure are favourable ... ," he and his coauthors recommended these as target values for seat design.

The substantial work of Andersson's research team led to recommendations that lumbar supports be constructed to preserve, to the extent possible, the standing lumbar lordosis in sitting, with the objective of reducing lumbar spine loads as measured by intradiscal pressure. These recommendations have been echoed by many others since (2, 3, 4). A lumbar support intended to preserve the standing lordosis will be located at approximately the apex of the standing curvature, around L3, and will be longitudinally convex to mate with the desired spine curvature.

ALTERNATIVE VIEWS – Porter and Norris (15), noting that the lumbar support specifications in the literature are based primarily on physiological rationales, constructed a wooden laboratory seat to compare the lumbar support specifications recommended by Andersson *et al.* (9) with sitter preferences. Plastic probes inserted from the rear of the seatback provided quantitative measurement of spine curvature. A total of 37 male and 25 female subjects sat in the experimental chair adjusted to three conditions: (a) seatpan horizontal, seatback 90 degrees to seatpan, (b) seatpan inclined 15 degrees from horizontal, seatback 30 degrees rearward of vertical, and (c) same as (a) but with the knees extended to simulate a driving position. The seatpan and seatback angles in conditions (b) and (c) were taken from the recommendations in Andersson *et al.* (9). The lumbar support could be adjusted to 0-, 20-, or 40-mm prominence, and adjusted to any vertical position. Porter and Norris found that people preferred the 20-mm prominence to either of the other prominences in all test conditions. They also found that the preferred lumbar support height was about 120 mm above the hip joint center, although there was considerable variation among subjects. These experiments show that the postures that Andersson produced with a 40- to 50-mm lumbar prominence are not those that are preferred in an experimental chair with both reclined and vertical back angles. In general, postures with substantially less lordosis are preferred.

Some researchers have also questioned whether a lordotic lumbar spine posture is in fact desirable when seated. Adams and Hutton (16) argue that the advantages of a flexed spine posture outweigh the disadvantages. They cite increased transport of disc metabolites with changing pressure levels as a factor in favor of flexed-spine postures.

The Porter and Norris research began to address an important issue in lumbar support design. Andersson and others have demonstrated apparent physiological advantages to sitting with substantial lumbar lordosis. Keegan has reported from clinical observations that patients treated for low-back disorders are more likely to be comfortable when sitting reclined with lumbar lordosis. However, an important question is whether lumbar support contours that are intended to produce or maintain lordotic spine postures are used by sitters in that way. Using a wooden laboratory chair generally unrepresentative of auto seating, Porter and Norris found that subjects preferred to sit with a maximum lumbar prominence about half of that found in standing. This is close to Keegan's neutral posture, but less than Andersson's recommendation for minimal disc pressure.

In the current study, an experiment is being conducted to determine if people sit with substantially different postures in a seat with a prominent, longitudinally convex lumbar support than they do when the seatback contour is flat. If a sitter does not use the convex support in the manner intended (that is, sitting with substantial lordosis), then the sitter may experience substantially less support at the lower levels of the lumbar spine than he or she would when sitting against a flat seatback (5). This paper presents preliminary results from 24 subjects tested in a one-hour driving simulation. The research is continuing with additional subjects participating in long-term sitting sessions as well as short-term testing of other seatback contours.

METHODS

OVERVIEW – Volunteer subjects participated in one-hour driving simulations using each of three lumbar support contours. At ten-minute intervals, a sonic digitizer was used to record the locations of body landmarks and the subject's longitudinal back curvature.

SUBJECTS – Six subjects were recruited in each of four stature-gender groups, as shown in Table 1. Subject age ranged from 19 to 72 years with a mean age of 40 years. Nineteen standard anthropometric measures were collected from each subject but are not reported here. Two measures of hip and spine flexibility were also recorded.

Table 1
Subject Anthropometry

Group	Gender	n	Stature Min-Mean-Max (in.)	Stature Min-Mean-Max (mm)	Stature Min-Mean-Max (%ile by gender)*
1	female	6	60.3–61.3–62.8	1533–1556–1595	6–14–29
2	female	6	62.8–63.5–64.6	1594–1613–1640	29–40–57
3	male	6	68.0–68.9–69.9	1727–1750–1776	33–46–62
4	male	6	71.9–72.3–72.7	1826–1836–1847	85–88–91

*Based on normal approximations to data from Gordon *et al.* (17).

SEATING BUCK AND DRIVING SIMULATOR – A laboratory seating buck was constructed to reproduce the seat, steering wheel, accelerator pedal, and brake pedal positions and orientations of a contemporary minivan. Figure 2 shows the seating buck. The seating reference point (SgRP) is located 781 mm rearward and 352 mm above the accelerator heel point (AHP), giving an H30 seat height of 352 mm. The center of the front surface of the steering wheel is located 465 mm rearward and 721 mm above the AHP. The instrument panel in the laboratory buck is located about 100 mm forward of its position in the vehicle to facilitate digitization of driver posture. The accelerator pedal, brake pedal, and steering wheel are instrumented and connected to a driving simulator program running on a Macintosh computer. The simulated road scene is projected onto a screen approximately 10 feet in front of the driver's eye point, providing a field of view measuring approximately 44 degrees horizontally and 20 degrees vertically.

Figure 2. Laboratory seating buck.

SEAT – A minivan seat was extensively modified for use in testing. All of the foam and covering material on the seatback were removed. Part of the metal frame that supported the headrest was cut away to reduce the prominence of the headrest. An adjustable lumbar support supplied by Schukra North America was installed in the seat. The front surface of the support frame was covered with a 2-mm-thick sheet of Teflon. A second layer of Teflon was cut to fit within the seatback frame and installed over the lumbar support. A soft, 15-mm foam sheet was laid over the outer Teflon sheet and covered with a thin fabric. A motorized lumbar support adjustment provides approximately 120 mm of vertical travel. The experimenter adjusts the lumbar prominence by hooking varying-length retaining rods between the top and bottom edges of the lumbar support frame.

Back-contour measurement rods were mounted in a frame attached to the seatback, after the manner of Porter and Norris (15). The 6.4-mm-diameter, 424-mm-long steel rods were installed on 25-mm pitch approximately 25 mm to the right of the seatback centerline. A central rib in the Schukra support prevented placement on the centerline. Sixteen rods are located at 25-mm intervals along the rack, although usually only 12 rods can be used because of lumbar support frame interference (depending on the vertical lumbar support position). During data collection, the rounded tip of each rod is pressed firmly against the seat foam, which is accessible through a slit in the Teflon

Figure 3. Rear view of seatback, showing adjustable lumbar support and contour-measurement rods.

sheet supporting the foam. The soft foam is readily compressed to a uniform thickness against the seated subject. Figure 3 shows a view of the seatback, including the adjustable lumbar support and the contour measurement rods.

SONIC DIGITIZER – Posture and contour data are collected using a Science Accessories Corp. GP8-3D sonic digitizer. This and similar systems have been used extensively at UMTRI and other biomechanics labs for collection of spatial data (18). In the current study, two sonic emitters were mounted collinear with the tip of a hand-held probe. The emitters produce a wide-band sound pulse when an electric current arcs across a spark gap. An orthogonal array of four microphones receive the sound. An interface unit calculates the sound transit time to each microphone, applies a conversion factor to obtain distance, and sends these values via a serial connection to a computer. Software written for this application applies a calibration factor to adjust for changes in temperature and humidity and calculates the three-dimensional location of the emitter using the three shortest microphone distances. The location of the probe tip is calculated from the locations of the two probe emitters.

TEST CONDITIONS – Three lumbar support contours were investigated. Each was characterized by the displacement of the most prominent point on the lumbar support (the point having the furthest forward location) relative to the supporting structure. In lumbar support (LS) Condition A, the support frame was allowed to flatten under loading from the subject to produce an approximately flat surface. In LS Condition B, a metal retaining rod was used to hold the top and bottom edges of the Schukra support such that the point of maximum prominence was 10 mm forward of its position in condition B. For LS Condition C, the point of maximum prominence was 25 mm forward of its position in Condition A.

These test conditions do not necessarily correspond to 0, 10, and 25 mm of lumbar support under the definition used by Andersson *et al.* and Porter and Norris because the reference plane cannot be determined without identifying a particular sitter and posture. Instead, these conditions represent relative levels of lumbar support. Condition C should provide the opportunity for supported spine postures that are substantially more lordotic than either condition A or B. In these preliminary data from one-hour driving simulation trials, there are few significant differences in posture and contour between the A and B conditions, while there are highly significant differences between the A and C conditions. Consequently, only results from LS Conditions A and C are discussed. Given the nature of the findings, no conclusions of substance are lost by neglecting the data from Condition B.

PROTOCOL – For each subject, each lumbar support condition is tested on a different day. At the start of testing, the subject changes into form-fitting tights and a loose-fitting shirt to facilitate access to body landmarks. The subject is trained to locate the pubic symphysis landmark. To digitize the point, the subject palpates the anterior-superior margin of the pubic symphysis and presses the digitizer probe tip firmly against that point.

Prior to testing, the experimenter fixes the lumbar support at the appropriate prominence (test condition), places the seat track in its full-rear position, locates the seatback recliner at a nominal 20 degree angle, and sets the steering wheel angle adjustment to a neutral position. The lumbar support is initially positioned at the center of its vertical travel.

The subject sits in the vehicle buck and manually adjusts the seat track, seatback recline angle, and steering-wheel tilt for maximum comfort. The vertical position adjustment for the lumbar support is motorized and controlled by the subject with a switch mounted to the right of the seat. The subject is encouraged by the experimenter to try a range of different positions to find the most comfortable combination of adjustments. When the subject has adjusted the seat and steering wheel satisfactorily, the experimenter activates the driving simulator. The lights are dimmed while the simulator is running to improve the visibility of the road scene. If necessary, the experimenter provides feedback and instruction on operating the simulator. In general, subjects readily follow instructions to keep the simulated vehicle in the right lane of a two-lane, winding road, and maintain an 88 km/h speed (displayed on the screen with a simulated head-up display as 55 mph).

After two minutes of operating the simulator, the experimenter instructs the subject to maintain his or her current posture while the simulator is paused and the lights are brought up. The experimenter uses the digitizer probe to record the subject's back contour and posture. First, each of the back contour probes in turn is pressed firmly against the subject's back, bottoming out the thin foam layer, and the location of the rear of the probe is digitized. Next, two fiduciary points on the contour-probe rack are recorded to define a projection plane perpendicular to the probes. These points also provide a precise measure of seatback angle. The seatback pivot is digitized to provide a reference point that is fixed relative to the seatpan. The body landmarks listed in Table 2 are then digitized. The subject locates the pubic symphysis landmark using the procedures learned previously. The contour and body landmark digitization typically requires two minutes. The simulator is then restarted, and the subject drives until 10 minutes have elapsed from the time the simulator was previously paused. The simulator is paused again and data collection performed as before. The total test time is one hour, providing seven data collection intervals (0, 10, 20, 30, 40, 50, and 60 minutes). The actual time the subject is seated is approximately 3 minutes greater, because of adjustment time and the two-minute initial drive.

Table 2
Body Landmarks

Landmark	Definition
GLABELLA	Undepressed skin surface point at the most anterior prominence on the brow on the midsagittal line.
TOP HEAD	Undepressed skin surface point at the most superior point on the head.
OCCIPUT	Undepressed skin surface point at the most posterior point on the occipital prominence.
C7	Depressed skin surface point over most posterior point on corresponding spinous process.
ASIS(L), ASIS(R)	Depressed skin surface point over anterior-superior iliac spine. Located by palpating at trunk-thigh junction to locate the most anterior point on the ilium.
PUBIC SYMPHYSIS (PS)	Anterior-superior margin of the pubic symphysis. Subject is trained, using a model skeleton, to locate point with probe. Subject is instructed to compress the tissue toward the bone to the extent comfortable.
TOP STERNUM	Undepressed skin surface point at the most superior margin of the jugular notch of the manubrium in the midline of the sternum.
BOTTOM STERNUM	Undepressed skin surface point at the most inferior margin of the manubrium in the midline of the sternum.
LATERAL FEMORAL CONDYLE (LFC)	Undepressed skin surface point at the most lateral prominence of the right femoral condyle.

ANALYSIS

For each test, the digitizer control software writes a data file that contains the buck coordinates of each point recorded. At each measurement interval, the contour and posture data are extracted and translated to an XZ-plane origin at the seat pivot point so that postures from different seat track positions can be directly compared. Linear interpolation is used to obtain 12 equally-spaced contour points from the unevenly spaced contour data (some probes are obstructed by the lumbar support mechanism). Posture data analyses presented here are restricted to the sagittal (XZ) plane. Two additional points used to define posture are calculated as described in Table 3. Shoulder position is estimated using the torso geometry for midsize males reported by Schneider *et al.* (19) as shown in Figure 4. The posture variables listed in Table 4 and shown in Figure 5 are calculated for each measurement interval. Two measures of back contour are also obtained that are similar to the definition of lumbar support discussed in the introduction. Figure 6 shows the calculation procedure schematically.

Table 3
Calculated Body Landmarks

Landmark	Definition
HJC	Sagittal position of mean of hip joint centers. Hip joint center locations calculated from ASIS(L), ASIS(R), and PUBICSYMPHYSIS using method of Bell as adapted by Manary *et al.* (18).
SHOULDER	An approximation to the location of the glenohumeral joint in the midsagittal plane. The relationships among TOP STERNUM, C7, and the glenohumeral joint for the midsize male in Schneider *et al.* (19) were used to estimate the shoulder joint location. See Figure 4.

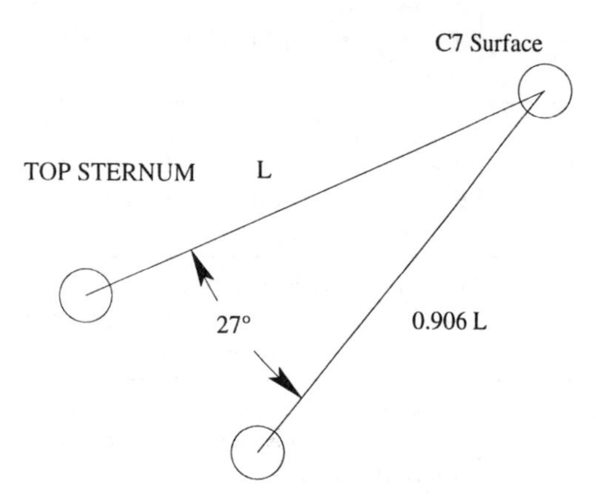

Estimated Glenohumeral Joint Center (SHOULDER)

Figure 4. Method of estimating shoulder joint location in the sagittal plane.

Table 4
Posture Variables

Variable	Definition*
Head Angle	Angle wrt horizontal of line formed by GLABELLA and OCCIPUT landmarks. Larger angles indicate greater neck extension.
Thorax Angle	Angle wrt vertical of line from TOP STERNUM to C7. Larger angles indicate more reclined thorax orientation.
Sternum Angle	Angle wrt vertical of line from BOTTOM STERNUM to TOP STERNUM. Larger angles indicate more reclined sternum orientation.
Pelvis Angle	Angle wrt vertical of line from PUBIC SYMPHYSIS to the mean of ASIS(R) and ASIS(L). Larger angles indicate more rearward pelvis rotation.
Torso Angle	Angle wrt vertical of a line from HJC to SHOULDER. Larger angles indicate more reclined torso orientation.

* All angles are measured in sagittal plane. Body landmarks are defined in Tables 2 and 3.

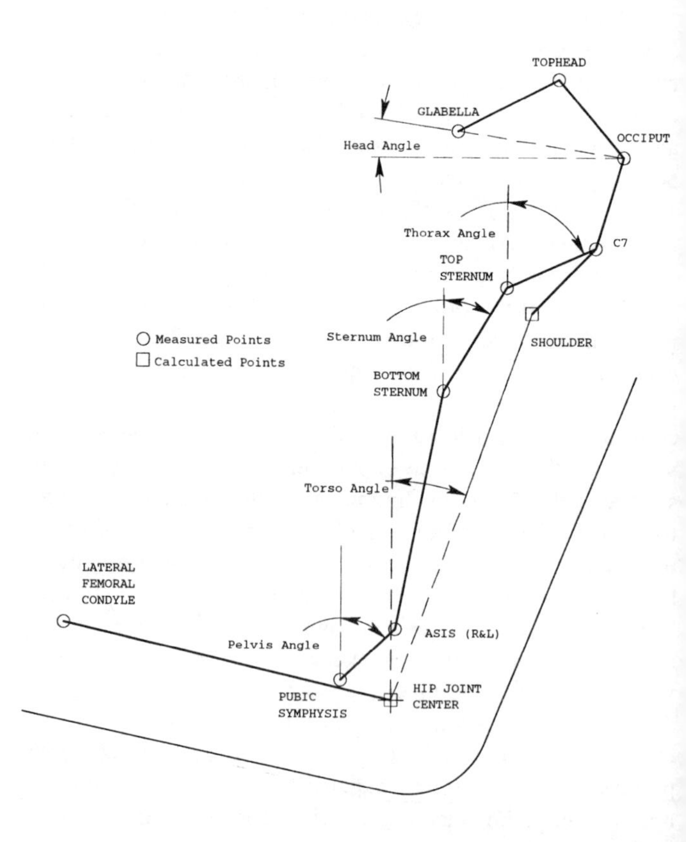

Figure 5. Body landmarks and posture variables.

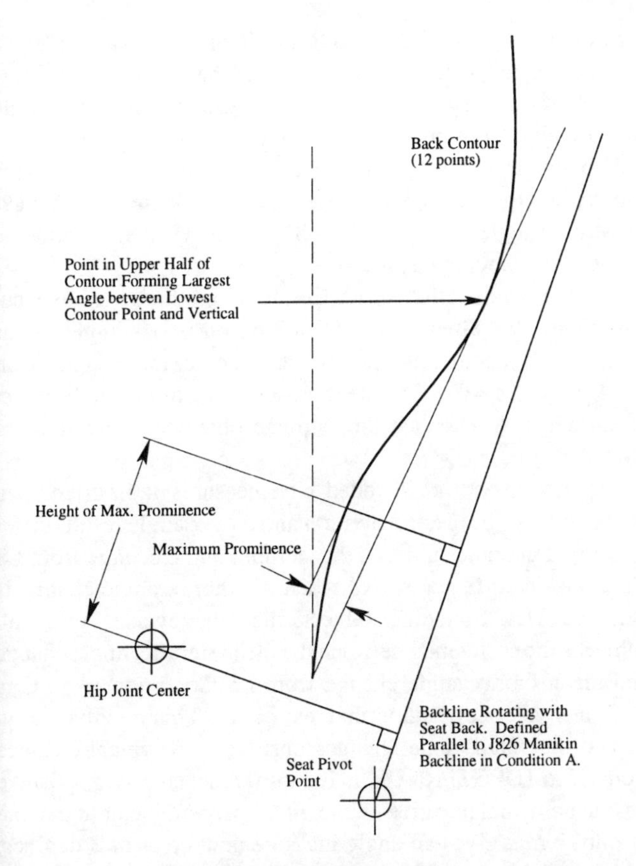

Figure 6. Schematic of calculation of maximum prominence and height of maximum prominence.

RESULTS

Large intersubject differences in sitting posture were found. Figure 7 shows seated pelvis angle by subject. All seven measurements from both the A (flat) and C (25-mm prominence) conditions are shown. Data for each subject are shown in a different position on the horizontal axis (14 data points for each subject). The mean pelvis angle by subject across conditions ranges from 39 to 72 degrees. The mean pelvis angle across all subjects and test conditions is 56.5 degrees. An important observation is that the points for each subject are fairly well grouped. The average standard deviation of pelvis angle within subject across test conditions is only 3.3 degrees, while the standard deviation of the mean pelvis angles by subject is 8.3 degrees. From Figure 7, it is apparent that any systematic, within-subject effect of test condition on pelvis angle is much smaller than the intersubject pelvis angle variance. Another observation is that there appear to be substantial differences between subject groups in pelvis angle. The group means are 58, 50, 64, and 54 degrees for the small females, midsize females, midsize males, and large males, respectively.

Figure 8 shows a similar plot of sternum angle by subject. As with pelvis angle, the data are well grouped within subject, and intrasubject differences between test conditions are generally smaller than intersubject differences. The overall mean is 24 degrees, with a range of subject means from 11 to 33 degrees. The standard deviation of subject means is 5.9 degrees, while the average within-subject standard deviation is 3.2 degrees. Head angle and thorax angle also show substantial intersubject variability.

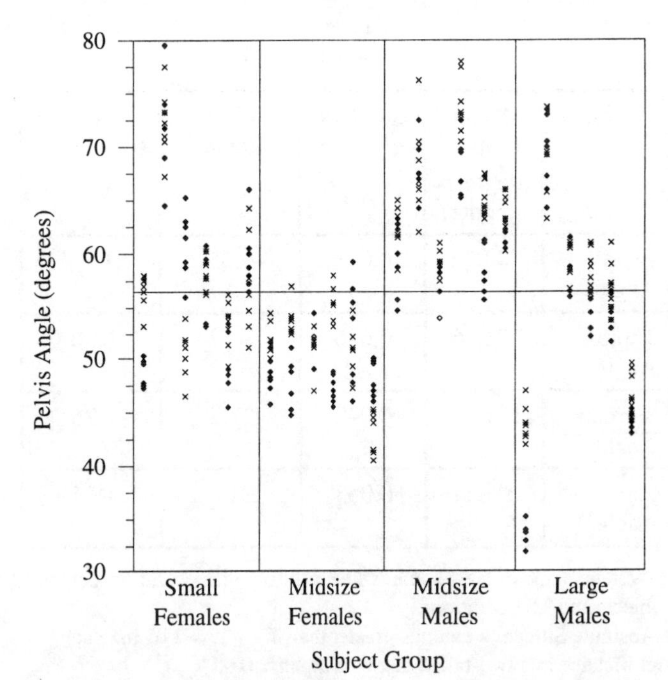

Figure 7. Pelvis angle by subject and group. Overall mean = 56.5°; Subject mean range = 39 to 72°. Condition A shown with **x** symbols, Condition C shown with ♦ symbols.

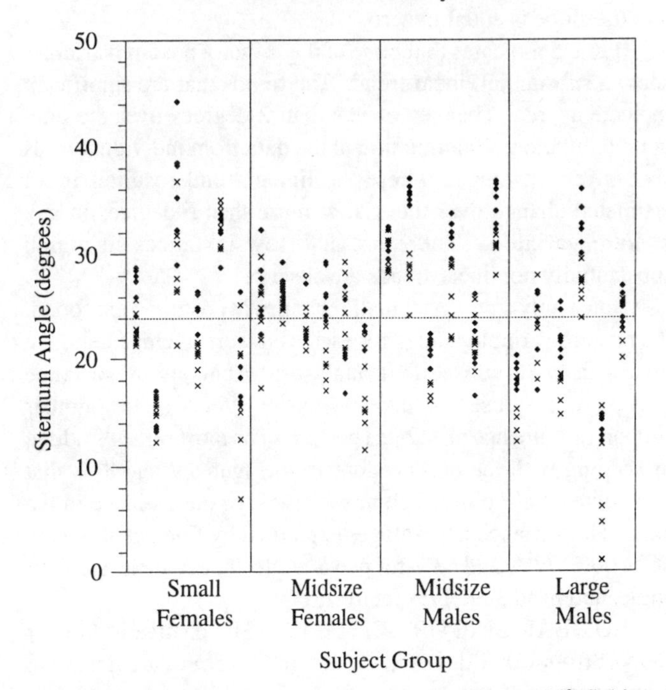

Figure 8. Sternum angle by subject. Overall mean = 24°; Subject mean range = 11° to 33°. Condition A shown with **x** symbols, Condition C shown with ♦ symbols.

TIME EFFECTS – The primary purpose of the one-hour driving simulation is to determine if there are any systematic changes in posture over time. Preliminary examination shows that, for all but a few subjects, there is little difference between subsequent measurements.

A least-squares line was fit to the seven data values of each variable for each trial. No difference was found between the mean slopes for the two LS Conditions, using a paired comparison, so the slopes for the two conditions were averaged within subject. The mean value of the average slope is significantly different from zero, or nearly significant, for sternum angle, thorax angle, and pelvis angle, but not for head angle.

Table 5
Test of Linear Time Effect

Variable	Mean Slope* (degree/ minute)	Std. Dev. of Slope	Student's *t* Value†	Change in 60 Minutes (deg)
Head Angle	-0.002	0.048	–0.25	--
Thorax Angle	0.016	0.036	2.19	0.96
Sternum Angle	0.021	0.049	2.05	1.26
Pelvis Angle	0.024	0.053	2.02	1.44

* Average slope of least-squares linear fit to variable vs. time across subjects and LS conditions.
† Absolute Student's *t* values greater than $T_{(0.025, 23)} = 2.07$ indicate significance in a two-tailed test with alpha = 0.05.

Table 5 shows the mean within-subject slope (degrees/minute), the standard deviation, and Student's *t* value testing the hypothesis that the slope is equal to zero.

Table 5 indicates that none of the primary posture variables show a substantial linear trend. The trends that are significant indicate average changes of less than 2 degrees over the one-hour simulation. Examination of the data from individual trials shows few instances where the linear trend resulted in an estimated change over the trial of more than 5 degrees in any posture variable. There are also few instances in which substantially nonlinear trends are observed.

Since only small systematic changes in posture were found, the posture variable values for each trial were averaged and the means used in subsequent analyses. Analysis of variance (ANOVA) was used to investigate the effects of the lumbar support prominence on subject posture. The most salient finding, in keeping with the observations from Figures 7 and 8, is that intersubject variability accounts for most of the variance in the data. The variance percentages explained by Subject alone are 56%, 60%, 73%, and 63% for pelvis angle, thorax angle, sternum angle, and head angle, respectively.

LUMBAR SUPPORT EFFECTS – The change in lumbar support from Condition A to Condition C, a nominal increase in prominence of 25 mm, produced significant differences in pelvis angle and sternum angle ($F_{(1, 20)} = 4.92$, $p = 0.038$; $F_{(1, 20)} = 33.3$, $p < 0.001$). The LS Condition effect approaches significance for thorax angle ($F_{(1, 20)} = 2.28$, $p = 0.147$), but is not significant for head angle ($F_{(1, 20)} = 0.28$, $p = 0.60$).

The least-squares estimate of the A–C LS Condition effect on pelvis angle is 2.0 degrees, indicating that, on average, subjects sat with more upright pelvis angles in Condition C. The effect of LS Condition (A–C) on sternum angle is –3.4 degrees, indicating that, on average, the subjects sat with their sternums more reclined in Condition C.

A measure related to net spine flexion can be obtained by subtracting sternum angle from pelvis angle. Larger values of pelvis–sternum indicate greater spine flexion. The effect of LS Condition (A–C) on pelvis angle–sternum angle is highly significant ($F_{(1, 20)} = 32.3$, $p < 0.001$). The least-squares estimate of the effect is 5.4 degrees, equivalent to the sum of the absolute effects of LS Condition on pelvis angle and sternum angle individually. In summary, one result of a nominal increase of 25 mm in lumbar support prominence is to reduce net flexion in the lumbar and thoracic spine an average of 5.4 degrees, if changes in sternum angle are assumed to reflect changes in the orientation of the sitter's thoracic spine.

Although the effect of LS Condition on thorax angle is not significant, the effect of LS Condition on pelvis angle minus thorax angle, another measure of net spine flexion, is significant ($F_{(1, 20)} = 12.1$, $p = 0.002$). The least-squares estimate of the effect is 3.4 degrees, less than the estimate obtained using sternum angle.

Although both are intended to be measures of the orientation of the sitter's ribcage, the sternum and thorax angle results differ slightly. For some subjects, the sternum angle change from LS Condition A to C is positive when the thorax angle change is negative. Data from other subjects show the reverse. Across all subjects, the correlation between the mean sternum angle change and mean thorax angle change from LS Conditions A to C is 0.72, a weaker correlation than expected. This may be due in part to the small angle changes observed. A typical distance from C7 to TOP STERNUM is 138 mm. A net error (*e.g.*, positive on one point and negative on the other) perpendicular to the line of only 7 mm gives an angle measurement error of 3 degrees, which is the estimated magnitude of LS Condition effect. So, while the trends are consistent, measurement error as a percentage of the true angle difference may account for the lower-than-expected correlation between the two estimates of ribcage rotation.

SUBJECT GROUP DIFFERENCES – There are significant differences in pelvis angle between subject groups. The group means are shown in Table 6. Comparing all pairs of groups, using Tukey-Kramer HSD with alpha = 0.05, the mean pelvis angle for midsize males is significantly higher than for midsize females. Other comparisons are not significant. The other posture variables do not show significant group differences.

Table 6
Pelvis Angles by Subject Group

Group	Mean Pelvis Angle (degrees)	Std. Dev.
Small Females	57.9	7.8
Midsize Females	50.1	3.4
Midsize Males	64.0	5.2
Large Males	53.9	10.5

OTHER EFFECTS OF LS CONDITION – The lumbar support prominence also affected the horizontal position of the subject's hips. The left and right hip joint centers (HJCs) were calculated from the pelvis posture data using the method of Manary *et al.* (18). A mean HJC was calculated by averaging the left and right HJCs and used as a measure of subject hip position in the XZ plane. The mean HJC is further forward on the seat for LS Condition C than for condition A ($F_{(1, 20)} = 6.26$, $p = 0.02$). The mean difference is estimated to be 11.4 mm. The mean HJC is also about 3 mm higher in the LS Condition C.

The fore-aft positions of the shoulder relative to the seat are not significantly different between subject groups or between the two LS conditions. However, shoulder joint height varies, as expected, with stature. Shoulder height also varies with LS Condition ($F_{(1, 20)} = 10.8, p = 0.004$). On average, shoulder joint locations are 6.8 mm higher in Condition C.

Torso angle, measured as the angle from the vertical of a line from mean HJC to the estimated shoulder joint location (in the XZ plane), does not differ significantly among the subject groups. Torso angle is significantly different between LS conditions ($F_{(1, 20)} = 8.97, p = 0.007$). On average, subjects sat about 1.8 degrees more reclined (larger torso angles) in Condition C than in Condition A. Means over all subjects are 23.3 degrees and 25.1 degrees for LS Conditions A and C respectively. Seventeen of the twenty-four subjects sat with a larger torso angle in Condition C. This observation is consistent with the 11-mm average forward shift in mean hip joint location with the more prominent support.

BACK CONTOUR – There are significant differences in back contour between the two LS conditions. The average maximum lumbar prominence for Condition A is 1.0 mm, indicating that, on average, subjects sat with a nearly flat low-back contour. For Condition C, the average maximum lumbar prominence is 8.9 mm. Standard deviations are 4.5 mm and 6.9 mm for LS Conditions A and C, respectively. The difference in prominence is highly significant ($t_{(23)} = 7.46, p < 0.001$). Some intergroup differences in lumbar contour prominence approach significance. Figure 9 shows the difference in maximum prominence between the two LS conditions for all subjects. Midsize female subjects show a larger effect of lumbar support prominence on back contour than do the other subjects. One small female subject sat with a flatter lumbar contour in Condition C than in Condition A. Only two subjects increased their lumbar contour prominence more than 15 mm in Condition C relative to Condition A.

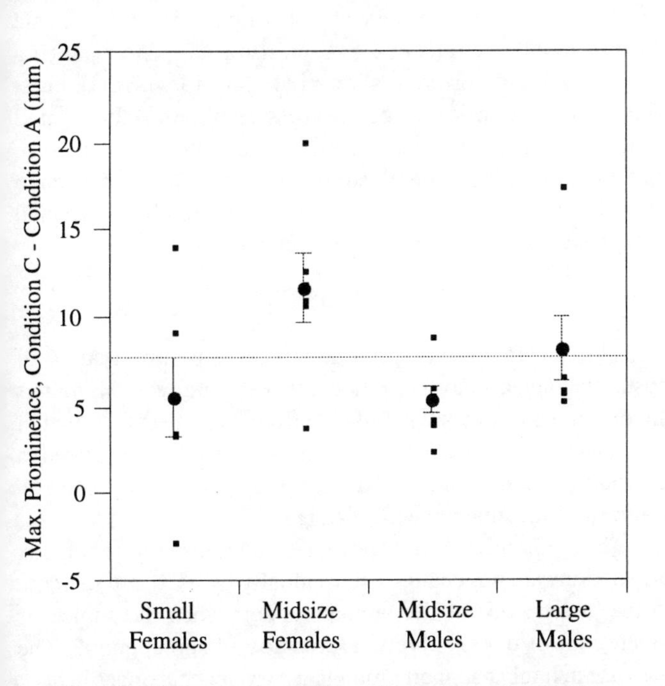

Figure 9. Difference in maximum lumbar contour prominence between LS Conditions A and C. Larger dots are group means. Error bars indicate ± 1 group standard deviation. Horizontal line is overall mean.

The height of the maximum prominence of the lumbar contour does not differ between lumbar support conditions, largely because the maximum prominence was small in LS Condition A. However, in the data from LS Condition C, the mean height of the maximum prominence for large males is significantly lower than the other groups ($t_{(23)} = 2.18, p = 0.04$), as shown in Figure 10. Data from one subject substantially increased the variance in the large-male group. Subject 401, the subject contributing the upright pelvis angles shown in Figure 7, is an outlier relative to his group on many measures. For all subjects, the height of the point of maximum prominence above the HJC along the backline is 139 mm with a standard deviation of 24 mm (see Figure 6 for definition of backline).

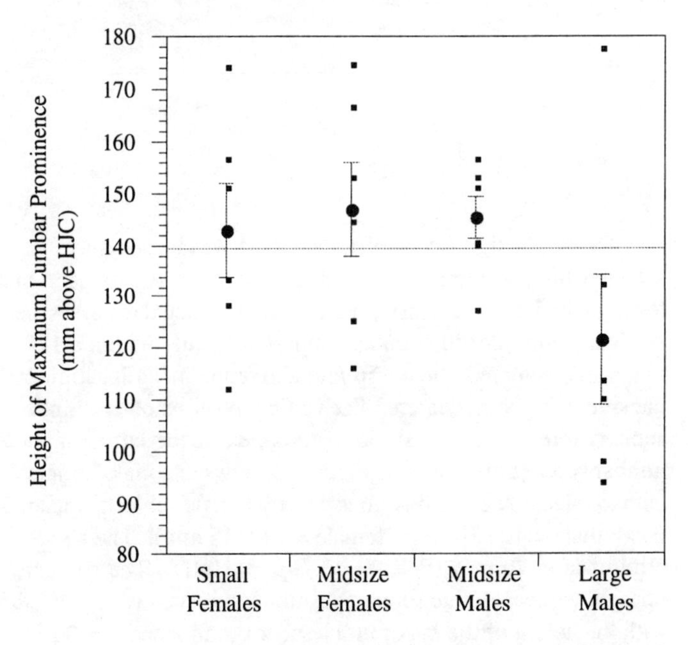

Figure 10. Maximum lumbar prominence height above HJC on back line for LS Condition C. Large dots are group means. Error bars indicate ± 1 group standard deviation . Horizontal line is overall mean.

SEATBACK ANGLE – Seatback angle was calculated from fiduciary points on the contour-probe frame. This reference plane is angled 18 degrees more upright than the SAE J826 manikin back angle when the manikin is placed in the seat with LS Condition A. Consequently, this differential was added to the angle measured from the contour-probe frame to obtain a manikin-referenced back angle. Note that this is not the same as the torso angle, which is defined as the angle relative to vertical of a sagittal-plane line connecting the mean HJC and shoulder point.

The subject-selected seatback angle varies with both subject group and LS Condition. Figure 11 shows a plot of seatback angle versus LS Condition by Group. The interaction approaches significance ($F_{(3, 20)} = 2.99, p = 0.055$). In general, larger subjects choose more reclined seatback angles than smaller subjects, but choose less reclined seatback angles with the more prominent lumbar support. Smaller subjects, in contrast, tend to choose slightly larger seatback angles with the more prominent support. The difference between groups was reduced in LS Condition C. The overall mean back angle across groups is 23 degrees. The seatback angle is not significantly different for LS conditions A and C.

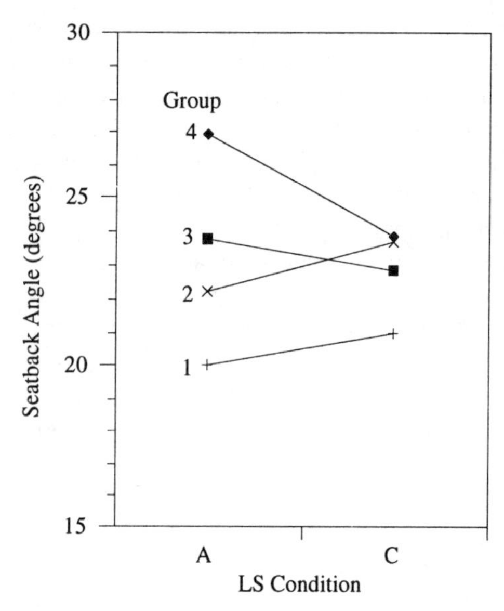

Figure 11. Seatback angle versus LS Condition by group.

There are significant differences between genders in the selected lumbar support position, as shown in Figure 12. In testing with LS Condition C, subjects are instructed to adjust the vertical position of the lumbar support vertically within a 120-mm range centered about 156 mm above the mean location of the sitters' hip joint centers. The vertical position of the lumbar support refers to the position of the apex of the curve of the lumbar support mechanism. Figure 12 shows that male subjects tend to select lower LS positions than do female subjects (means ± s.d: male = 121±18 mm, female = 147±15 mm). The gender difference is 26 mm ($F_{(1, 20)} = 13.7$, $p = 0.001$). The subject-selected vertical lumbar support position is moderately correlated with the height of the maximum lumbar prominence ($r = 0.58$). There is no significant difference between the mean lumbar support position (134 mm above HJC) and the mean height of the maximum lumbar contour prominence (139 mm).

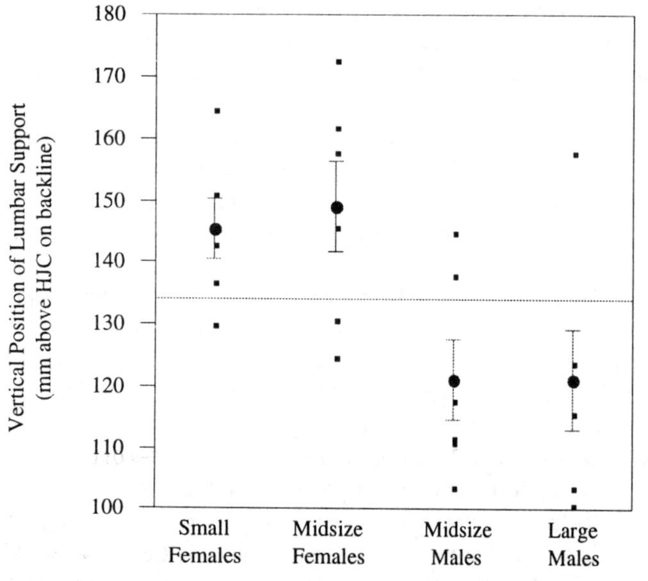

Figure 12. Subject-selected vertical lumbar support position, in millimeters above mean HJC location, measured along back line.

SUMMARY OF RESULTS – Although the analyses above show that the experiment was powerful enough to detect small within-subject differences in posture between lumbar support conditions, these differences were minor, particularly compared with the large intersubject variability. In this preliminary analysis of data from a larger study, the average effects of adding a vertically adjustable 25-mm lumbar support to a flat backrest contour are to:

- decrease pelvis angle by 2 degrees (more upright),
- increase sternum angle by 3.4 degrees (more reclined),
- decrease flexion of the thoracic and lumbar spine by 5.4 degrees,
- increase the maximum prominence of the lumbar lordosis by 8 mm,
- increase torso angle by 1.8 degrees, and
- shift the hip joint center forward by 11 mm.

These findings can be partially visualized by use of a kinematic model of the torso. A planar representation of the thorax, lumbar spine, and pelvis were developed based on the interpretation by Haas (20) of anthropometric data from Schneider *et al.* (19), Snyder *et al.* (21), and Reynolds *et al.* (22). The model is intended to represent the spine linkage of a midsize male, and consists of rigid links connected by pin joints located at the intervertebral joint centers from L5/S1 to T8/T9. The model was initially adjusted to produce a flat lumbar-spine posture, shown with light lines in Figure 13. The hip joint center of the model lies at the origin of the plot. Lines connect joint centers and spinous processes below T8. The lumbar spine was then extended 5.4 degrees, the mean spine extension produced by the 25-mm support, using the even distribution of lumbar motion recommended by Hubbard *et al.* (23). The pelvis of the model was rotated around the hip joint center to produce a pelvis angle 2 degrees more vertical than the flat-spine model. The model was translated forward 11.4 mm and upward 3 mm, to account for the difference between the two LS conditions in mean HJC location. The torso angle was also adjusted to approximate the mean torso angle in LS Condition C. The resulting representation of posture is shown in Figure 13 with dark lines. The lumbar extension of 5.4 degrees results in only a small lumbar lordosis, with a prominence close to the 9-mm average measured in testing. The illustration in Figure 13 is intended to assist in visualizing the findings, but should not be interpreted as representing any individual subject's postures.

DISCUSSION AND CONCLUSIONS

This paper presents preliminary results from a study of postural adaptation to seatback contour in auto seats. Findings are presented from analyses of data from 24 subjects. A total of 48 subjects will participate in testing similar to that described in this paper. Additional seatback contours, some with subject-controlled adjustments, will be tested.

The preliminary results show that preferred pelvis and thorax orientations are not changed substantially by a 25-mm increase in the prominence of the lumbar support when the subject is given control over the vertical location of the support. The increase in lumbar support prominence produces changes in back contour that are about one-third the magnitude of the support prominence increase, on average. A schematic visualization of the torso skeletal linkage shows that increased lumbar support

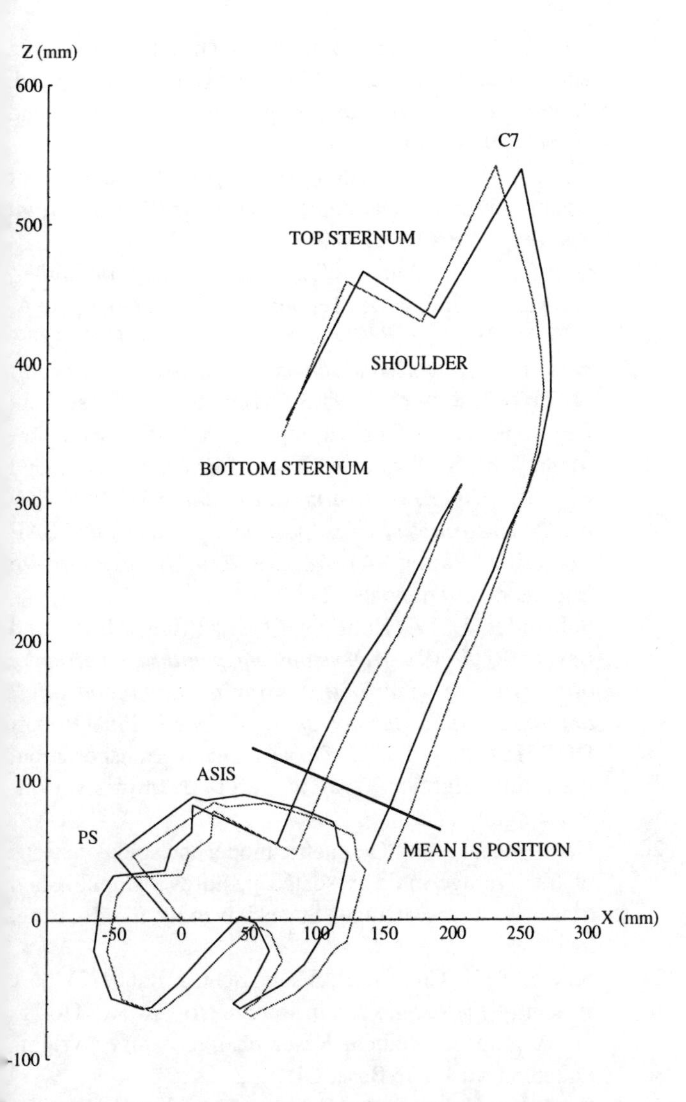

Figure 13. Schematic representation of the torso using a planar kinematic model. The sagittal-plane profile of the pelvis and sacrum are shown, along with a line connecting the sternum points, shoulder joint center, and spinous process surface points. Another line connects spine joint centers from T8/T9 to L5/S1. Light lines show a flat lumbar spine posture, illustrating the mean pelvis, thorax, sternum, and torso orientations obtained with lumbar support Condition A. The dark lines show the model translated forward to match the Condition C HJC location with the pelvis angle, sternum angle, thorax angle, and torso angle adjusted to match the mean data from Condition C. The dark line perpendicular to the spine indicates the mean subject-selected lumbar support position. The distance between the model back contours on that line is 15 mm.

prominence causes the subject to sit forward on the seat with only a small increase in lumbar lordosis.

It should be emphasized that these findings are preliminary and that more study of postural adaptation in auto seats is needed. The current data are limited by sample size and by the static laboratory conditions used in testing. Differences between postures chosen in an on-road environment and those measured in the laboratory are expected to be small, however, since the primary physical constraints affecting posture selection (vision requirements and control placement) are included in the laboratory seating buck design.

These findings are in agreement with earlier work. The flat lumbar spine profiles recorded by Schneider *et al.* (19) in a study of driver anthropometry have been criticized as unrepresentative of driving postures in present-day seats because of the lack of lumbar contour in the test seats. However, the preliminary results of the present study show that the addition of a 25-mm lumbar support does not substantially alter drivers' lumbar spine postures.

As noted in the introduction, current lumbar support design recommendations are strongly influenced by the goal of reducing lumbar disc pressure, which is related to the loads on the disc annulus and the surrounding ligaments. As Andersson and others have shown, lumbar lordosis generally results in reduced disc pressures. However, seated postures with substantial lumbar lordosis were not selected by the subjects in this study, even when the seat was configured to provide support for lordotic postures.

Among the factors that may contribute to the prevalence of flat-spine postures is the influence of the posterior thigh muscles in restricting forward pelvis rotation. The hamstring muscles connect the pelvis and leg across the knee and hip joints and produce a restriction on pelvis orientation that varies depending on knee angle (24, 25). When the knees are extended beyond 90 degrees, as is typical of automotive postures, the relatively erect pelvis angle necessary to produce substantial lumbar lordosis with a reasonable thorax orientation may not be possible for many sitters without discomfort in the backs of the thighs. Consequently, more reclined pelvis orientations may be selected, resulting in greater spine flexion and a flat lumbar spine contour. In the current study, a relatively high seat height was used (H30 = 352 mm), potentially reducing the influence of the hamstring muscles on pelvis posture by allowing less-extended knee angles than would be required with a lower seat height. For vehicles with lower seats, including most passenger cars, lumbar lordosis may be even less likely than for the seat height tested, although further research is necessary to verify this assumption.

The apparent physiological benefits of lumbar lordosis cannot be realized if sitters do not select such postures. If lordotic lumbar curvatures are not prevalent even when the seat is designed to accommodate them, then the purpose of lumbar supports in auto seats should be reconsidered. The preliminary findings from the present study suggest that seatbacks with fixed lumbar supports should provide support for nearly flat spine profiles, rather than for the standing spine curvature associated with lower disc pressure. Those people who prefer to sit with substantial lordosis can be accommodated by providing an adjustable-prominence support located approximately 140 mm above the hip joint center.

These investigations are continuing with a larger group of subjects and several additional test conditions. Subsequent reports will describe a general model of the postural effects resulting from changes in seatback contour. This research is expected to contribute to the formulation of a new rationale for lumbar support design.

ACKNOWLEDGMENTS

This research was sponsored by Lear Seating Corporation. The authors extend their thanks to Carol Flannagan for statistical direction, to Brian Eby for equipment fabrication, and to Leda Ricci for assistance in the preparation of this paper. Thanks also to Joseph Benson of Schukra North America for providing the adjustable lumbar support used in this study.

REFERENCES

1. Åkerblom, B. (1948). *Standing and sitting posture with special reference to the construction of chairs.* Doctoral Dissertation. A.B. Nordiska Bokhandeln, Karolinska Institutet, Stockholm.
2. Chaffin, D.B. and Andersson, G.B.J. (1991). *Occupational Biomechanics, 2nd Ed.* Wiley and Sons, New York.
3. Reynolds, H.M. (1993). Automotive seat design for sitting comfort. In W. Karwowski and B. Peacock, eds. *Automotive Ergonomics*, pp. 99–116. Taylor and Francis, New York.
4. Reed, M.P., Schneider, L.W., and Ricci, L.L. (1994). *Survey of auto seat design recommendations for improved comfort.* Technical Report UMTRI-94-6. University of Michigan Transportation Research Institute, Ann Arbor, MI.
5. Reed, M. P., Saito, M., Kakishima, Y., Lee, N. S., and Schneider, L. W. (1991). *An investigation of driver discomfort and related seat design factors in extended-duration driving.* SAE Technical Paper 910117. Society of Automotive Engineers, Warrendale, PA.
6. Andersson, G. B. J., Örtengren, R., Nachemson, A., and Elfström, G. (1974). Lumbar disc pressure and myoelectric back muscle activity during sitting. I. Studies on an experimental chair. *Scandinavian Journal of Rehabilitation Medicine*, 6(3), 104-114.
7. Andersson, G. B. J. and Örtengren, R. (1974). Lumbar disc pressure and myoelectric back muscle activity during sitting. II. Studies on an office chair. *Scandinavian Journal of Rehabilitation Medicine*, 6(3), 115-121.
8. Andersson, G. B. J. and Örtengren, R. (1974). Lumbar disc pressure and myoelectric back muscle activity during sitting. III. Studies on a wheelchair. *Scandinavian Journal of Rehabilitation Medicine*, 6(3), 122-127.
9. Andersson, G. B. J., Örtengren, R., Nachemson, A., and Elfström, G. (1974). Lumbar disc pressure and myoelectric back muscle activity during sitting. IV. Studies on a car driver's seat. *Scandinavian Journal of Rehabilitation Medicine*, 6(3), 128-33.
10. Andersson, G. B. J., Murphy, R. W., Örtengren, R., and Nachemson, A. L. (1979). The influence of backrest inclination and lumbar support on lumbar lordosis. *Spine*, 4(1), 52-58.
11. SAE J826 Handbook (1991). *Volume 4, Devices for use in defining and measuring vehicle seat accommodation – SAE J826*, pp. 34.52-56, Society of Automotive Engineers, Warrendale, PA.
12. Keegan, J. J. (1953). Alterations of the lumbar curve related to posture and seating. *Journal of Bone and Joint Surgery,* 35-A(3), 589-603.
13. Keegan, J. J. (1964). *The medical problem of lumbar spine flattening in automobile seats.* SAE Technical Paper 838A. Society of Automotive Engineers, New York.
14. Keegan, J. J. and Radke, A. O. (1964). Designing vehicle seats for greater comfort. *SAE Journal, 72(9)*, 50-55.
15. Porter, J. M. and Norris, B. J. (1987). The effects of posture and seat design on lumbar lordosis. In E.D. Megaw, ed. *Contemporary Ergonomics,* pp. 191-196. Taylor and Francis, New York.
16. Adams, M. A., and Hutton, W. C. (1985). The effect of posture on the lumbar spine. *Journal of Bone and Joint Surgery*, 67-B(4), 625-629.
17. Gordon, C. C., Churchill, T., Clauser, C. E., Bradtmiller, B., McConville, J. T., Tebbetts, I., and Walker, R. A. (1989). *1988 anthropometric survey of U.S. Army personnel: Methods and summary statistics.* Final Report NATICK/TR-89/027. U.S. Army Natick Research, Development and Engineering Center, Natick, MA.
18. Manary, M. A., Schneider, L. W., Flannagan, C. C., and Eby, B. H. (1994). *Evaluation of the SAE J826 3-D manikin measures of driver positioning and posture.* SAE Technical Paper 941048. Society of Automotive Engineers, Warrendale, PA.
19. Schneider, L. W., Robbins, D. H., Pflüg, M. A., and Snyder, R. G. (1985). *Development of anthropometrically based design specifications for an advanced adult anthropomorphic dummy family, Volume 1.* Final Report DOT-HS-806-715. U.S. Department of Transportation, National Highway Traffic Safety Administration, Washington, D.C.
20. Haas, W. A. (1989). Geometric model and spinal motions of the average male in seated postures. Unpublished Master's thesis. Michigan State University, East Lansing, MI.
21. Snyder, R. G., Chaffin, D. B., and Schutz, R. (1972). *Link system of the human torso.* Report No. AMRL-TR-71-88. Aerospace Medical Research Laboratory, Wright-Patterson Air Force Base, OH.
22. Reynolds, H. M., Snow, C. C., and Young, J. W. (1981). *Spatial geometry of the human pelvis.* Memorandum Report AAC-119-81-5. Federal Aviation Administration, Civil Aeromedical Institute, Oklahoma City, OK.
23. Hubbard, R. P., Haas, W. A., Boughner, R. L., Conole, R. A., and Bush, N. J. (1993). *New biomechanical models for automobile seat design.* SAE Technical Paper 930110. Society of Automotive Engineers, Warrendale, PA.
24. Stokes, I. A. F. and Abery, J. M. (1980). Influence of the hamstring muscles on lumbar spine curvature in sitting. *Spine,* 5(6), 525-528.
25. Boughner, R. L. (1991). A model of average adult male human skeletal and leg muscle geometry and hamstring length for automotive seat designers. Unpublished Master of Science thesis. Michigan State University, Department of Metallurgy, Mechanics, and Material Science, East Lansing, MI.

950142

Evaluation of Objective Measurement Techniques for Automotive Seat Comfort

Janilla Lee and Ted Grohs
Lear Seating Co.

Mari Milosic
Wayne State University

ABSTRACT

This paper is the second in a two-part series that provides an overview of four objective methods for measuring Seat Comfort. It provides a detailed description of three of the four concepts, Electromyography (EMG), Spinal Loading, and Body Motion. Each of these methods is discussed in terms of Theory, Test Technique, and Test Data that are strictly linked to measurable biological parameters. Graphs are included to show the nature of the specific biological activity being measured and how these activities change in response to environmental stimuli acting on the human body. A comparison of the four techniques is provided. Suggestions are made to help future efforts in developing these objective measures of Seat Comfort.

INTRODUCTION

Almost everyone agrees that comfort is a very subjective matter. To understand what car owners define as a comfortable seat requires market research, and most of these studies use statistical methods to analyze the data.[1,2] Usually the seat designs are aimed to satisfy the normal population. However, this kind of study is often mixed in with some psychological factors, current appearance and style trends. The result may shift as time goes by.

If we can come up with a scientific method to quantify seat comfort, we can eliminate these kinds of expensive studies and subjective argument. This objective, scientific method should be focused on human body comfort requirements, i.e. human biological (Physiologic and Orthopedic) requirements, in terms of seating posture. Understanding this requirement will allow seating suppliers to offer "healthy" seats to the OEM market. Our research project has been concentrating on searching for good objective comfort measurables for automotive seating. This paper is the second part of a two-part Seat Comfort paper. Our first paper[3] was published two years ago. It had an extensive discussion of a DOE (Design of Experiment) project used to correlate subjective evaluations with Pressure Distribution and Electromyography (EMG) which were the 2 objective measurables that we introduced at that time. Although we used EMG data, we did not discuss the technique. We will do so in this paper. In addition, we will also introduce two more objective measurables: Spinal Loading and Body Motion. To summarize the entire project, we will compare all four measurables studied by pointing out the advantages and disadvantages of each technique. All studies were conducted in the static condition.

OBJECTIVE MEASURABLES

The 3 objective measureables that we will discuss here are all based on human body Physiological or Orthopedic phenomenon. Their theories and techniques are similar in many ways, so our discussion will be grouped by Theories, Test Technique and Test Data.

BASIS FOR STUDY

Electromyography (EMG) - The postural muscles of the back function to protect the spinal structures and help maintain the body in a comfortable configuration during normal activity. If the body is forced into an inefficient position, these muscles will have to work harder to re-establish a normal un-stressed posture.[4] The result will be fatigued muscles due to high levels of activity. It would be advantageous to reduce activity in the postural muscles to avoid fatigue and discomfort. However, if the activity of these muscles is too low, numbness could occur due to lack of blood flow. It can therefore be deduced that some low value of muscle activity is required to both eliminate numbness, yet avoid fatigue.

Electromyography (EMG) signals can be used to measure the myoelectrical activity of muscles. Muscle contractions send a measurable electric potential that can be recorded by EMG. In order to establish a correlation between muscle activity and levels of comfort or discomfort in an automobile seat, EMG measurements of the postural muscles have been investigated. It is proposed that seat comfort, based on reduced muscle fatigue and avoided numbness, can be predicted by EMG measurements. Reduced fatigue and decreased muscle activity also tend to reflect the potential for injury prevention. It is therefore an additional goal of this study to develop EMG measurements as a design guideline to establish seats that improve physiological conditions for the passenger.

Spinal Loading - The intervertebral disks of the spinal column act as important shock absorbers when the spine is loaded in the axial direction. This loading condition is inherent in everyday activity, as the human body is subjected to the forces of gravity during normal standing and walking situations. Kazarian reports that with the application of a compressive load onto an isolated intervertebral disc, there occurred an immediate compressive distortion. When the load was maintained at a constant value, a gradual flow or "creep" was noted to occur in the height of the disk over time. This suggests that a compressive load on the spine causes a slow decrease in the height of the disks over time as they react to absorb the load. Hence, an overall decrease in body stature results.[5,6]

When the axial load on the spine is removed, the spinal disks exhibit a recovery behavior in which a portion of the "loss" due to creep is regained. The spinal disks, in essence, grow at a continuously decreasing rate as they recover. This behavior occurs in the same exponential trend as that of creep.

Althoff has shown that sitting produces a decreased stress on the spine in comparison to standing. Therefore, if a loaded posture (standing) is followed by a more relaxed posture (sitting), the spine will tend to recover and produce a positive change in stature. The result is a growth in spinal length, which is a measurable, physiological quantity. It is therefore proposed that seat comfort levels can be predicted by the extent of spinal growth that occurs in a seated phase following a pre-loading of the spine.

Body Motion - It has been claimed that uncomfortable chairs cause frequent and intensive body motions.[7] If a person feels uncomfortable, he/she will tend to adjust his/her posture to improve the situation. Since body position and orientation can be measured by digital motion tracking systems, these phenomena can potentially be utilized as an objective measurement of comfort. It is proposed that a greater incidence of trunk motion over a given time will indicate an attempt to relieve discomfort. Hence seat comfort can be quantified by the measurement of the extent of body movements that occur in a seated phase.

METHODS OF MEASUREMENT - In conjunction with the Orthopedic Biomechanics Department of Wayne State University, an experiment was designed to evaluate measurement techniques for the quantification of seat comfort. Three seats with known different levels of seat foam firmness were supplied by an OEM for this experiment. The experimental set-up and methods of data acquisition described here were conducted at the Wayne State University Orthopedic Biomechanics Laboratory. Through extensive pilot studies conducted and evaluated by the Biomechanics Dept., the methods described here were determined to be most effective and efficient for the lab capabilities.

Electromyography - EMG data was measured by use of self adhesive, tri-contact surface electrodes. The surface electrodes were attached to clean, abraded skin at the following locations: cervical paraspinal (neck), upper trapezius (shoulder), mid, and lower erector spinea (mid and lower back). The electrode cables were fed into a physical filter (band width = 5 - 500 Hz) which was connected to a digital Tektronix data acquisition system. Total recording length was 2000 points with a sample clock of 500 μs over a 1 second acquisition time. Data was collected via a hand trigger and saved digitally to a hard disk.

In order to quantify muscle activity, the raw EMG data had to be processed into a more comparable form. The digital data was first transformed into ASCII format and amplified. Data was then full-wave rectified to generate an absolute value of the EMG with positive polarity. This rectified data was run through a low-pass filter (cut-off frequency = 6 Hz) to produce a linear envelop that closely follows the trend of muscle activity, and hence the shape of the tension curve created by the muscle contraction. The final processing integrates the EMG data to establish the area under the curve, and to produce a single, quantifying value for the amount of muscle activity in that particular muscle group of the body

Testing was conducted according to a pre-determined procedure developed from the pilot studies. Following an initial pre-loading of the spine, EMG measurements were taken at 15 minute intervals for 1 1/2 hours with the test subject in a seated posture. The initial loading phase involved the test subject assuming a standing posture for a duration of 30 minutes. Once pre-loaded, the test subject was instructed to sit in the pre-adjusted seat with his/her feet flat on the floor and arms resting in lap. This was an arbitrary posture adopted for control and comparison purposes. Upon initial seating, an EMG reading was taken to establish a submaximal reference contraction to which all other readings would be compared.

All three seats involved in the study were tested with the same protocol for each test subject. Subjects were tested during the same time period on alternating days in order to recreate the same testing conditions for each seat. The entire EMG test protocol was conducted in conjunction with the Spinal Loading test.

Spinal Loading - Stature measurements were acquired with the use of a stadiometer developed by Althoff (Figure 1). This apparatus consists of a frame with a back tilt of 10 degrees off the vertical. Exact position of the test subject is controlled by three probes adjusted to meet the apex of the spinal curves, a head and buttocks rest, hip positioners, and glasses with a 45 degree, angled-mirror mount for head control. With additional modifications for control of knee flexion and foot position, reproducibility can be closely monitored, and stature measurements can be accurately taken to within 0.1 mm.

Measurements are taken once the subject has been marked and the stadiometer adjusted. Marker placement for stature measurement is positioned at the joint space between cervical vertebrae C6 and C7 of the neck. This was determined by the pilot studies to be the best placement for measurement reproducibility.

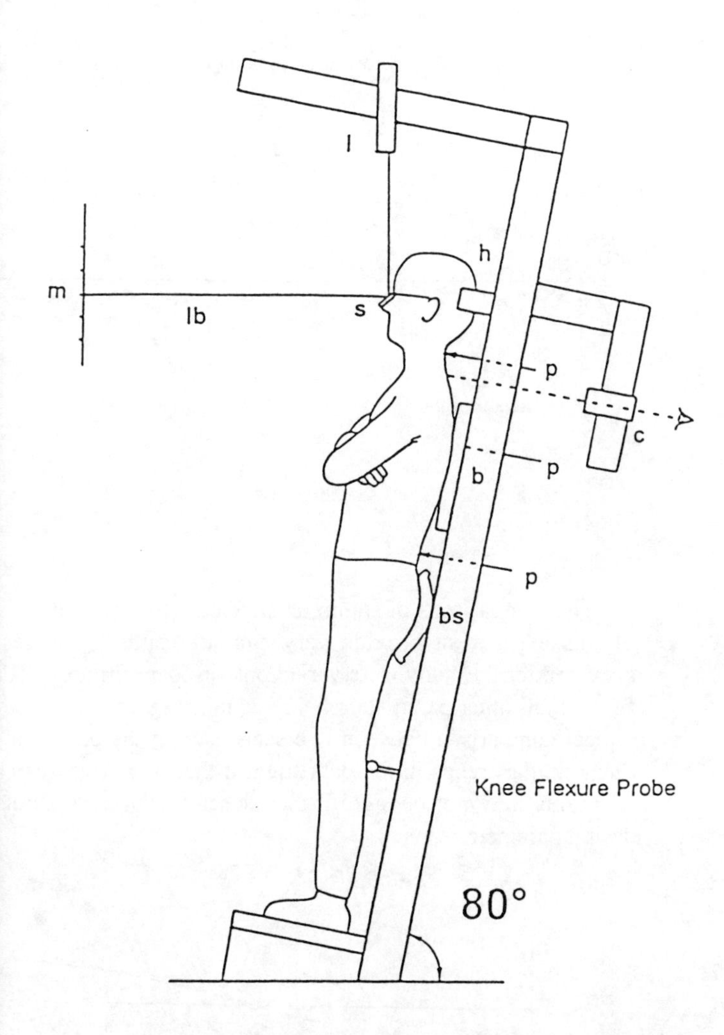

Figure 1. The apparatus used for precise measurement of stature. h: head support, bs: buttock support, p: probes to control spinal contour, b: back support plate, c: camera, l: laser, s: spectacle frame with mirror, lb: laser beam, m: marker to locate the reflected laser beam

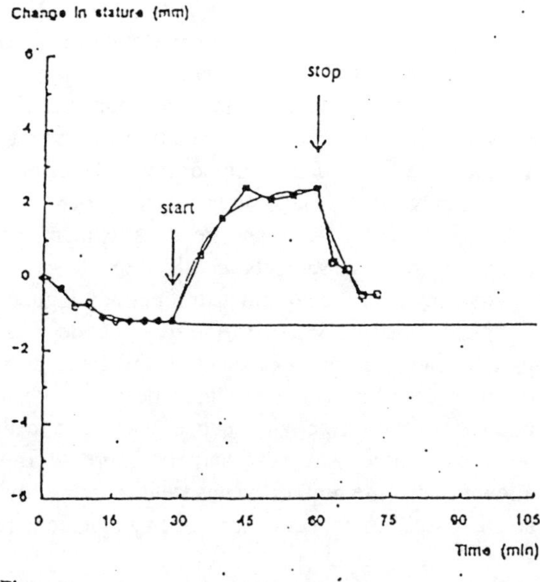

Figure 2

The test protocol was conducted in two phases. First, the test subject was pre-loaded by standing for a duration of 30 minutes. This procedure, adopted from Athoff's pretest protocol, subjects the spine to axial loading, and forces the "creep" trend of the spinal disks. Stature measurements were taken every 3 minutes throughout this phase to establish the creep trend. For the second phase, the test protocol adopts the same procedure as that of the EMG tests, and the two tests are run simultaneously. The subject adopts a seated posture for a total duration of 1 1/2 hours. Stature measurements are taken every 15 minutes to establish the recovery trend.

Data acquisition was accomplished via a Mitutoyo Digital Indicator / IBM XT Interface. Digital data output was fed into a PC and converted to ASCII readable format via a BASIC program. For each test, this data could then be curve fitted and extrapolated to a given point in time to establish the relationship between the creep and recovery trends (Figure 2). The result is a quantifying value which can be used for comparison purposes.

For more accurate comparison between the three experimental seats, stature change data was normalized according to spinal disk size. The change in stature is dependent on the size of the spinal disks, since larger disks will displace more fluid and produce more prominent changes. Diskal area was calculated with anthropometric measurements using the methods described by Colombini et al. This normalization helps eliminate the dependence of the data on the variation in test subject anthropometric factors.

Body Motion - This measurement technique was not experimentally pursued at the time of this study due to equipment problems. The motion tracking system that was utilized (The BIRD, interfaced with an IBM PC) allowed for digitization of X, Y, and Z coordinates of an object or point deformation. The plan was to use this system to measure seat deflection at contact points of the subject's back in order to determine the amount of movement or re-adjustment that occurred during a seated phase. However, the BIRD did not produce statistically repeatable results during initial checking of its measurement capabilities. No alternative motion tracking system has been investigated for reliability at the time of this study.

TEST RESULTS

EMG - Upon review of the test data obtained from this protocol, it is evident that EMG signals are not readily comparable among test subjects. EMG is governed by the body's physiological mechanisms. Therefore, a specific subject's muscle geometry will influence signal values, as will the events that preceded the acquisition of the signal. In other words, an EMG signal can be affected by glucose levels, diet, variation in sleep patterns, and levels of activity prior to the test. These parameters make EMG signals very difficult to control in a test environment.

In order to best relate the correlation of EMG signals to seat comfort, the change in signal data on a given day for a single electrode site per seat was studied. This method provides the most realistic simulation of an in-car situation without the introduction of considerable error due to the dependency on the large number of variables. From this analysis, a very important trend is evident. The signal intensity for test subjects in all three seats shows some consistency over the 90 minute test period. Figure 3 illustrates a typical consistency pattern for all electrode sites at each 15 minute data interval. According to this trend, it is likely that defining comfort with EMG may be related to a consistent intensity of muscle activity coupled with some median activity level over time. A seat may be more "forgiving" (i.e., allowing of some high or inconsistent muscle activity) as long as the muscle activity is congruous in all the body regions.

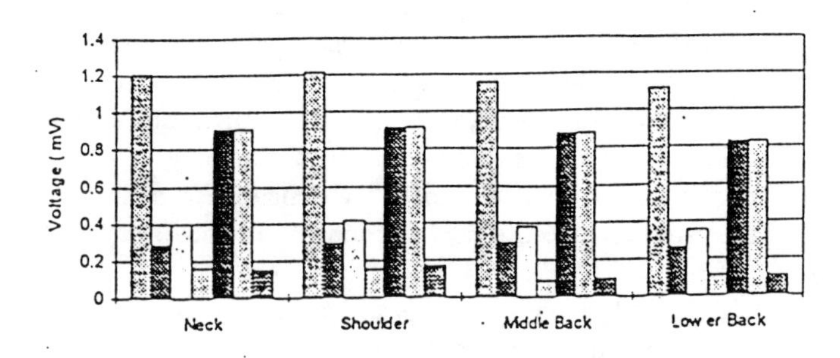

Figure 3

Since subjective surveys do not provide information regarding the physiological condition of a person at the time of the test, EMG data cannot be successfully correlated to subjective comfort data. Overall ratings in this experiment also provide little correlation due to the similar number of consistencies per seat. An additional study involving two seats with distinctively dissimilar parameters and a revised subjective survey would be required to provide more insight into the effectiveness of EMG as a measurement of comfort.

Spinal Loading - Figure 4 shows the typical creep and recovery trends for the stature change data of a single test subject for a given seat. It has been verified by this experiment that the curve fits to these trends produce a distinctive, measurable difference at some terminal time. This difference yields a quantifying number that is comparable among seats.

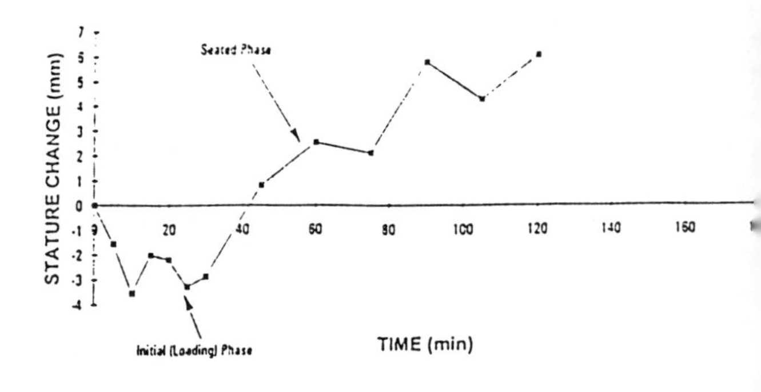

Figure 4

The comparisons of stature change data for the three seats of this experiment provide very similar results. The seats provided for the study were varied only in foam firmness (ILD #), with all other parameters being equal. Figure 5 shows the typical comparison between two seats used in the pilot study. Such similar trends in the data suggest that this measurement technique may not be significant for seats with very similar design parameters.

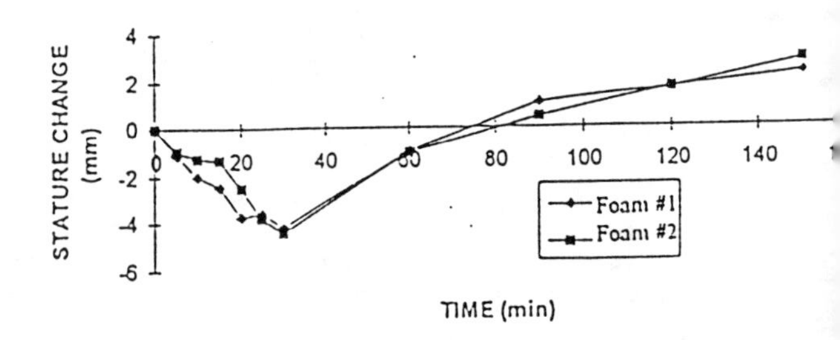

ILD(Foam #2) > ILD(Foam #1)

Figure 5

Stature change data did not directly correlate with subjective comfort data obtained in this experiment. However, the spinal loading method should not be discounted as an objective comfort measurable. The subjective surveys used in this experiment provided a discrete scale by which test subjects would rate comfort (Ref. 1). This type of scale does not distinguish the most critical or more forgiving subjects in a comparative analysis. Also, due to time constraints, the subject base was not large enough to provide average comfort ratings over a significant population. Therefore, a more relative scaling, with additional information regarding the physiological condition of the person at the time of the test, would be more appropriate for correlation.

Spinal load experimental data seems to provide the most straight forward, comparable data of all the techniques studied. However, there are some factors that have been found to affect the data. Since intervertibral disks are influenced by pressure, load, and physiologic concerns, the possibility for erroneous data exists for test subjects who may have disc damage or pre-loaded spines due to activity prior to the test. Data collected in these situations cannot be directly related to the comfort of a seat. For future investigation with this measurement technique, the initial creep trend of the spinal disks should be forced on the subject, perhaps by use of a weight vest worn during pre-loading. This would help discount the effects of any prior activity of the test subject.

COMPARISON

SEAT COMFORT OBJECTIVE MEASURABLES COMPARISON

	Pressure Distribution	EMG (ELECTROMYOGRAPH)	Spinal Loading	Body Motion
Test Time	10 min.	30 Min.	4-6 Hrs.	30 Min.
Static	YES	YES	YES	YES
Dynamic	YES	YES	N/A	YES
Short Term	YES	YES	N/A	YES
Long Term	YES	YES	YES	YES
Advantages	. Test procedure is simple and fast. . Pressure mats are transparent to occupants. . Test equipment is portable.	. Test procedure is more repeatable than pressure distribution. . Test equipment is portable.	. Test data is straight forward, no interpretation or base data required. . Test is repeatable.	. Test is simple and fast. . Test equipment can be made portable.
Disadvantages	. Human anatomy differs greatly. . Sitting postures are not totally repeatable.	. Each test subject needs to establish his own base EMG data and muscle activity pattern. . The sensors may add discomfort to occupants. (Test equipment interferes with test itself)	. A complete test takes 4-6 hours. . Test is limited in the lab only.	. Many people have habitual motions that will confuse the data.

TABLE I

CONCLUSION

From the table above and the data we gathered, we believe that spinal loading is the most promising objective measurable among the four in this study. It provides direct data, needs less interpretation or data manipulation to predict comfortable seats. However, to establish the design criteria and optimize the test procedures, further research is needed to develop this technique. The limitation of confining this technique within the lab environment will not be a concern once dynamic simulation is available in the lab. Due to the termination of this project, the dynamic phase of this development is not currently available.

ACKNOWLEDGMENTS:

Many thanks to Dr. King H. Yang and Mari Melosic of the Biomechanical Engineering Department of Wayne State University for their work in conducting this research and development project with us. Also many thanks to Joe Conti of Lear Seating for his contribution.

REFERENCES:

1. Reed, M.P., *An Investigation of Driver Discomfort and Related Seat Design Factors in Extended-Duration Driving,* SAE Technical Paper 910117, 1991.

2. Meada, M., *Studies on Evaluation of Vehicle Seats and Static-Ride Comfortability,* Japan Auto Parts Industries Association, Seat Technology Committee, 1989.

3. Lee, Janilla, and Ferraiuolo, Paul, *Seat Comfort,* SAE Technical Paper 930105, 1993.

4. Theysohn, H. and Greiff, H., *Study of the Reaction of Several Muscles While Driving A Car,* 1990.

5. Althoff, I., Brinckmann, P., et al, *An Improved Method of Stature Measurement for Quantitative Determination of Spinal Loadings.,* Spine 12:682, 1992.

6. Sullivan, A., McGill, S.M.,*Changes In Spine Length During and After Seated Whole-body Vibration,* Spine 15:1257, 1990,

950143

The Use of Electromyography for Seat Assessment and Comfort Evaluation

Tamara Reid Bush and Frank T. Mills
Michigan State Univ.

Kuntal Thakurta
Johnson Controls, Inc.

Robert P. Hubbard and Joseph Vorro
Michigan State Univ.

ABSTRACT:

A need to develop methodologies to obtain objective measurements of the effects of different seat contours on people is evident. In an effort to monitor muscle activity during static seated postures, electromyography (EMG) was employed. In an experimental setting, fatigue was induced in back extensor muscles for different seated postures. The resultant EMG signals were then sampled bilaterally for three different vertebral levels and the effects of the different seating systems on posture were evaluated.

In preliminary tests involving 4 subjects of similar size and build, utilizing three differently contoured seats, findings support the use of EMG to quantify muscular fatigue as a viable means of objectively measuring the effects of different seat contours.

INTRODUCTION:

In the past, the evaluation of seat design utilized subjective methods such as questionnaires filled out by the seated occupants. Recently, the automotive seating industry has begun to examine and use seating evaluation methodologies which provide objective, quantitative data [1]. The predominating philosophy is to combine objective and subjective measurement systems for the seat evaluation process, with this, it is anticipated that the task of designing "comfortable" seats will become easier and more efficient.

When an individual is in a seated posture, many physiologic events occur. These events include the flow of blood to transport metabolic products to and from localized areas; and muscular activity to maintain posture and permit movement. A properly designed seat is one in which the body would be supported with anatomical integrity and, in addition, promote normal body functions.

Johnson Controls Product Evaluation Group and a team composed of researchers from the Department of Anatomy and the Department of Materials Science and Mechanics at Michigan State University are exploring the use of surface electromyography (EMG) for seat evaluation. This modality monitors muscle activity in a non-invasive fashion. Specifically, surface EMG is being used to monitor the element of fatigue in paraspinal muscles (back extensors) during static seating situations involving three differently contoured seats. The fatigue data are then being evaluated between seats, and pressure mapping values, obtained by Johnson Controls Inc. (JCI), from the same seat contours are being used to aide in interpretation.

Several seat studies have been conducted employing the use of electromyography. In one such study performed by Lee and Ferraiuolo [2], EMG activity levels and pressure mapping data were correlated with subjective data for a static seated test. Their correlation between the subjective data and the electromyographic data was stated as being inconclusive.

Reed et al. [3] implemented EMG in a study pertaining to driver discomfort. They analyzed the electrical signals in a dynamic situation (a simulated driving test). These researchers were unable to come to a clear conclusion regarding total quantitative muscle activity levels during driving. Due to their low data sampling rate, however, a frequency analysis was not conducted on the data.

A third group studied the duration of muscle activity levels during a driving evaluation [4] and attempted to correlate these data with comfort. Greiff and Güth measured the occurrence of lasting tension (which they defined as more than 3% of a maximum tension), the occurrences were then counted if the duration was longer than 12 seconds. They hypothesized that low muscle activity levels or activity levels for short amounts of time meant the driver was comfortable. They found that lower activity levels occurred while individuals were driving with the cruise control on and therefore concluded the driver was more comfortable.

For the study discussed in this paper, static tests were performed evaluating muscle fatigue which was indicated by a shift in median frequency [5]. Muscle fatigue is not a new parameter of study. Many fatigue studies have been performed on a variety of muscle groups [5,6,7] including the paraspinals. From studies such as these, the shift in median frequency has been established as a viable means of measuring muscle fatigue.

METHODS:

TEST PREPARATION - The electromyographic data were recorded with pairs of disposable pregelled, self-adhering, silver/silver chloride electrodes, measuring 20 mm wide, 35 mm in length and 1 mm thick. These specific electrodes were chosen because of their thinness and ability to maintain adhesion for extended periods of time. The electrode pairs were attached in a bipolar configuration with a 10 mm separation [5].

The skin was prepared by shaving all visible hair, cleansing with an alcohol swab, lightly abrading the area with a gauze pad, and thoroughly drying the region. The electrode pairs were placed bilaterally at three vertebral levels, approximately 25 mm lateral to the spinous processes, and parallel to the muscle fiber direction. The three vertebral levels chosen for this study were the 5th and 6th thoracic vertebral level (T5-6), the 9th and 10th thoracic vertebral level (T9-10) and the 3rd and 4th lumbar vertebral level (L3-4). Prior to the experimental protocol, a brief summary of the testing procedures was given to each subject. After this summary, all subjects were asked to read and sign an informed consent agreement.

EXPERIMENTAL PROTOCOL - Each seat was placed in the design position, and electromyography was used to monitor muscle activity during contractions of the back extensors. Two different sets of tests were performed on separate days, one set involved seats 1 and 2 while the second set involved seats 2 and 3.

Seat 1 promoted a slumped posture with little lumbar support, and was termed a low contour seat. Seat 2 promoted an erect posture with a large amount of lumbar support and no regions of high shoulder pressure. This seat was termed a highly contoured seat. Seat 3 exhibited a small amount of shoulder pressure, a neutral amount of lumbar support and prominent lateral support in the torso region. This seat was termed as neutrally contoured. These classifications were derived from schematics of the seats and pressure mapping data provided by Johnson Controls.

The first test compared the low contour seat with the highly contoured seat, seats 1 and 2 respectively. The second test compared the same highly contoured seat with the neutrally contoured seat, seats 2 and respectively.

For each test, 4 subjects (2 men and 2 women) wi standing heights ranging from 1.68 to 1.73 meter sitting heights ranging from 0.85 to 0.94 meters ar weights ranging from 54 to 77 kilograms were teste Ideally the same four subjects were to be tested in eac of the two groups. Due to availability however, only 3 the 4 subjects were the same between the two te groups.

The seats were set in their design positions, and th subjects were allowed to move the seat forward and a to accommodate placement of the feet on the foot-plat No other adjustment of the seat position was allowe During testing, the pelvic region of each subject wa secured in the seat by the use of a seat belt.

DATA ACQUISITION - Load cells were used t monitor the amount of back extension force. Thre maximum voluntary contractions (MVC) tests wer performed on each subject for each seat to ascertain th individual's maximum potential extension force. T accomplish this, the subject extended into the seat wit a maximum force, and this force was then monitored b a load cell. The highest force value of the three tests fo each seat was selected as the representative MVC value for that specific seat.

For the first group of tests between seats 1 and 2 these values were used as the MVC values and 80% o each value was selected as the target testing value fo that seat.

For the second set of tests between seats 2 and 3, a comparison was made between the representative MVC values for that pair of seats. In an attempt to magnify the differences due to contour changes between the seats, one force value was chosen for both seats. Eighty percent of the lowest of the two representative values was selected as the target testing value for that pair of seats. The reason 80% of the lower MVC value was chosen instead of the overall higher MVC value, was to assure the subject would be able to maintain a contraction for 45 seconds for all six trials.

Each subject performed a 45 second fatigue test with the raw EMG data being sampled for 6 muscular sites at 2000 Hz. All data were collected using the Noraxon Myosystem 2000 and the Noraxon Spectral Analysis software. Data were collected when a subject reached and stabilized at the desired target MVC value. The starting seat for testing was alternated between subjects; so two of the subjects began testing in seat 1, and two of the subjects began testing in seat 2 for the first group of tests. The seats were alternated throughout the six test trials. It is noted that a random pattern of testing seat assignments would have been better, and future testing protocols will be conducted in this manner. After the first trial, subjects rested for 10 minutes, the seats were switched and then another trial was performed.

For the second set of tests between seats 2 and 3, a protocol identical to the one described above was followed.

DATA ANALYSIS - A dynamic median frequency analysis was performed on the first 40 seconds of the myoelectric data from each trial. This produced a median frequency value for approximately each second of data. Therefore the results from each test consisted of approximately 40 individual median frequencies. A linear regression and standard error analysis were then performed using the Excel software package.

RESULTS:

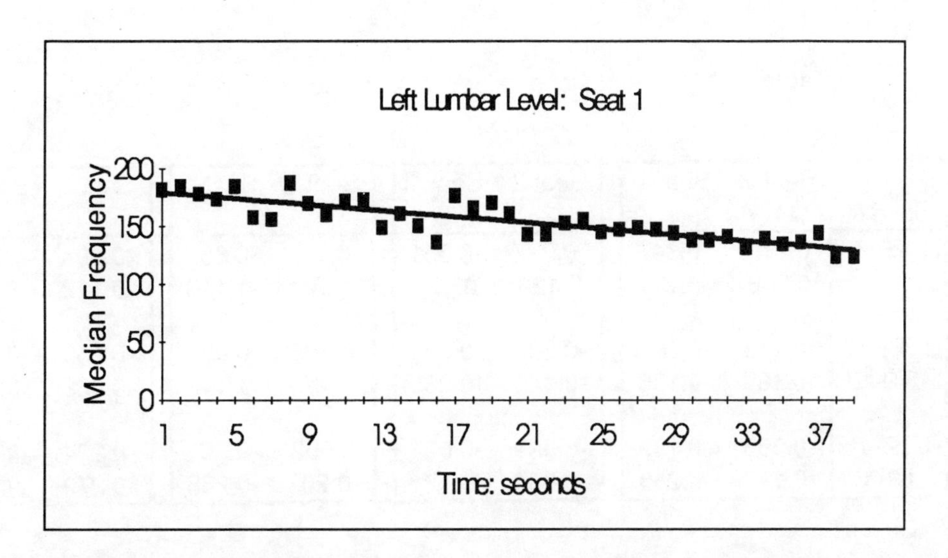

Figure 1: Linear regression line through median frequency data. Each frequency was obtained by a dynamic analysis of a window of data, with each window being approximately equal to one second.

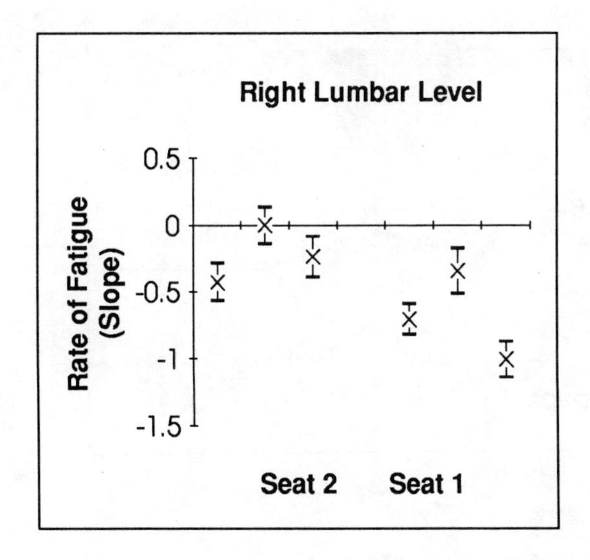

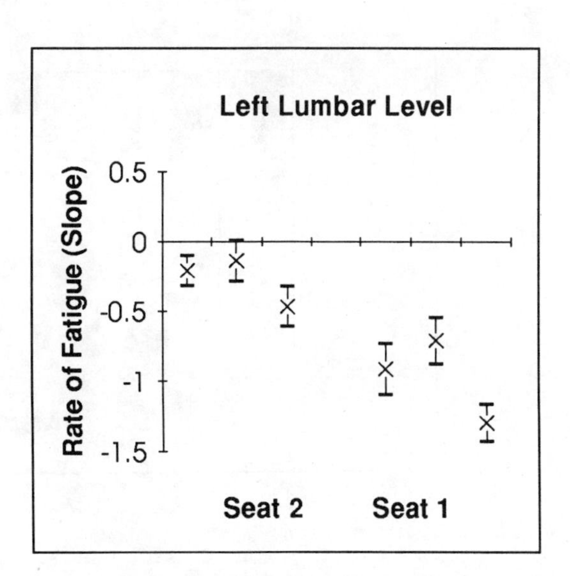

Figure 2a and 2b: Rate of fatigue or slope value of the regression line for one subject (all 6 trials) at the lumbar level, with ± standard error.

EXPLANATION OF THE SAMPLE DATA PLOTS - Figure 1 shows an example of median frequency data with a best fit line calculated by linear regression. This plot represents data acquired from the paraspinal muscles of the left 3rd and 4th lumbar vertebral levels. The downward slope indicated a shift in median frequency corresponding to muscle fatigue.

Figures 2a and 2b are examples of bilateral contraction profiles of the 3rd and 4th lumbar vertebral levels showing overall slopes of the median frequency (rate of fatigue) with a deviation range of plus and minu one standard error. The more negative the slope valu the faster the rate of fatigue. The first three points a the data for the three tests in seat 2 and the secor three points are the data for the three tests in seat The data points are grouped in order from the first to la trial for each seat. Each chart contains data from or muscle group for all six trials.

OVERALL ANALYSIS:

Table 1: Average of slope values for each subject for Seats 1 & 2.

Subject	Seat 2 1	Seat 1 1	Seat 2 2	Seat 1 2	Seat 2 3	Seat 1 3	Seat 2 4	Seat 1 4
R T5-6	0.270	-0.347	-0.227	-0.381	-0.709	-0.837	-0.095	-0.321
L T5-6	0.213	-0.121	0.126	0.071	-0.741	-1.119	-0.112	-0.109
R T9-10	-0.113	-0.367	-0.336	-0.324	0.303	-0.035	-0.158	-0.250
L T9-10	-0.187	-0.106	-0.141	-0.265	-0.221	0.337	-0.082	-0.539
R L3-4	-0.093	-0.139	-0.082	-0.092	0.002	-0.051	-0.224	-0.687
L L3-4	-0.147	-0.250	-0.064	-0.171	-0.203	-0.268	-0.270	-0.973

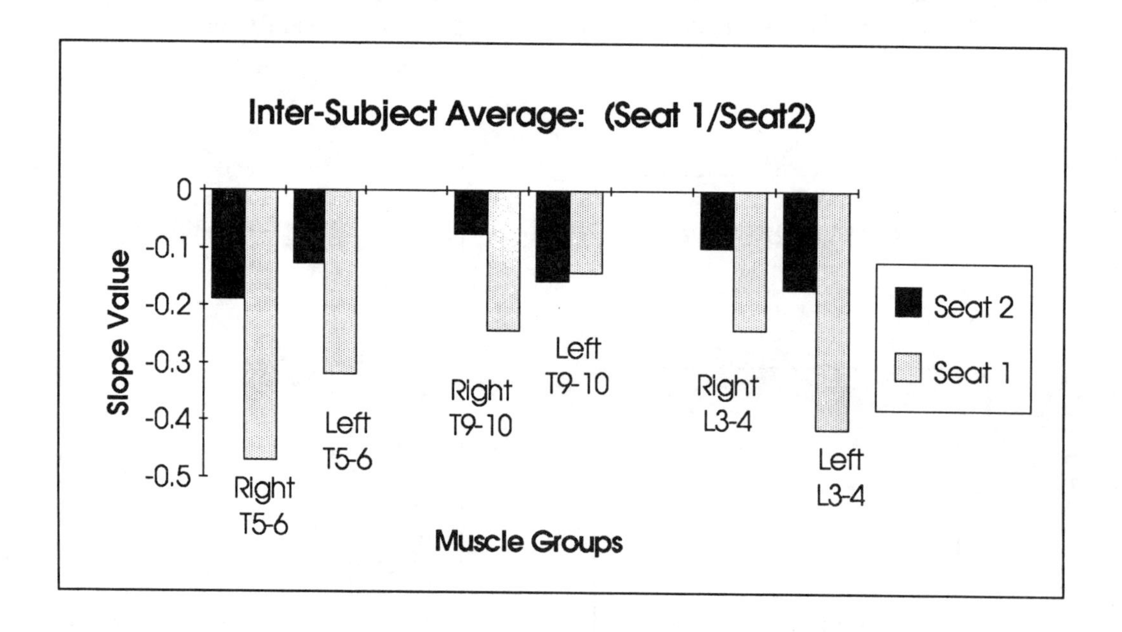

Figure 3: Combined average slope values of all subjects for each muscle group for seats 1 & 2.

TEST 1 (SEAT 1 AND SEAT 2) - The region which showed the largest effect of fatigue was the upper thoracic region (5th and 6th thoracic vertebral levels) and is demonstrated in Figure 3. For the right side of the body, all four subjects typically displayed a higher fatigue rate in seat 1 (low contour) as compared to seat 2 (high contour). When an overall average was taken, the subjects fatigued in seat 1 more than two times faster than seat 2.

The second largest seat related difference of the data occurred in the lumbar region of the subjects. Overall, the fatigue rate was more than two times higher in seat 1 than 2 and is demonstrated as such in Figure 3. This trend was seen bilaterally for all subjects, (Table 1).

Muscular activity in the mid-thoracic region (9th and 10th thoracic vertebral levels) showed the least amount of fatigue difference between seats. Figure 3 shows that when fatigue rates were examined as a whole, they did not differ appreciably on the left, but did on the right side. When examining individual subject data for the T9-10 region, it can be noted that the subjects did not consistently show a higher rate of fatigue in one seat, and were not consistent bilaterally as to the muscle response in each seat.

The inconsistencies between the left and right sides may be attributed in part to an unequal load application throughout the entire back region and/or to right handed dominance (consistent in all subjects)[8].

A two sample paired t-test of the means was performed on each vertebral region (right and left side values were combined). Both the T5-6 vertebral region and the L3-4 vertebral region were found to have significant differences between seats at a significance level of 0.05. For a one-tailed analysis, the values were as follows: T5-6 region: $p=0.01$, L3-4 region: $p=0.03$.

However, for a two-tailed analysis only the T5-6 vertebral region showed significance: T5-6 region: $p=0.01$, L3-4 region: $p=0.06$.

Table 2: Average of slope values for each subject for Seats 2 & 3.

Subject	Seat 2 1	Seat 3 1	Seat 2 2	Seat 3 2	Seat 2 3	Seat 3 3	Seat 2 4	Seat 3 4
R T5-6	-0.088	-0.176	0.053	0.091	0.012	0.148	-0.562	-0.228
L T5-6	-0.293	-0.295	-0.014	-0.212	-0.188	-0.199	-0.713	-0.233
R T9-10	-0.007	-0.110	-0.040	-0.120	-0.076	-0.050	-0.543	-0.316
L T9-10	-0.209	-0.028	-0.026	-0.059	-0.085	-0.193	-0.320	-0.304
R L3-4	-0.010	-0.051	0.020	0.156	0.004	-0.019	-0.013	0.058
L L3-4	0.033	-0.087	-0.004	0.019	-0.130	-0.043	0.169	0.132

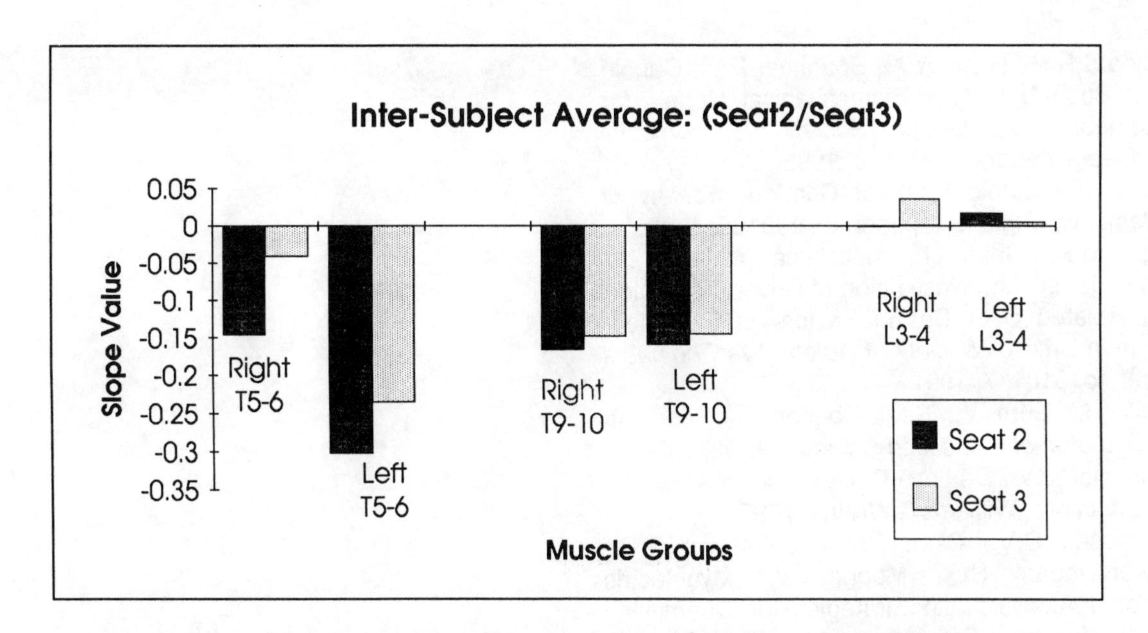

Figure 4: Combined average slope values of all subjects for each muscle group for seats 2 & 3.

TEST 2 (BETWEEN SEATS 2 AND 3) - A trend was not well defined in the data for seats 2 and 3 (highly contoured vs. neutrally contoured). When examining the individual subject data (Table 2) for the upper thoracic region, it can be seen that of the four subjects, one fatigued faster in seat 3, one fatigued faster in seat 1 and the other two had no consistent bilateral muscular performance.

For the mid-thoracic region, the overall average shows an equivalent distribution between seats with slightly lower fatigue rate exhibited in seat 3 (Figure 4). The lumbar region did demonstrate fatigue when the overall subject data was analyzed.

A two sample paired t-test of the means was performed on the three vertebral regions and no significant difference was shown at a significance level of 0.05.

CONCLUSIONS:

To develop objective measures of automotive seat design, a protocol involving induced muscle fatigue, measured by electromyography, was devised to examine effects of the seating system on posture. An analysis of the resultant spectral components of the myoelectric signals was made to compare with different seat designs. From these preliminary results it is evident that gross changes of the seating system, specifically contours affect posture, which in turn affect muscle activity and are detectable by EMG analysis. Further statistical analysis and a larger scale study will be necessary to verify this trend. It is possible, however, to theorize from this information that similar results would occur in a natural driving environment, over an extended time period.

REFERENCES:

1. Hubbard R.P., Haas W.A., Boughner R.L., Canole R.A., Bush N.J.: New Biomechanical Models for Automobile Seat Design. *Society of Automotive Engineers*. paper no. 930110, 1993.
2. Lee J., Ferraiuolo P.: Seat Comfort. *Society of Automotive Engineers*. paper no. 930105, 1993.
3. Reed M.P., Saito M., Kakishima Y., Lee N., Schneider L.: An Investigation of Driver Discomfort and Related Seat Design Factors in Extended-Duration Driving. *Society of Automotive Engineers* paper no. 910117, 1991.
4. Greiff H., Güth V.: Seat Comfort Studies with EMG. *Automotive Seating Review*. :38-42, 1994.
5. Basmajian J.V., De Luca C.J. : *Muscles Alive*, 5th ed. Baltimore, Williams & Wilkins, 1985.
6. Kondraske G.V., Carmichael T., Mayer T.G., Deivanayagam S., Mooney V.: Myoelectric Spectral Analysis and Strategies for Quantifying Trunk Muscular Fatigue. *Archives of Physical Medical Rehabilitation* 68:103-110, 1987.
7. Merletti R., Lo Conte L.R., Orizio C.: Indices o Muscle Fatigue. *Journal of Electromyography an Kinesiology*. 1: 20-33, 1991.
8. Biedermann H.J., Shanks G.L., Inglis J. : Mediar Frequency .Estimates of Paraspinal Muscles Reliability Analysis. *Electromyography Clinica Neurophysiology* 30:83-88, 1990.

950144

Evaluating Short and Long Term Seating Comfort

Kuntal Thakurta, Daniel Koester, Neil Bush, and Susan Bachle
Johnson Controls, Inc.

ABSTRACT

This paper reports the results of a study comparing the subjective assessment of short and long duration sitting comfort associated with an 80 mile highway drive. Thirty-six subjects evaluated five cars selected from the small car market segment. Subjects completed a comfort assessment questionnaire and were pressure mapped before and after the ride and drive. Subjective evaluation of several seat zones were analyzed and compared to pressure readings. The results for four zones including shoulder, lumbar, ischial tuberosity, and thigh areas are discussed.

INTRODUCTION

The design of automotive seating for improved occupant comfort is one of the primary goals for seat system engineering teams. Comfort measurement is difficult because of such factors as user subjectivity, occupant anthropometry, seat geometry, and amount of time spent sitting. The methodology for evaluating subjective comfort is achieved through highly structured questionnaires that direct the occupant to assign feelings of discomfort to specific regions of the seat.

Objective measures, such as seat interface pressure, are also used to give additional insight into the occupant/seat interface. However, there are limitations in the use of this type of objective measurement for determining occupant seating comfort. Two major drawbacks of the systems for recording human-seat interface pressures are hysteresis and creep [1].

These shortcomings affect the accuracy of the results obtained during long duration sitting pressure measurement. In a study involving 100 participants and seven different seat design parameters, the authors reported no correlation between subjective (human) and objective (pressure) measurements [2]. Another study involving 140 participants showed strong correlation for the averaged vehicle data, but showed weak correlation between individual respondent's scores and their pressure distributions [3]. In both studies, subjective occupant comfort was evaluated with seat pressure distributions under showroom conditions, hence no conclusions were derived regarding long term or dynamic seating comfort.

A primary goal of this project was to investigate whether the current pressure measurement tools can be used in conjunction with subjective questionnaires to evaluate and predict long term comfort.

METHODS and MATERIALS

SUBJECTS: Thirty six subjects were recruited from within Johnson Controls to participate in this study. The subjects consisted of nine females and twenty-seven males. Basic anthropometric measurements were made for each subject. The table below shows the mean subject anthropometries and demographics.

Table 1: Subject information including means and standard deviations.

	Mean	S.D.
Age (years)	30.4	9.2
Standing Height (mm)	177.2	10.7
Weight (kg)	75.6	16.3

EQUIPMENT: Five vehicles were selected from the small car market segment. For the purpose of this paper they are going to be denoted by numbers 1 through 5. Each of the test vehicles were mid-level equipped with manual seat adjusters and cloth seats. Three vehicles of each type were used such that three subjects were on the road at one time such and four sets of subjects could be run in one day. Subjects were scheduled to drive at the same time and day of the week for each vehicle in order to minimize any effect in their subjective comfort ratings due to time and environment variations.

PRESSURE MEASURING SYSTEM: The pressure distribution system consists of an array of 44 rows by 48 columns of force sensitive thin flexible sensors which measure pressures at 10 mm intervals. By attaching these thin flexible mats to seat backs and seat cushions, information regarding pressure magnitudes and locations can be obtained both graphically and numerically. Figure 1 shows a typical map of the pressure distribution that is created in real time during the test. The system advantages include data resolution, high speed data collection, real time displays, and portability. With the help of this pressure distribution system, Johnson Controls has developed an objective tool, which not only determines body pressure magnitudes at pre-defined zones of the seat, but also evaluates body pressure distributions at occupant-seat interface areas. Figure 2 is a load distribution pie chart that indicates the percentage of load at defined zones at the occupant-seat interface.

For the current study, two pressure maps of each subject with the seat adjusted to the subject's driving comfort position were taken, one before and one after the ride and drive. The mats were removed after showroom data collection and care was taken so that the mats were consistently placed in the same locations after the 80 mile drive to record subject's post ride pressure distributions. The pressure mats were slid under the occupant while in the vehicle to minimize occupant movements and changes in anatomical posture after the 80 mile ride and drive.

QUESTIONNAIRE: A two-part subjective questionnaire was used to guide subject evaluation before and after an 80 mile ride and drive for each vehicle seat. The questionnaire utilized a ten point scale along with descriptive words. The first part of the questionnaire consisted of a total of 42 questions involving different attributes of the seat for determining showroom comfort. The second part of the questionnaire consisted of 48 questions involving different seat attributes to determine occupant long term comfort after the 80 mile drive.

PROTOCOL: The schedule of subject activities took less than two hours for each vehicle as follows :

Step 1: Anthropometric measurements (taken one week prior to the ride and drives)
Step 2: Pressure map with seat in driving/comfort position
Step 3: Initial seat comfort questionnaire completion inside the vehicle
Step 4: 80 mile highway ride and drive
Step 5: Post drive seat comfort questionnaire
Step 6: Post drive pressure map in driving/comfort position

DESIGN OF EXPERIMENT: To understand the effect of different variables on occupant seat comfort, multiple variables were explored; specifically lumbar support, thigh support, ischial support and shoulder support. Subjective ratings obtained from the thirty-six participants, in the four categories for both pre and post evaluations, were compared to the corresponding distribution of pressure in these four categories. Comparisons were also investigated for overall seat comfort for each of the five vehicles.

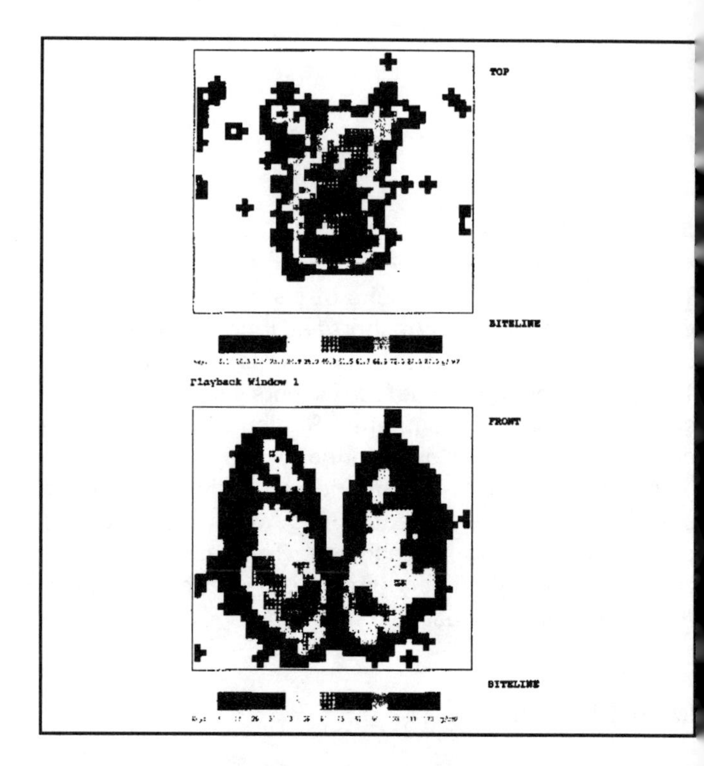

Figure 1: Pressure Distribution. Reprinted [4].

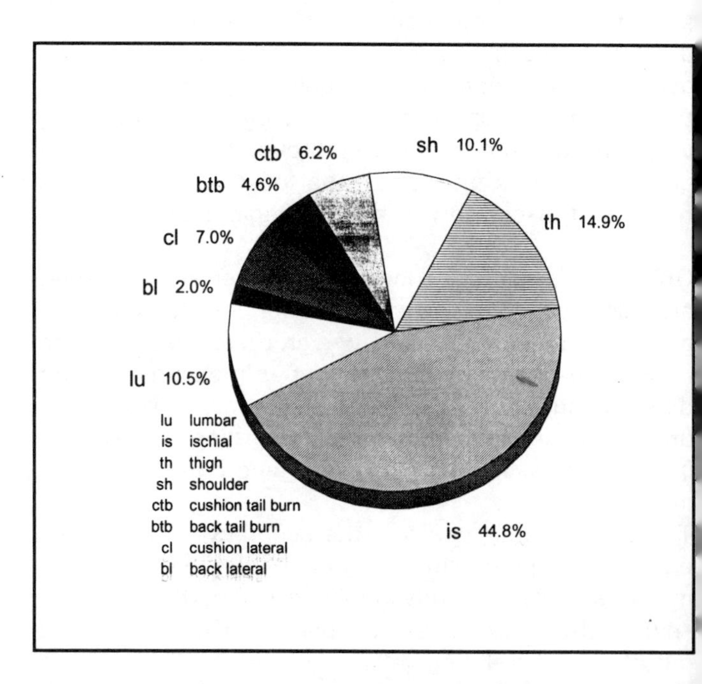

Figure 2: Load Distribution. Reprinted [4].

RESULTS

Analysis of the subjective data before (Pre) and after the ride and drive (Post) was done comparing overall comfort with the comfort within each of the body areas being considered. Figure 3 shows the percent of subjects who either reported improved comfort, no change, or decreased comfort.

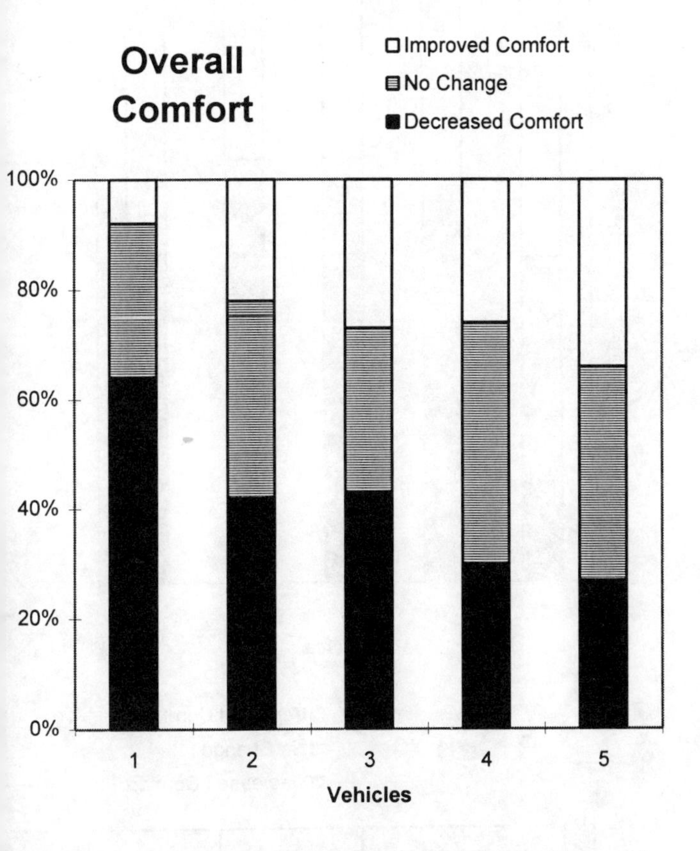

Figure 3: Overall comfort change before and after the ride and drive as a percent of total subjects.

The chart shows that the largest percentage of the subjects 64% experienced decreased comfort in Vehicle 1 while the lowest percentage experiencing decreased comfort was in Vehicle 5 at 27%. It was interesting to note that the subject's percentage of decreased comfort showed a decrease from Vehicle 1 to Vehicle 5 while percentage of improved comfort showed a gradual increase over the same time span from Vehicle 1 to Vehicle 5. This may be due to the fact that with increased time spent in the vehicles, occupants have a tendency to be less critical about their perceived comfort. The percentage of subjects who reported no change was relatively constant across all vehicles ranging from 28 % in Vehicle 1 to 44 % in Vehicle 4. It is somewhat surprising that a percentage of the subjects, from 8% to 34%, rated the comfort of the seats higher after the ride and drive. Figure 4 shows the change in comfort for the seat zones being analyzed.

Analysis was done in order to compare the correlation of lumbar and shoulder support to occupant's overall back comfort and similarly to compare ischial and thigh support to occupant's overall cushion comfort. Tables 2 & 3 show the highest and lowest correlation values (r^2) obtained in each category for the five different vehicles.

Table 2 - Correlation values (r^2) for showroom and long term, highest subjective comfort within seat zone

	Showroom Comfort Pre		Long Term Comfort Post	
	Veh.	Value	Veh.	Value
Shoulder Support	5	0.77	2	0.53
Lumbar Support	3	0.82	3	0.79
Ischial Support	5	0.89	5	0.77
Thigh Support	4	0.67	4	0.59

Table 3 - Correlation values (r^2) for showroom and long term, lowest subjective comfort within seat zone

	Showroom Comfort Pre		Long Term Comfort Post	
	Veh.	Value	Veh.	Value
Shoulder Support	2	0.43	5	0.10
Lumbar Support	2	0.00	5	0.34
Ischial Support	2	0.32	2	0.39
Thigh Support	2	0.10	2	0.04

No strong correlation was found for any specific vehicle for either showroom or long term comfort in the four categories evaluated. However, it was interesting to observe that vehicles 3, 5 and 4 showed the highest correlation for both showroom and long term across the factors of lumbar support, ischial support and thigh support respectively. Likewise, Vehicle 2 appeared to show the lowest correlation for both showroom and long term regarding ischial and thigh support. Even though vehicle 2 showed the lowest correlation between the above factors and overall back and cushion comfort, it was rated as best for overall showroom comfort.

PRESSURE DISTRIBUTION: In order to correlate subjective responses for each of the four variables to the distribution of pressures, the pressure mat was divided into zones. These zones were based on analysis of pressure distributions for different occupants and on occupant anthropometric dimensions. The data obtained from the pressure mats were recorded in an ASCII format and percentage distributions of pressures in each area were compared to the subjective data. For the statistical analysis, two additional variables were considered, type of vehicle and the time of measurement i.e showroom vs. long term. Table 4 shows the means and standard deviations of the results from the pressure distributions after the 80 mile drive.

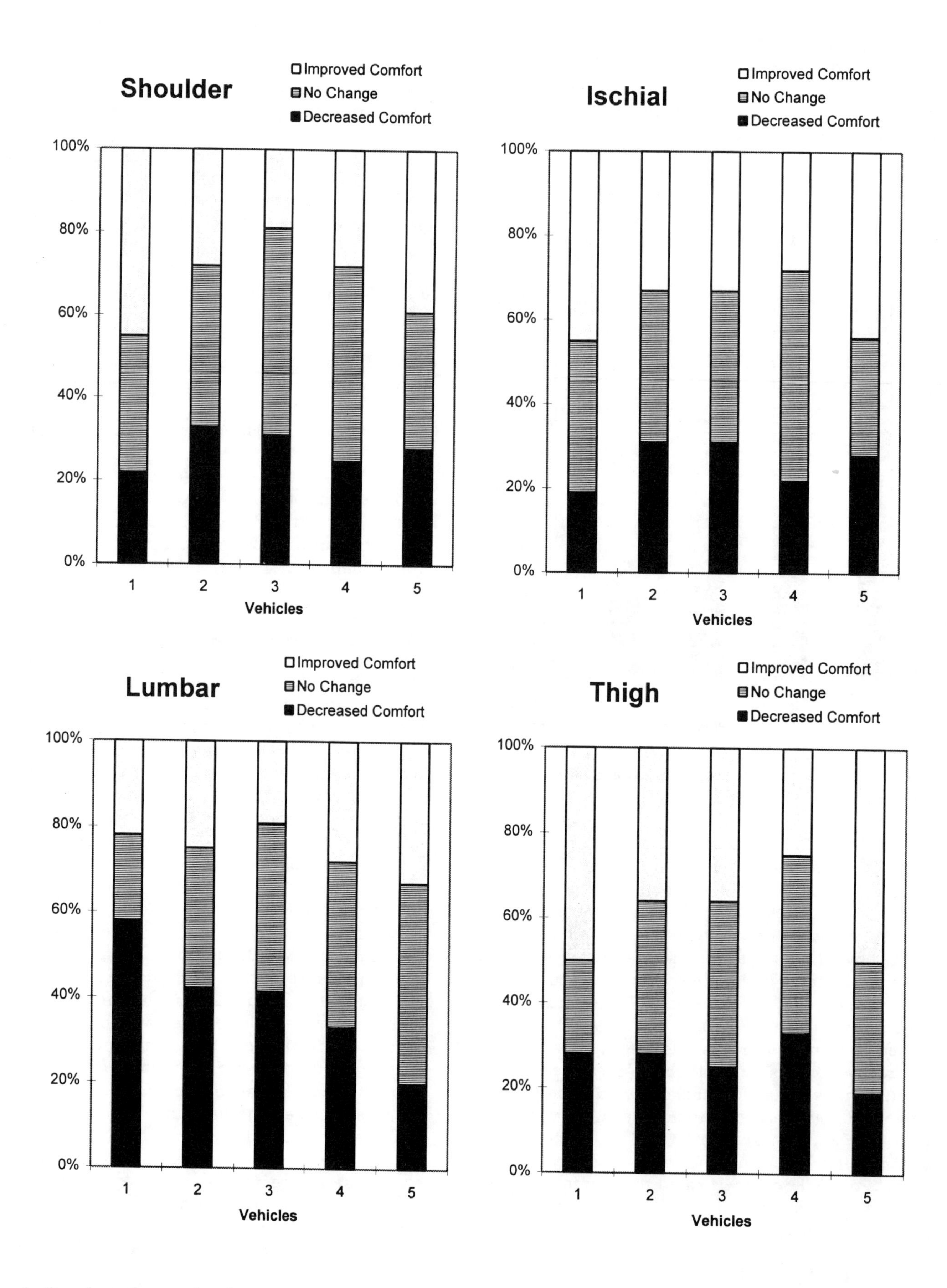

Figure 4: Comfort change for four seat zones before and after ride and drive as a percent of total subjects.

Table 4 - Summary of pressure distribution results after the 80 mile drive

Veh.	Lumbar Mean (SD)	Shoulder Mean(SD)	Thigh Mean(SD)	Ischial Mean(SD)
1	7 (3)	7 (3)	15 (5)	45 (10)
2	9 (3)	10 (3)	13 (5)	45 (10)
3	10 (3)	10 (3)	10 (4)	46 (5)
4	9 (3)	8 (3)	12 (5)	47 (10)
5	10 (4)	9 (4)	11 (5)	42 (7)

The mean values shown above are percentages of occupant's load distribution, while the numbers in the brackets show the standard deviations. The results obtained from the pressure distributions were quite consistent for all the five vehicles as shown in the table above. However, subsequent analysis of the results from the pressure distributions, the type of seat (vehicle), the time of measurement and occupant's subjective response yielded the following results :

- The effect of distribution of pressure to subjective comfort is significant. Lumbar support and Ischial support appeared to be more significant than shoulder and thigh support.

- The effect of time of measurement is significant.

- There is significant difference between the five seats.

It should be observed that these results are preliminary. However, they suggest that seat comfort is not linearly dependent on any variable, but is in fact a complex non-linear interaction of different variables. The study also showed the importance of measuring overall seat comfort in both showroom and long term sitting conditions. Using only one factor may lead to premature and incomplete conclusions about occupant seating comfort. Better understanding of the different variables and their interactions is needed in the future.

CONCLUSIONS

The current study was designed to explore occupant differences in comfort for short term and long term. The results provided in this paper are from a small part of the study and preliminary in nature. These results suggest the non-linearity and complexity of determining occupant comfort. At this time further research is under way at Johnson Controls, Inc. to determine the inter-relationships between different variables like body pressure distribution, posture, time of measurements, occupant demographics, seat attributes and features, etc to develop better mathematical models to measure seat comfort.

ACKNOWLEDGEMENTS

The authors wish to acknowledge the following people for their help during the course of this study: Thomas Moco and Baso Shelton for conducting the drives and helping in the data collection and subject recruitment and measurements; Joe Woelfel and Scott Danielson from Terra Research for helping with the statistical analysis; To all the thirty-six participants from Johnson Controls, Inc. for spending the time and effort to help out with study.

REFERENCES

1. Ferguson-Pell, M., Cardi, M.D. "Prototype Development And Comparative Evaluation Of Wheelchair Pressure Mapping System", Applied Research. Assist Technol. Vol 5, Pg 78-91, 1993.

2. Lee, J., Ferraiuolo, P.G. "Seat Comfort", Society Of Automotive Engineers, 930105, (1993).

3. Frusti, T.M., Hoffman, D.J. "QuantifyingThe Comfortable Seat Developing Measurable Parameters Relating To Subjective Comfort". Automotive Body Interior And Safety Systems. IBEC 1994.

4. Kumar, A., Bush, N., Thakurta, K. "Characterization Of Occupant Comfort In Automotive Seats. Automotive Body Interior And Safety Systems. IBEC 1994.

950146

Seat System Fatigue Test

Russ Davidson, Janilla Lee, and Ed Pan
Lear Seating Corp.

ABSTRACT

Currently one OEM requires seat systems to pass a fatigue test consisting of a varying load applied to the seat back for a large number of cycles. The test was derived from a seat occupant study documented in SAE paper 840509, Passenger Car Seating Loads; A Human Factors Engineering Problem. This test takes an average of two weeks to complete for a given program. The purpose of our study is to develop an equivalent fatigue test that shortens the required testing time and provides the same results and confidence.

INTRODUCTION

All materials and parts have a life. A part that fails after a period of time or during the part's functional use is known as a fatigued part. The part was structurally sound to perform the part design function once but not throughout the expected life. Fatigue life depends on material and the part environment. An improper design can induce fatigue when there is excessive stress at some points within the part. Fatigue is a complicated subject, one of the gray areas in engineering.

In automotive seating, there have been more recalls due to fatigue failure than any other reason. The rationale for the failure is the difficulties in determining fatigue criteria for a complicated system like an automotive seat and in providing proper testing. The testing criteria should be correlated to field data to provide the proper product validation.

In working with an OEM on seat design, there are several different fatigue tests in the product validation process. The fatigue tests were developed like most tests to duplicate past design failures. New programs naturally pick up the developed test as a requirement to prevent the past design failure in the new program. As more programs have been developed, a larger number of test requirements have also developed. Since the new requirement has history, known or unknown, there is a natural reluctance to delete or change any test requirement.

As stated, changing an OEM required test can be a difficult task. For the majority of tests required for seats, there is not a desire to change because the tests are simple and direct. Fatigue tests, on the other hand, are very time consuming. Compressed program development has made time a valuable commodity. Every step is being scrutinized for added value. If non-value added testing can be identified and determined redundant, there should exist some willingness to change that test requirement.

BACKGROUND

Current OEM system fatigue testing was developed from SAE paper Passenger Car Seating Loads; A Human Factors Engineering Problem[1]. In this study a typical bucket seat was instrumented to interpret back loads applied to the seat. While the vehicle was stationary and during driving, test occupants were asked to use the seat and do various non driving tasks (i.e. get ones wallet, straighten coat, etc.). Resultant seat back loads were tabulated for a variety of occupant sizes doing the various tasks. The study derived extraordinary loads applied to a seating system through various occupant movements. The current OEM cycle test used the tabulated information to develop an extraordinary load fatigue test for seating systems.

The fatigue test consists of 28 different rearward loads. Each load is applied to the seating system for a specific number of cycles to reproduce the loading observed during the SAE study. The repetitive pattern, defined as one block, consisted of 54 cycles made up of the 28 different rearward loads. The test consists of 2264 blocks totaling 122,256 cycles. The variety of loads range from 156 N to 1250 N each having a total cycle count anywhere from 2264 to 9056 as outlined in Table 1.

The loads are applied to the seat back 406 mm above 'H' point and 45 degrees off vertical. Considering that most seat back angles are 20 to 25 degrees off vertical, the load angle is 65 to 70 degrees to the back.

The current difficulties with the existing test are the duration of the test and non-perpendicularity of the loading angle. The current test takes 34 hours to complete one sample. The typical sample size is six, resulting in 8.5 continuous test days.

The primary reason for the long test time is the large number of cycles. Since the cycles are not constant amplitude, the question arises of how significant are the loads that are less than the maximum? For example the highest load of 1250N is applied once per block for a total of 2264 cycles. Similarly, the minimum load of 156N is also applied once per block for 2264 total cycles. The question that needs to be answered is how much damage does the 156N load causes compared to the 1250N load? If we can equate the damage caused by the 156N load at 2264 cycles to 1250N at 'X' number of cycles (with 'X" << 2264), a great deal of time could be saved. If this comparison holds for the extreme loads, all other loads and corresponding cycles should be able to be equated to the maximum load.

The non-perpendicularity causes difficulty in load application. Large clamps are required to hold the load cylinder to the seat back. Due to the oscillatory load and the 20 to 25 degree component of the load due to the non-perpendicularity of the loading, the clamps tend to slide down the seat back during testing (especially on tubular seat backs). The goal of this study is to develop an equivalent test with fewer number of cycles and constant perpendicular loading which will significantly improve product development time and cost.

THEORY

The following relationships are going to yield some specific values relating the current test to fatigue theories. Current fatigue analysis is not an exact science and should not be used as specific design practice without extensive verification testing and/or repeated inspection. The specific values were obtained from what information is readily available and is intended to reflect the general intent of this study.

All fatigue failures can be related to static strengths through a variety of experimental curve-fit relations (S-N Diagrams). Data obtained from rotating specimen tests show steel has a typical cycle life of 10^3 at a stress loading equaling eight-tenths ultimate strength (see Figure 1). The data curve continues to a cycle life of 10^6 at a stress loading equaling one-half ultimate strength. Stress loadings lower than the fatigue strength at 10^6 have infinite life for ferrous materials. Assuming complete reversed bending and ignoring the effects of stress concentrations, this is a very generalized relationship of fatigue behavior in steels,. The focus is that high stress levels require fewer cycles to cause failure. Likewise, lower stress levels require extremely larger cycle counts to cause failure.

The S-N diagrams work well for repeated sinusoidal stress of constant amplitude. However, constant amplitude for the life of an element is a rarity. Oscillatory (amplitude) loadings can be related using Miner's Law to predict failure[2,3].

$$n_1/N_1 + n_2/N_2 + n_3/N_3 \dots = K$$

Where n_n is the number of cycles at a particular load causing a particular stress, and N_n is the number of cycle at that particular stress to cause fatigue failure. K is a constant ranging from .5 to 2.5. Therefore, each stress load can be treated independently as a percentage of the total fatigue life or lost life (relating K to unity i.e. $n_1/N/K$).

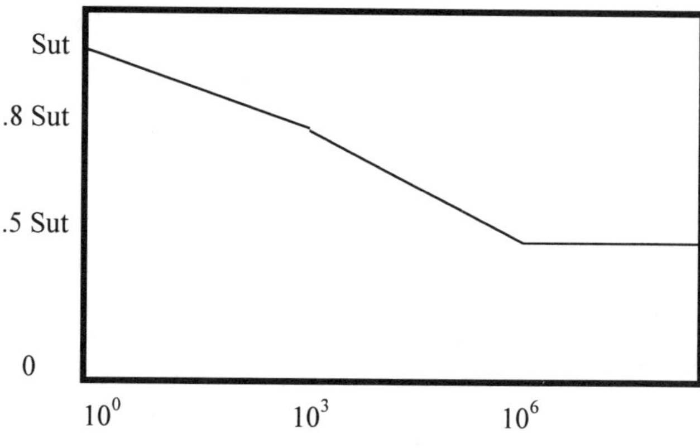

Figure 1 Stress versus Cycles

To analyze any structure directly, one needs to quantify the fundamental characteristics that define stress: section moment of inertia, moment at that section caused by the loading, and material structural strength. Given a particular loading condition, the weakest point in a structure can be identified. The structures fatigue life is dependent on the strength of the weak section and the magnitude of the applied load. The weak section will remain the weak section relative to the remaining structure as long as the attitude and locations of the load remains the same. The magnitude the load directly affects the life of the structures.

For the seats in this study assume the weak point has been identified. Let the section have the following properties and loading:

Section Moment of Inertia(I) = $1.09 \times 10^{-8}\ m^4$
Material AISI 4340 S_{ut}=1035 MPa
Section distance from neutral axis to outer edge(c) = 12 mm

Load(F) = 156 N to 1250N (See Table 1)
Perpendicular Distance from section neutral axis to Load (d) = 494 mm
Bending Moment = F x d

Therefore the stress in the section is:

$$\sigma = Fdc/I$$

$$\sigma = 680 \text{ MPa for a load of } 1250 \text{ N}$$

Using these assumptions, stress levels can be calculated for all corresponding loadings of the fatigue test (See Table 1). From the stress levels developed, percentage lost life can be calculated for each loading (see Figure 2.) to find cycle to failure at a particular load (N). Fatigue data for 4340 material was chosen purely due to information availability. The vertical line is used for 'R ratio' equal to zero, since the stress returns to zero between each loading ($R = \sigma_{min} / \sigma_{max}$). The percentage of lost life is tabulated for each load of the subject fatigue test, assuming the seat will fail on the last cycle of the test (see Table 1).

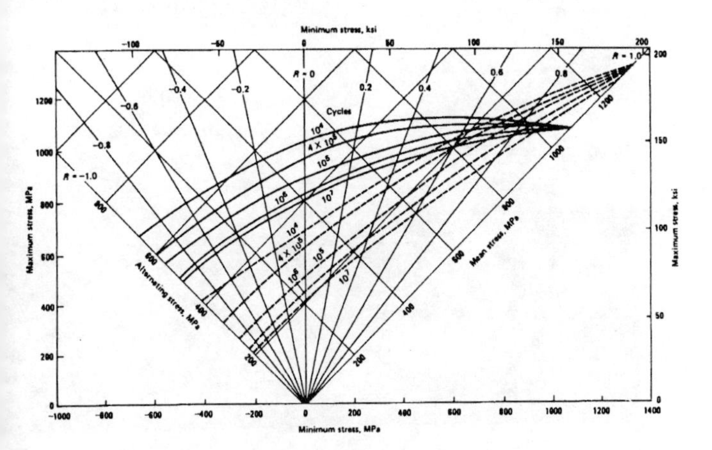

Figure 2. [4] Master Diagram for AISI 4340
Dashed lines represent data obtained from notched specimems, Kt=3.3

In Table 1 all insignificant loads are shaded. Insignificant loads are defined as their sum equaling less than 3 % lost life. Summing the remaining load cycles, now defined as significant loads, yields 20,376 significant cycles, less than twenty percent of the total cycle count. In theory a test run using only the significant loads and corresponding cycles would be .97 equivalent to the current OEM fatigue test. The significant loads still total six separate loads. The goal is to have a test with one load and cycles equaling the current test.

Therefore, if the maximum load case (1250) is used, as the only loading to be applied to the seat, the number of cycles can be calculated from the percentage of lost life shown in Table 1. If 2264 cycles depleted 26.26% of the life of the part, then 100% of the part's life will be reached at 8621 cycles as shown below.

$$2264 \text{ cycles}/.2626 = 8621 \text{ cycles}$$

Our hypothesis is that a test applying a load of 1250N for 8621 cycles should duplicate the effect of the current test.

Table 1 Percentage Lost Life of Variable Fatigue Loading

Load #	Cycles/ Block	Total Cycles	Sigma (MPa)	N 10^4	n/N	%Life	% Life
668	3	6,792	363	1,000	6.79E-03	0.079	
757	4	9,056	412	100	9.06E-02	1.050	
690	2	4,528	375	1,000	4.53E-03	0.053	
1,104	1	2,264	601	5	4.53E-01		5.252
521	1	2,264	283	1,000	2.26E-03	0.026	
748	2	4,528	407	1,000	4.53E-03	0.053	
165	3	,792	90	1,000	6.79E-03	0.079	
530	1	2,264	288	1,000	2.26E-03	0.026	
788	1	2,264	429	100	2.26E-02	0.263	
1,250	1	2,264	680	1	2.26E-00		26.261
672	2	4,528	366	1,000	4.53E-03	0.053	
369	4	9,056	201	1,000	9.06E-03	0.105	
156	1	2,264	85	1,000	2.26E-03	0.026	
610	1	2,264	332	1,000	2.26E-03	0.026	
783	2	4,528	426	100	4.53E-02	0.525	
743	3	6,792	404	1,000	6.79E-03	0.079	
966	4	9,056	526	8	1.13E-00		13.130
525	1	2,264	286	1,000	2.26E-03	0.026	
1,233	1	2,264	671	1	2.26E-00		26.261
427	3	6,792	232	1,000	6.79E-03	0.079	
160	2	4,528	87	1,000	4.53E-03	0.053	
752	1	2,264	409	1,000	2.26E-03	0.026	
432	3	6,792	235	1,000	6.79E-03	0.079	
1,206	1	2,264	656	2	1.13E-00		13.130
676	2	4,528	368	1,000	4.53E-03	0.053	
374	2	4,528	203	1,000	4.53E-03	0.053	
223	1	2,264	121	1,000	2.26E-03	0.026	
961	1	2,264	523	2	1.13E-00		13.130
Total	54	122,256			0.86	2.836	97.16

EXPERIMENT

Four OEM production seat test specimens were obtained. Seat I was tested to failure using the current OEM fatigue test with a live floor. Seat II was tested to failure just as Seat one except on a hard bed. Seat III was tested to failure using the maximum load (1250 N) in the same setup as Seat I except on a hardbed. Seat IV was tested to failure using the component of the maximum load (1113 N perpendicular to the seat back) on a hardbed. Failure is define as the first cycle not to take full load.

RESULTS

Test Sample	Cycles to Failure	Observations
Seat I Variable Load Live Floor	81,140	The outboard rear track to floor mounting bracket fractured completely at the section through the attaching rivet (See Figure 3.). No apparent damage to the front outboard track bracket.
Seat II Variable Load Hardbed	24,948	Test halted due to excessive seat back deflection not allowing maximum load. The front outboard track bracket had radial cracks of 6 mm long starting from the slot forward and of 8 mm long starting from the slot rearward (See Figure 4.). The outboard rear track to floor mounting bracket had cracks 15 mm long inboard and 18 mm long outboard approaching the attachment rivet (Figure 5.).
Seat III 1250 N Load Hardbed	6,061	The front outboard track bracket separated at the mounting bolt similar to the cracks found in Seat II (See Figure 6). The outboard rear track to floor mounting bracket had cracks 13 mm long inboard and outboard approaching the attachment rivet (See Figure 7).
Seat IV 1113 N Perpendicular Load Hardbed	7,463	The front outboard track bracket separated at the mounting bolt. The outboard rear track to floor mounting bracket had cracks 13 mm long inboard and outboard approaching the attachment rivet.

Figure 3. Seat I Rear Track Bracket Separation

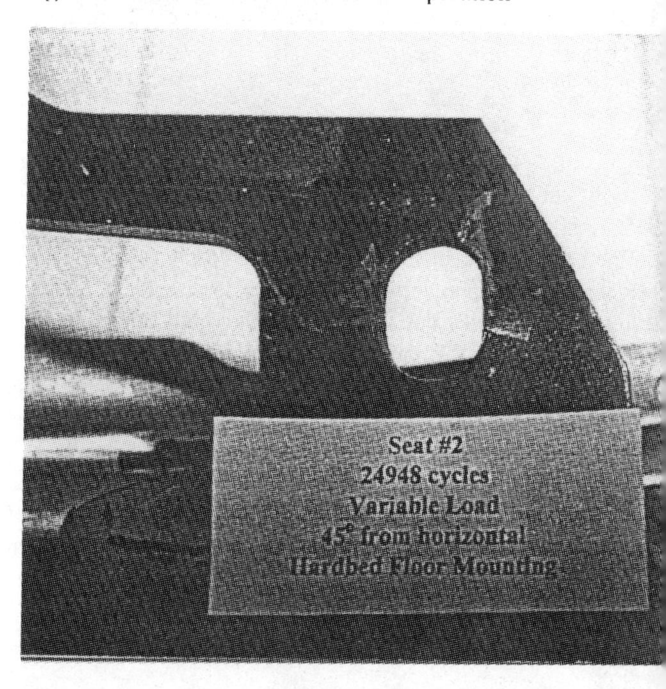

Figure 4. Seat II Front Track Bracket Damage

Seat I and II did not complete the full cycle count. The cycle count for both Seat I and Seat II can be converted to equivalent cycles at maximum loading similar as before by taking actual cycles at maximum loading and dividing by percentage lost life.

Seat I
81,140 cycles/54 cycles = 1502.6 blocks
1250N occurs once per block for 1502.6 cycles
1502.6 cycles/.2626 = 5722 cycles

Seat II
24,948 cycles/54 cycles = 462 blocks
1250N occurs once per block for 462 cycles
462 cycles/.2626 = 1759 cycles

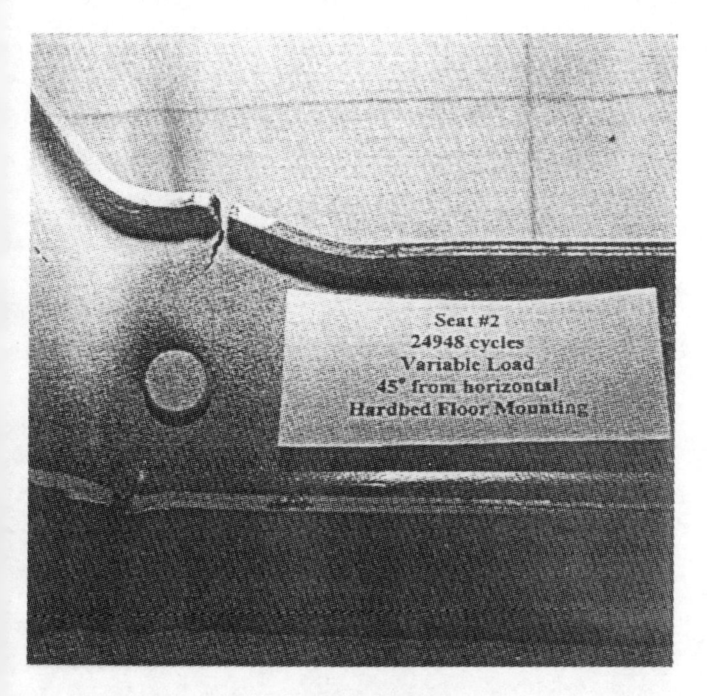

Figure 5. Seat II Rear Track Bracket Damage

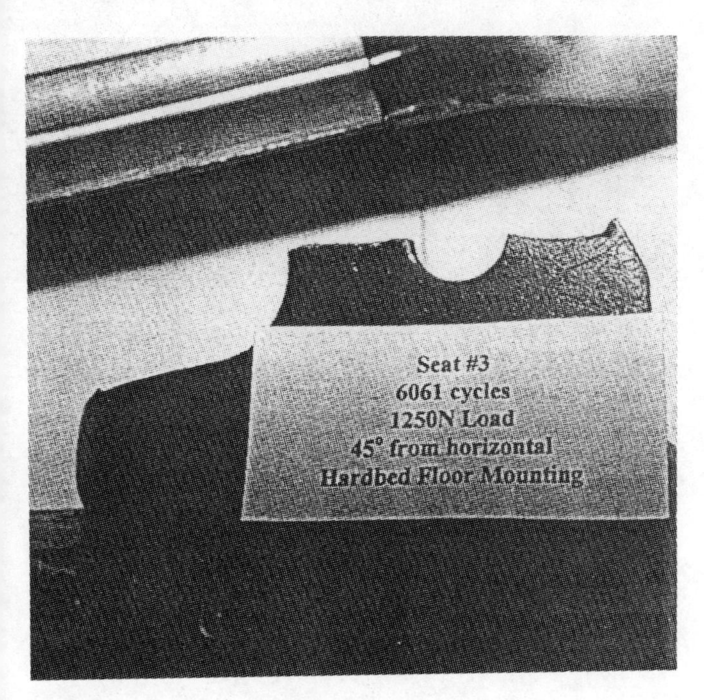

Figure 6. Seat III Front Track Bracket Damage

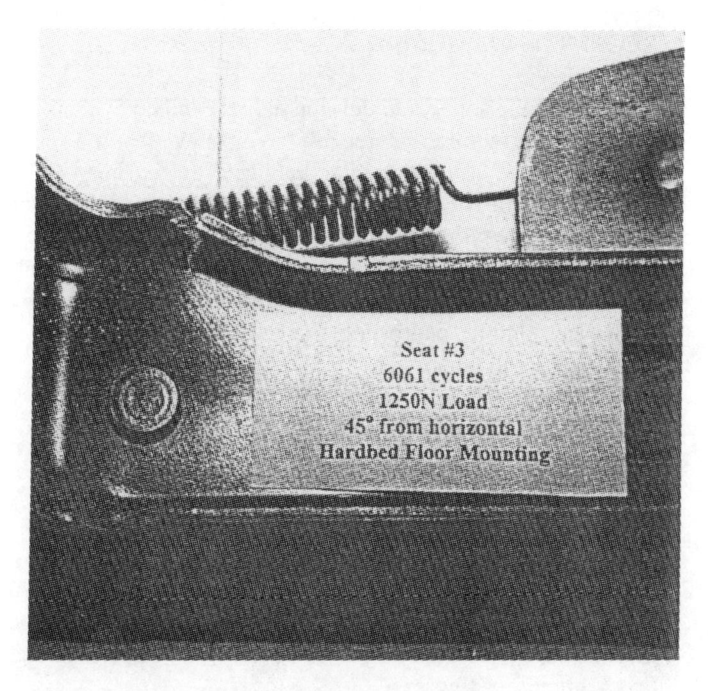

Figure 7. Seat III Rear Track Bracket Damage

CONCLUSIONS

Seat I experienced failure at 81,140 cycles (5722 equivalent maximum load cycles). Seat II experienced failure at 24,948 cycles (1759 equivalent maximum load cycles). The lost life observed in Seat II most likely is due to the increased stiffness (rigid fixture versus production floor pan) at the respective mounting points.

Seat III experience failure at 6061 equivalent maximum load cycles while Seat IV saw failure at 7463 perpendicular maximum load cycles.

The results are inconclusive of indicating whether our hypothesis was endorsed or denied. Seat II and Seat III should have had comparable outcomes. With the limited data recovered and the variability of fatigue testing, the outcome can not be explained. Seats IV and III did have comparable results; however, without further data points it is inconclusive whether a true correlation exists or not.

There does, however, exist enough information indicating further study. Seat II possesses drastically reduced life not witnessed in the testing of Seats III and IV. This occurrence may suggest an abnormality either in the manufacturing process or detailed components of Seat II's configuration.

Because the initial seat specimens chosen for the study are complex both in design, structure, and manufacturing variability, the tests should utilize test specimens of a lesser complexity to reduce the number of variables. Please be reminded that the systems chosen were known to produce failure prior to the successful completion of the test under study.

RECOMMENDATIONS

The goal of the study is to develop a direct and equivalent test to the current required fatigue test. We believe that the single amplitude load can be used to replicate the existing OEM fatigue test, but we still need to verify the equivalence. The seat structure with the applied load is basically a cantilevered beam. Replacing the seat structures with bar stock will enable us to increase the sample size and eliminate the seat variables. With the simple approach, true evaluation and definite conclusions will be possible.

ACKNOWLEDGMENTS

Thanks to Ted Grohs and John Dodson for actual specimen testing. George Betsistas for specimen procurement. Joe Conti, Jim Masters and Kathy Palazzolo for paper editing. Eric Cambell for figures.

REFERENCES

1. Libertiny, G. Z., (1984) Passenger Car Seating Loads; A human Factors Engineering Problem, SAE Technical Paper Series, 840509.

2. Shigley, J. E. and Mitchell, L. D., (1983) Mechanical Engineering Design, McGraw-Hill, New York, pp. 336-337.

3. Collins, J. A., (1981) Failure of Materials in Mechanical Design, John Wiley & Sons, New York, p. 216, 198.

4. Metals Handbook, 9th Edition, Volume 1, Properties and Selection: Irons and Steels, American Society for Metals, Metals Park OH, p. 667.

950147

The Effects of Fabric Backcoatings on Automotive Interior Manufacturing Processes

John Olari
Lear Seating Corp.

ABSTRACT

While fabric backcoatings comprise a relatively small part of the cost of a fabric, it is believed they have a profound affect on interior trim manufacturing processes. This study used the results of tests done on 105 fabrics to determine the effect that backcoatings have on a fabric's processing characteristics. Five tests were used to determine the processing characteristics of the fabric. The results showed that backcoatings do indeed have an affect on a fabric's processing capabilities. The results also suggest that the interaction between fabric construction and backcoating needs to be explored in greater detail.

INTRODUCTION

Backcoatings have been used on fabrics for many years and are necessary to meet many durability, flammability and sewability standards. Backcoatings are used on velour fabrics to keep the tufts from pulling out during sliding entry situations and on knit fabrics to give the fabrics dimensional stability during sewing. Backcoatings are also used in a variety of applications to impart fire resistance to the fabric especially in the automotive, airline and home furnishings industries.

Backcoatings not only have a considerable effect on sewing processes but they also affect the manufacture of molded seating, headliners, and door panels. The backcoating must allow the fabric to stretch when it is molded and also not interfere in the many different bonding processes used to manufacture automotive parts. It must support the fabric while not adversely affecting the hand and it must allow the fabric to breathe. The intent of this study was to determine how backcoatings affected the properties desired in the manufacture of automotive interior trim.

BACKCOATING COMPOSITION

The largest component of most backcoatings is the latex emulsion. Acrylic latexes are by far the most widely used. Other polymers that are used include urethane, butyl rubber, polyvinyl acetate and polyvinyl chloride. The second largest component of a backcoating is usually filler. Fillers are inorganic materials that are used to extend the latex thus reducing the cost of the backcoating. The most common fillers used today are calcium carbonate and clay. Alumina trihydrate is also used as a filler and it also gives the backcoating some fire retardant properties.

While a backcoating's primary purpose is to impart durability and physical stability it is also used to impart fire resistance to the fabric. The most widely used fire retardant additives are antimony trioxide and decabromo diphenyl oxide. Other halogenated and non-halogenated compounds are also used and certain polymer types such as those made from melamine also impart fire resistance.

The rest of the ingredients in a backcoating are used mainly as processing aids. Surfactants, froth aids, thickeners and catalysts are all used to aid in the production of the backcoating or to aid in the application of the backcoating to the fabric.

Backcoatings are usually applied to the fabric by roller coating or knife coating. The backcoating is sometimes frothed before application to reduce drying times and to allow less coating to be applied to the fabric than could be achieved without frothing. Frothing is also used to keep the backcoating from penetrating too far into the fabric.

EXPERIMENT

Five objective tests were used to evaluate the effect of backcoatings on a fabric's processing characteristics. The tests used were Bondability [1], Breathability [2], Trimability [3], Hand [4], and Creasing. Bondability tests the bond strength of a fabric adhered to polyurethane foam using Lear's Sure-Bond process. Breathability tests the porosity of the fabric using a Teledyne Gurley densometer. Trimability tests a fabrics ability to conform to molded seating contours by measuring the elongation of a fabric in each of 4 directions. The Hand test measures the force required to compress a fabric bonded to a foam pad by 5% of its original thickness. The Creasing test measures the length of creases formed when a fabric bonded to a foam pad is deflected with a point load.

BACKCOATING IDENTIFICATION

105 fabric samples were tested from a variety of suppliers. The samples were tested using IR spectroscopy to determine the main polymer structure of the backcoating. The IR scan was also used to determine what if any fillers were used in the backcoating.

The IR scans identified 16 different base polymers in the backcoatings that were used on the fabrics tested. (See table 1.) The IR scans also identified 4 different fillers used in these back coatings.

Polymer Type	Filler Types	# of Times Used
Acryl/Ester	N,K,A	10
Acrylate-1	N,K,A,C,KA,KC	23
Acrylate-2	N,K,A	11
Acrylate-3	K	2
Buta Copoly	N	6
Buta/PVA	A	1
Nylon	N	5
Polyester-1	N	4
Polyester-2	A,C,AN	7
Polyester-3	A	1
Polyester-4	C,KA	5
Polyglycol	N	9
PVA/VC	A	8
PVC/Acrylate	N	2
Urethane	N	2
Vinyl Acetate	N,A	5
None		4
		105

N = None
A = Alumina Trihydrate
C = Calcium Carbonate
K = Kaolin Clay
AC = Alumina + Carbonate
KA = Kaolin + Alumina
AN = Antimony Trioxide
KC = Kaolin + Carbonate

Table 1: Backcoating Identification and Filler Types

PROCESSING CHARACTERISTICS

After identifying the polymers and fillers used in the backcoatings the samples were then subjected to the five tests described above. The data collected was then analyzed using various statistical methods to determine if backcoating types affected the processing characteristics of the fabrics. The fabrics were analyzed as a whole and were also split into groups by fabric construction. (see table 2)

BONDABILITY - The bondability test requires that samples exhibit foam failure or an adhesive strength of at least 9N. Only 9 out of the 105 fabrics tested did not exhibit foam failure. Therefore the bond strength numbers are more a matter of the foam physical properties than the effect of backcoating. Of the nine that failed three were acryl/ester backcoatings, three were acrylate-1 backcoatings, two were acrylate-2 backcoatings, and one was polyester-4 backcoating. When bond failure versus fabric construction was analyzed the results were also spread out with two being flat woven, three being weft insert, three woven velour and one 32 Gauge knit. The results show that backcoating type and fabric construction are not factors in bondability with the exception of the weft insert construction.

HAND - When the fabrics were analyzed as a single group a positive correlation was made between hand and backcoatings containing calcium carbonate. When the fabrics were sorted by fabric construction and then analyzed several correlations were discovered. For 32 Gauge knit fabrics there was a negative correlation for Acrylate-1 backcoatings and for woven velour fabrics there was a negative correlation for Acryl/Ester backcoatings. Hand measures the force required to compress a seat cushion 5% of its original thickness therefore lower numbers are more desirable.

TRIMABILITY - The fabrics were analyzed as a single group and as groups sorted by fabric construction. No correlation was found between trimability and backcoating. Trimability measures the amount of stretch in a fabric and higher numbers are more desirable.

BREATHABILITY - The fabrics were analyzed as a single group and a negative correlation was found between breathability and PVA/PVC backcoatings. The fabrics were then sorted by fabric construction and analyzed. Several correlations were discovered. For 32 Gauge knit fabrics there is a positive correlation between breathability and Acryl/Ester backcoatings and a negative correlation between breathability and Polyglycol backcoatings. Breathability measures the amount of time required to pass 200cm^3 of air through the fabric therefore lower numbers are more desirable.

Fabric Construction	# of Fabrics Teste
32 Gauge Knit	22
Circular Knit	4
Tricot Knit	9
Knit	6
Flat Woven	25
Woven Velour	36
Weft Insert	3

Table 2: Fabric Construction

CREASING - The following test method was developed to objectively measure a fabrics tendency to crease under loading. This creasing test measures the length of creases created in a fabric when a seat cushion is indented by a point load. The indentor is a rod with a 4.75mm radius and the cushion is subjected to a 30N force. The load is held and the pad is evaluated for crease lines. The longest visible line is measured with values rounded to the nearest 10mm therefore lower numbers are more desirable. The fabrics were analyzed as a single group and as groups sorted by fabric construction. No correlation was found between creasing and backcoating.

CONCLUSIONS

This study shows that there are correlations between backcoatings and fabric processing characteristics. Especially for hand and breathability characteristics. It was also assumed that backcoatings would have a significant affect on bondability although that assumption did not hold for this study. The results also suggest that the interaction between fabric construction and backcoating needs to be explored in greater detail. The fabrics for this study were randomly selected and the results give a good starting point for further investigation on the effect that backcoatings have on a fabric's processing characteristics.

REFERENCES

1. ASTM Test method D-751
2. General Motors-Fisher Body test Method 46-10
3. SAE Test method J855
4. SAE Test method J815

950148

Magnetic Induction Heating for Automotive Seat Trim Bonding

Clarice Fasano
Lear Seating Corp.

ABSTRACT

Magnetic induction heating can be used to bond automotive seat trim covers to foam pads. A thermoplastic film doped with ferromagnetic particles is placed between the trim cover and foam pad. An induction coil can be designed for specific seat contours in a tool press. When current is applied to the induction coil located in the tool, magnetic flux is induced. Eddy currents generated in the ferromagnetic particles activate the hot melt film. Heat is delivered directly to the bondline. The process does not damage heat sensitive trim materials (i.e.. leather, vinyl) as the tool surface remains cool. Process reversibility permits design for recyclability and a reduction in production scrap rates.

INTRODUCTION

Automotive interior trim designers can select a wide array of materials in order to style contoured seats. However, the adhesives used to bond the trim cover to the cushion and the methods used to affect an open time on the adhesive, can limit the seat trim material selection process. For example, some adhesives which require high temperatures to melt an adhesive may not be chosen for heat sensitive applications such as leather or vinyl. Moreover, the use of spray adhesives to bond trim cover materials to foam have several drawbacks. Plant issues such as the level of volatile organic compound (VOC) emissions. long cycle times, labor intensity, non-reversible bonding and high energy consumption are factors which detract from spray adhesives applications.

The challenge for seating manufacturers is to introduce new trim bonding technologies which permit design freedom. A process which delivers heat directly to the bond line can be achieved through magnetic induction heating. The process does not damage heat sensitive trim materials as the tool surface remains cool.

However, conventional magnetic induction heating applications are used to bond substrates such as metal, glass and thermoplastic parts. The magnetic induction heating technology has been adapted to bond polyurethane foam cushions to trim cover fabrics. Magnetic induction heating for trim bonding applications will be evaluated in terms of the bonding process, adhesives and tooling developments.

MAGNETIC INDUCTION HEATING PROCESS

The magnetic induction heating process utilizes a thermoplastic film, doped with ferromagnetic particles. An induction coil is designed into the seat tool press (Figure 1). A trim cover is placed over the seat tool press. The adhesive film is placed onto the trim cover. A vacuum is applied to fit the trim cover to the seat contours. A foam bun is placed over the adhesive film and trim cover. Next a platen is lowered in order to apply pressure to the seat assembly. When current is applied to the induction coil, by a radiofrequency generator, the magnetic flux generated induces currents in the ferromagnetic particles in the adhesive. The eddy currents induced in the ferromagnetic particles generate heat, thus melting the thermoplastic matrix. The process activates the hot melt adhesive, thus bonding the trim cover to the foam pad.

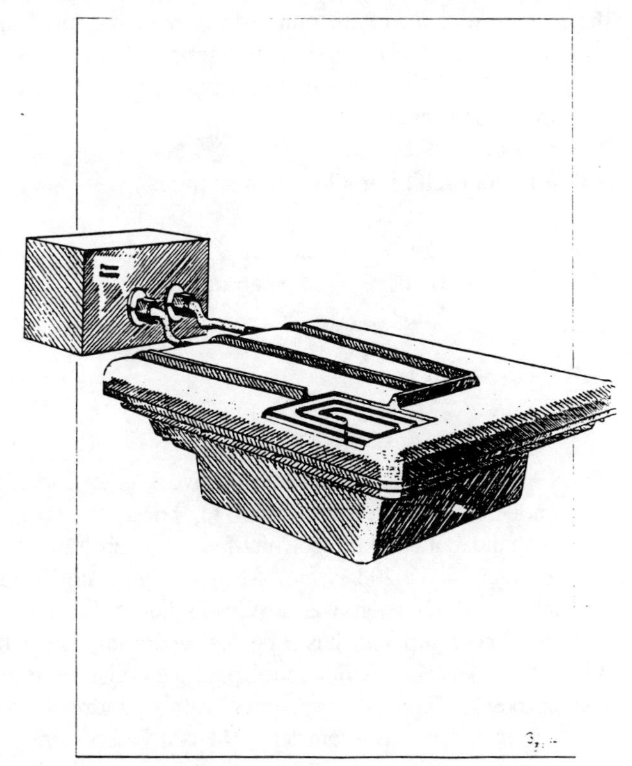

Figure 1: Seat tool with embedded induction coil, connected to a radiofrequency generator.

ADHESIVES DEVELOPMENT

BACKGROUND - The first phase of evaluating the efficacy of magnetic induction heating for trim cover to foam cushion bonding was to develop a viable adhesive for the process. Lear Seating Corporation (Southfield, Michigan) and Bostik Adhesive Company (Middleton, Massachusetts) worked in conjunction to develop an adhesive for this application.

The film medium was selected in order to vacuum form the trim covers to the seat tool contours. The adhesive film acts as a bladder so that a vacuum will pull the trim set to meet the contours of the seat tool geometry. The vacuum forming was incorporated into the magnetic induction bonding process from prior art developed by the Surebond (Lear Seating) process.

The ferromagnetic particles were required to emit heat during the magnetic induction bonding process. The heat dissipated during the induction process was transferred to the thermoplastic matrix. The ferromagnetic particles were supplied by Hellerbond Division (Columbus, Ohio). The powders selected were iron, magnetic stainless steel and magnetic iron oxide. Iron was chosen in the study as it is a ferromagnetic element. The magnetic stainless steel and iron oxide powders were chosen based on the research performed by Alfred Leatherman (Hellerbond Division). His research had shown that a significant improvement in the heating rate occurs when the metal particles and oxide particles are mixed together. *The use of the smaller oxide particles replaces the large iron particles while imparting similar heating efficiencies* [1].

Three thermoplastic hot melt films were selected based on the melt flow temperature ranges from dynamic mechanical analysis (Table 1). The melt flow temperatures ranged from 60 to 120 ^{o}C. The adhesive films were loaded with a varied mixture of powders.

Table 1: Hot Melt Film Flow Temperatures.

Hot Melt Film	Melt Flow Temperature (C)
A	78
B	60
C	120

TEST OBJECTIVE - The test objective was to find a thin film adhesive which was able to bond to both the polyurethane foam and the polyolefinic foam substrates.

Equipment - The 20 kW radiofrequency generator supplied by Hellerbond Division was a Westinghouse Electric type 20K65. The equipment has a normal operating range from 450 kHz to 4 MHz. A flat spiral pancake coil was used for test purposes. The coil was made with cylindrical copper conductors, 4.5 mm in diameter. The coil had 5 turns. The overall coil diameter was 12 cm. Spacing between the coils was approximately 4 mm. Water cooling through the copper coils was used during testing.

Materials - The adhesive film formulas were varied by changing the hot melt film matrix (A, B, C) and the powder loading (iron, magnetic stainless steel, iron oxide, mix). Films were developed in 0.05 and 0.13 mm thicknesses.

The bonding substrates chosen were a polyurethane slab stock foam (approximately 12 mm thick) and cloth trim sets. The trim sets had a polyolefinic foam backing (approximately 8 mm thick foam + 1 mm thick body cloth material). The bond interface was between the polyurethane foam and the polyolefinic foam.

Procedure - A 0.5 x 100 x 100 mm ^{3} Teflon sheet was placed directly on top of the pancake coil. The trim set material was the first material placed on the fixture, followed by the adhesive film, then the polyurethane foam slab stock. Pressure was applied pneumatically at 0.14 MPa during the test procedure. Samples were tested at two current levels at frequencies below 10 MHz. Trials were made on the samples in order to assess the amount of time needed to produce a bond. If the sample did not bond within 180 seconds then the test was halted. The samples were allowed to cool for 5 minutes before examining the bond. The bond was examined by a hand tear for cohesive or adhesive failure.

Evaluation of Adhesives Development Testing - The samples were subjectively evaluated for bond quality. In order to rate the adhesive bond quality , the type of bond failure (cohesive or adhesive) and bond area were rated. Bond area A designates bonding occurring in the 1 inch diameter center of the coil. Bond area B designates bonding beyond the perimeter of area A. Bond area was used as a multiplier for bond quality. Each sample was evaluated at the two interfaces. One interface was defined between the adhesive and the polyurethane (PUR) foam and the other between the adhesive and the polyolefinic (PO) foam. Quality points were deducted if burning occurred through the foam or trim cover (Table 2) The maximum number of quality points to be attained by this system is 20.

Table 2: Sample Calculations for Adhesive Bond Quality

	Interface PO & Film	*Total Quality Points*
Adhesive Failure to Foam	0	
Cohesive Failure to Foam	5	
Bond Area A	0	
Bond Area B	2	
Burn through Trim Cover	0	
Burn through Foam	0	
Total Quality Points Case I	**10**	**20**
Adhesive Failure to Foam	0	
Cohesive Failure to Foam	5	
Bond Area A	0	
Bond Area B	2	
Burn through Trim Cover	-2	
Burn through Foam	-1	
Total Quality Points Case II	**7**	**8**

Results - Bostik hot melt adhesive C (sample 17) with a blend of the Hellerbond iron, iron oxide and magnetic stainless steel powders, exhibited the best adhesion to the polyurethane and the polyolefinic foam backing. The bond was reversible when the adhesive was reactivated on the coil. Bostik hot melt adhesive B did not achieve bondability to the substrates under the aforementioned test conditions. Bostik hot melt adhesive A had more of an affinity for the polyurethane foam substrate than the polyolefinic foam substrate.

Some of the samples exhibited small burn holes in the foam or through the body cloth after the test sequence. The burning phenomena is due to overheating of the large micron sized particles; this can be attributed to the close proximity of the larger micron sized particles to the work coil. When the same samples were tested without being compressed no burning occurred.

The main objective of the study was to find an adhesive which would work in this application. Further studies are ongoing to determine the effects of particle type, particle size, particle density and resin binder as factors in developing the adhesive film.

MATERIALS TESTING FOR ADHESIVE SAMPLES - Bostik hot melt adhesive film C (sample 17) was used to bond samples of leather and vinyl trim sets. The leather and vinyl trim sets had a polyolefinic foam backing. Twenty leather samples and three vinyl samples were prepared for materials testing. The 100 mm x 100 mm samples were subjected to the following materials tests as shown in Table 3 below. Specifications following the General Motors Materials Test Procedures are designated by the GM prefix. A Gurley Densometer was used to evaluate the breathability of the coupon samples. The test measures the time it takes for 200 cc of air to pass through the sample. Materials are considered breathable if the time period to pass air through is less than 60 seconds.

Table 3 : Materials Testing for Adhesive Trim Samples.

Test	Specification
Flammability	GM 9070 P
Fogging	GM 9305 P
Oven Aging	90 C for 168 hours
Environmental Cycling, Interior Trim, Cycle D	GM 9505 P
Hydrolysis	GM 9231 P
Breathability before bonding	24 seconds
Breathability after bonding	1 second

TOOL DEVELOPMENT

BACKGROUND - Once a suitable adhesive film for the magnetic induction heating process was developed, the next phase of the project was to design a tool which could be used in this unique bonding application. A Ford Probe front seat cushion was selected as it is a current leather program. The Probe seat cushion has deep trenches in which the salvage from the trim cover is recessed (Figure 1).

OBJECTIVES - The prototype tooling phase of the project required coil design iterations and the tool material selection.

WORK COIL DESIGN - The work coil design is an integral part of the project. The adhesives development was performed using a simple spiral pancake coil. The surface was flat and the coils were spaced along the same plane. A higher level of complexity was involved when trying to fit the coils in the seat tool press based on the limitations of the seat geometry.

The work coil acts an inductor. A simple model would be an inductor (L) - capacitor (C) circuit . The theoretical model assumes that the inductor is resistanceless. The work coil is connected between the terminals of a charged capacitor from the radio frequency generator. *When the connection is made the capacitor starts to discharge through the inductor. Current going through the inductor establishes a magnetic field in the space around the inductor. The current persists until the magnetic field has disappeared and the capacitor has been charged in the opposite sense to its initial polarity. The process repeats itself in the reverse direction. In the absence of energy losses the charges on the capacitor surge back and forth indefinitely. This process is called electrical oscillation and is illustrated in Figure 2* [2]. *The internal oscillator tube grid circuit of the radio frequency generator must tuned to make the oscillating frequency match the fixed frequency of the tank circuit* [3].

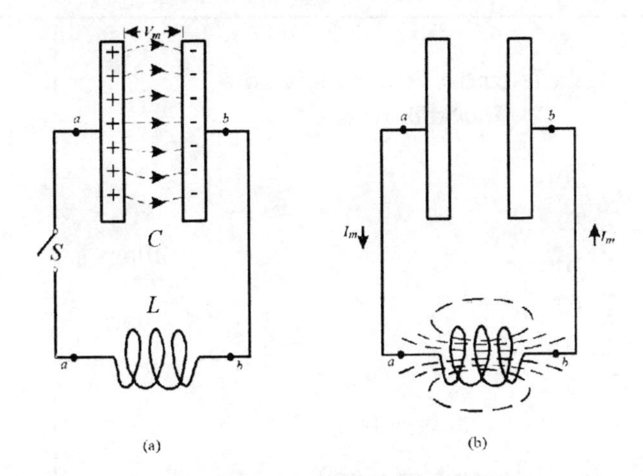

Figure 2: Schematic of Oscillation between the Electrical and Magnetic Fields (Reproduced from University Physics, 5th Edition, Sears, Zemansky, Young, June 1980)

The output frequency of the radio frequency generator is dependent on the inductance of the work coil. The inductance of the work coil is dependent of the number of turns in the coil, the spacing between the turns, the diameter of the coil conductors and the magnetic particle density in the adhesive. Changes in the work coil geometry affect the current and hence the output voltage of the generator [4].

The deep trench areas of the Probe seat pose a design challenge for the coil development phase of the project. The distance of the adhesive film from the center conductor plays a critical role in the temperature seen by the adhesive. The decrease in magnetic density with increasing distance from the conductor affects the differential temperature changes in the adhesive. The electric and magnetic fields close to the conductor can vary rapidly with space. *The mathematical expressions for near fields generally contain the terms* $1/r$, $1/r^2$... $1/r^n$, *where r is the distance from the source to the field point* [5]. A test was designed to study the effects displacement of the adhesive from the coil have on temperature variability.

TEST OBJECTIVES - The two variables which could be changed in the coil design study, without physically reconfiguring the coil, were the current and the cycle time for bonding. Coil design # 3 (zone 10) was chosen for the study.

Procedure - Trim cover material was plied over zone 10 of coil design # 3 at a fixed distance. A layer of adhesive film (sample 17) was placed on top of the trim cover material. A thermocouple tape was placed over the adhesive in order to observe the temperature changes over time. The adhesive needed to reach a temperature of 115 ^{0} C before melting would occur.

The goal was to see how far the adhesive could be displaced from the conductor and still meet the 115 ^{0} C melt flow temperature. The parameters varied were the current, cycle time and trim set thickness. The maximum current which could be attained by this coil design is 240 amps.

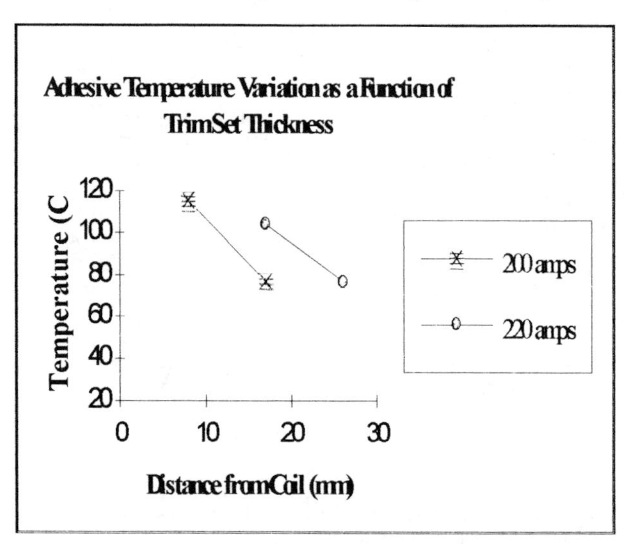

Figure 3: Adhesive Temperature Variation as a Function of Trim Set Thickness

Results - The temperature required to melt thermoplastic film A is 115 ^{0} C. Figure 3 exhibits that at a distance of 8 mm from the conductor the melt flow temperature of 115 ^{0} C for the adhesive film is reached in 30 seconds. Increasing the thickness of the trim material from 8 to 17 mm results in a decrease in adhesive temperature to 76.7 ^{0} C. Increasing the current from 200 to 220 amps increases the adhesive temperature to 104.4 ^{0} C. Yet 104.4 ^{0} C is still below the melt flow temperature. Increasing the trim set thickness to 26 mm decreases the adhesive temperature to 76.7 ^{0} C.

Leaving the trim set thickness at 26 mm and increasing the current from 220 to 240 amps raises the adhesive temperature to 98.9 ^{0} C (Figure 4). Again this is still below the melt flow temperature required by the adhesive in order to bond the substrates.

As cycle time is increased (Figure 4) at the highest current (240 amps) the adhesive temperature remains at 98.9 ^{0} C.

These results are significant as the adhesive film draping over the salvage area of the Probe trim set can be displaced 23 mm from the class "A" surface of the seat tool (13 mm for the tear strip and 10 mm for the salvage). Although a vacuum will pull the trim set over the contours of the tool, the process is being designed for melting the adhesive when the vacuum is not applied. After viewing these results a the coil was redesigned in order to meet the process criteria.

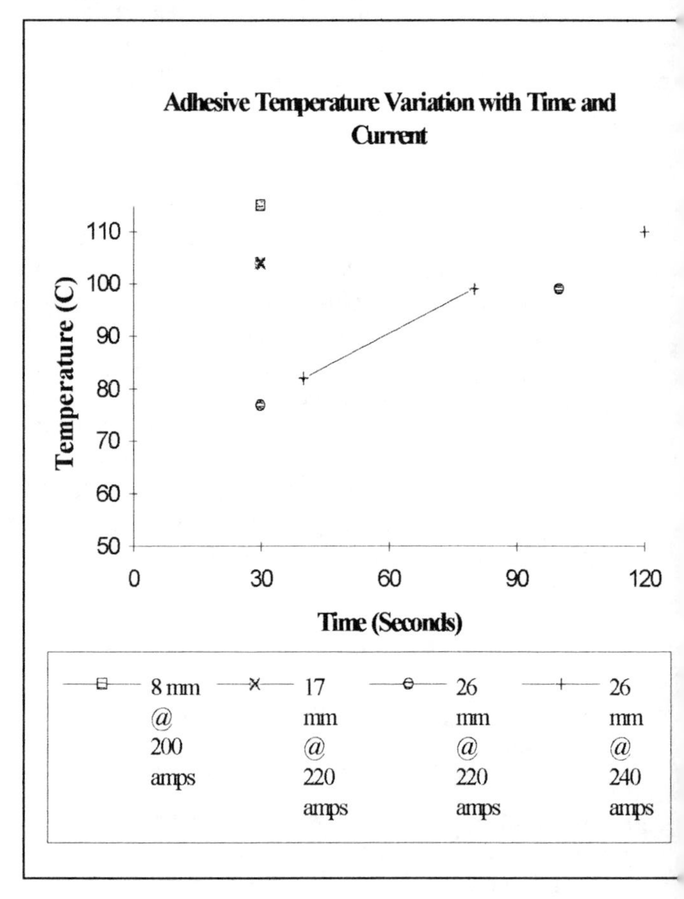

Figure 4: Adhesive Temperature Variation with Time and Current.

TOOL MATERIAL SELECTION - The material chosen for the seat must be selected on the basis of minimal deflection under vacuum and its dielectric strength. A finite element model was developed in order to evaluate the seat tool design. *The dielectric strength is critical as it is the maximum electric field that an insulator or dielectric can withstand without breakdown. At breakdown considerable current passes as an arc, usually with more or less decomposition of the material along the path of the current*

[6]. Given the high current and voltage requirements of this system different grades of epoxy and some polypropylene resins are being evaluated for use in this application.

CONCLUSIONS/SUMMARY

Magnetic induction heating can be used to bond polyolefinic backed trim set materials, consisting of cloth, leather or vinyl to polyurethane foam cushions. Bostik hot melt film adhesive, C, doped with a mixture of ferromagnetic particles, was used to bond the materials in this unique application. The bond is reversible when reactivated by the induction coil. A design of experiments is being implemented to study the effects of ferromagnetic particle loading factors in the hot melt adhesive film applications.

The magnetic induction heating process is dependent on the induction coil geometry. Complex seat geometry with deep trench areas require specialized coil designs. Prototype testing of the coil design for a Probe seat tool is in progress. Once a final design has been completed then assemblies will be made for validation testing.

Lear Seating Corporation has patents pending for the magnetic induction bonding process and the adhesive films developed for the process.

ACKNOWLEDGEMENTS

The work performed on this project was done in collaboration with Bostik Adhesives personnel Ron Kelley (Group Leader, Film and Laminating Adhesives), Jamie Donovan (Applications Specialist), Phil Souza (Product Manager) and Chris White (Automotive Business Development Manager), and Alfred Leatherman, President and General Manager of Hellerbond Division. The prototype tools were made by Mike Stein, Tool and Pattern Maker and Fred Self, Wood Pattern Maker from Progress Pattern Corporation. Maintenance and enhancements to the radiofrequency generator were performed by Progress Pattern Electricians Scott Doherty, Jim Korte and Ray Bailey. The rendition of the Probe seat tool (Figure 1) was a courtesy of Bob Pizio, Graphic Artist for Progress Pattern Corporation. I would like to extend my appreciation to the Surebond Foam Laboratory Technicians Don Allen, Jason Kingsley, Jim Hecker, Chad Linders, Brian Chadra and Mike Skibinski, Advanced Engineering Laboratory Technician, for their assistance and contributions during the development project.

REFERENCES

[1] Leatherman, A., "Induction Bonding of Plastics: Improvements in Versatility and Economics", ANTEC, pp. 1780 - 1783, 1992.

[2] Sears, Zemansky, Young, University Physics, 5th edition, Addison-Wesley Publishing Company, ISBN 0-201-06936-9, Reading, Massachusetts, 1980.

[3] Correspondence from A. Leatherman to C. Fasano, August 10, 1994.

[4] Correspondence from A. Leatherman to C. Fasano, March 15, 1994.

[5] Durney, C., Massoudi, H., Iskander, M., Radiofrequency Radiation Dosimetry Handbook, 4th edition, USAF School of Aerospace Medicine, USAFSAM-TR-85-73, Brooks, AFB, Texas, 1985.

[6] Hawley, G., Condensed Chemical Dictionary, 10th edition, Van Nostrand Reinhold Company, ISBN-442-23244-6, New York, New York, 1981.

950590

Computer Assisted Headlight Design and Research

Han-Wen Tsai and Chien-Ping Kung
OES, ITRI

Abstract

This report describes research done on actual automobile headlights with an aim to developing a means of using personal computers to simulate the flow of light from auto headlights. We were also able to establish optimal geometric shapes for reflectors and coverglasses. Afterwards, computer-aided ray tracing and ray attenuation operations were performed to calculate illuminance distribution results. Moreover, comparisons were made to determine the computer simulation analysis results and the matching of different parts, as well as the relative distribution of light rays. In this way, the accuracy and reliability of the computer simulation analysis could be verified.

Introduction

In the last few years, owing to the ongoing refinement of CNC simulation processing technology , research and design engineers have increasingly relied on concepts from the science of optics to design special geometrical patterns for headlight reflectors, as well as matching patterns for headlight coverglasses. This has led to raised standards in terms of actual road illumination, and there is now a need for enhanced control over the flow of light rays.

Nonetheless, errors remain no matter what concepts are used, and no matter how successful the design of specially shaped reflectors or parabolic reflectors with complicated etched patterns. This is especially true in common difference analysis and the result is that it is hard to accurately predict the effectiveness of product design. Hence, the question of how to cope with ever rising expectations for road illumination is even more important, and computers must naturally be employed to simulate the headlight design function.

In this article we will discuss the appropriate matching of designed reflectors and coverglasses with precise shapes of light bulb emitted ray patterns, and computer-aided ray tracing so as to accurately simulate actual road conditions.

Constructing Models of Light Bulb Function

The first step in the creation of computer-aided headlight simulation analysis is to set up accurate models of light radiation from the bulbs to be used. Often, the lack of precise bulb radiation models is the main reason behind poor on road performance.

Measurements cited in this article were made using a goniophotometer to measure the intensity profile of light bulb emissions. This was helpful in building emission models on the computer. Figure 1 shows various geometrical shapes of incandescent bulbs (cc-2v filament) built on the basis of computer simulation analysis of our own computer generated headlight designs. The area enclosed by the solid line in Figure 2 is the candela strength profile of the incandescent bulbs measured using the goniophotometer in x-z plane(refer to the coordinate of Figure 1), and table 1 show the candlepower data. Because the goniophotometer data is discrete and varied, after you establish a good geometric form for the bulb, choose an appropriate filament illuminance pattern, such as Lambertian Emitters. Then, after undergoing extensive Raytrace ray trace simulation, calculate the division of the filament light intensity and compare it to the goniophotometer data. The main cause of disparity between the two sets of data is the shape of the bulb, which can effect the flow of light. This can be corrected by adjusting the parameters for the light bulb shape.

The dotted line section of Figure 2 shows the results of intensity profile analysis of the computer simulation. Having a correct bulb intensity profile simulation provides a good basis for later analysis of the entire headlight system.

Constructing Reflector and Coverglass Models

Figure 3 concerns geometric forms of reflector and coverglass modules built with reference to computer simulation data . Using the computer also gives each module its own optical characteristics, such as the reflectance of reflectors, or the light transmittance of the coverglasses. As to the reflectors, a simple rectangular shape is used, which is divided into upper and lower boarders.

On the concave, reflective surface (such as the area marked in the illustration) the use of a smooth, parabolic surface in the simulation, makes a correct parabolic curve where the light rays meet the reflector. when you do the ray tracing. In addition to this, if the left and right division boarders in the reflector curve are not symmetrical, use facetted parabolic surfaces in the simulations.

The results of this may nonetheless be limited by the profile of the light unit, due to the shape of the reflector, and the use of facetted surfaces to construct a bordered curve may be helpful in trimming the extruded part of the reflector. Also, because of the change in the periphery of the curve, the use of a faceted surface simulation arc reduces the discrepancy of the results with the ray tracing.

The results of the computer generated coverglass patterns shapes simulation study can be also seen in Figure 3. For coverglass patterns that are relatively more complicated, the time it takes to build such an elaborate structure and endow it with needed optical characteristics is naturally much longer.

Raytrace Simulation

In order to analyze computer generated headlight systems, including the bulb, the reflector and the coverglass, after illuminance distribution simulation study, several million rays from the filament need to be traced and calculated to determine the amount of light energy lost. Then, a statistical estimation can be made of the illuminance distribution reaching the screen(show as the Figure 4). In this way, the extensive ray tracing work and calculation of loss can improve the effectiveness of computer calculated simulations studies. In our project, the CAD/LITE 3D computer program was used to complete the analysis work.

Figure 5 shows computer simulation analysis results of the headlight systems such as figure 4. The contour shape and illuminance intensity of the radiated light was ascertained on a ten meter distant screen. On our chart, the y-axis represents the vertical view angle of luminaire, and the x-axis charts the horizontal view angle. The numerical values marked on the square screen are for the headlight's horizontal and vertical view angles of luminaire, figured on one degree intervals and covering the area with even lux. If even more accurate results for the headlight optical simulation system are desired, you can compare your findings with actual accompanying light systems by using a photodiode sensor. Compare the results with your simulation findings, and calculate parameters for enhanced accuracy. Figure 6 shows the results of an illuminance distribution study done with the verticai and horizontal angle of illuminance set at 0.2 degree, and the distance from the headlight to the screen set at 10 meters (the small screen in Figure 4).

Analysis of Computer Simulation Errors

The simulation studies cited in this report were based on actual headlight designs, but we limited the scope of the study to include only headlights from the JIS regulation c1 series. The illuminance required value in lux for each point is shown in Table 2. Table 2 also marks the optics design values and the headlights of the computer simulation calculation on the basis of the real accompanying light value determined in our study. Under the Table 2, are the calculations for error in the design value and headlight accessory accompanying light value . The common difference was calculated to be less than eight percent. Moreover, the shape of the contour line representing the accompanying light of actual automobile headlights can be seen in Figure 7. Comparing with the contour line shown in Figure 5, it can be clearly seen that the two lines come quite close.

Common Tolence Analysis

In using computers to do headlight optical simulation anysis, besides determining on road performance, problems in design can be dealt with by using common tolence analysis, as, for instance, when dealing with the location of the bulb filament.

As seen in Figure 1, the incandescent bulbs used in this report were those formed in a V shape. Finding the optimal location of the V shaped filament within the reflector was the problem covered in Figure 3. The locus of greatest intensity was near the center of the V shape. To find the optimal location in the reflector, use the optics system in figure 3. First figure without the coverglass and set the intensity center within the reflector. After this, do a computer-aided ray trace simulation, calculate the set distance of the illuminance distribution, and select the point of highest value to determine the optimal reflector and bulb locations. As can be seen in Table 3, when the distance between the filament intensity center and the parabolic focus of the reflector is zero, the illuminance value is at its highest -- hence, this is naturally the best setting. When setting the optimal design setting, it can easily be seen that even subtle shifts in the bulb during the assembly process can have tremendous consequences. To deal with this problem, common tolence analysis can be done and the placement of the bulb can be altered accordingly to get the best results.

Conclusion

Illuminance distribution results from computer simulation studies can be used as reliable information in the design of automobile headlights. Computer-aided headlight design studies such as described in this article can speed the design process and reduce costs.

Reference

1. Gerhard Lindae, "Improvements of Low-Beam Pattern by use of Polyellipsoid Headlamps(PES), " SAE Tech. Paper No. 850228, 1985

2. Arira Kaneko, and Naohi Nino, "New Optical System for Low Profile Headlamps, "SAE Tech. Paper No. 890690, 1989

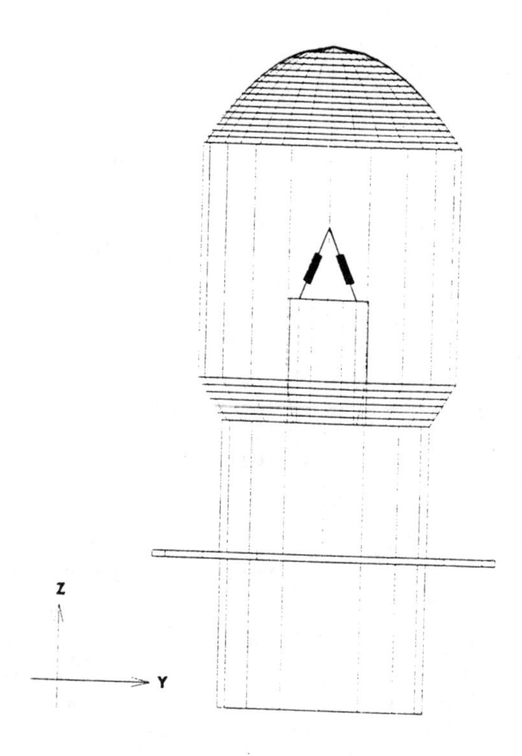

Figure 1: The geometrical shapes of incandescent bulbs(cc-2v filaments)

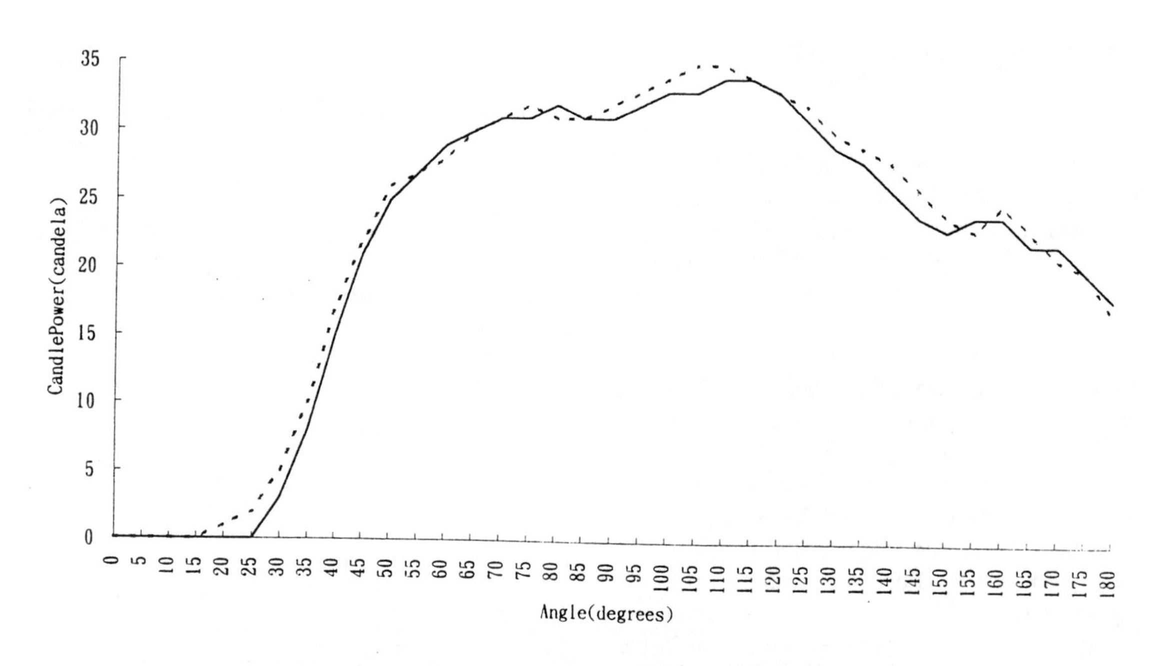

Figure 2: The candlepower strength profile of the incandescent bulbs measured using the goniophotometer and the results of intensity profile analysis of the computer simulation.

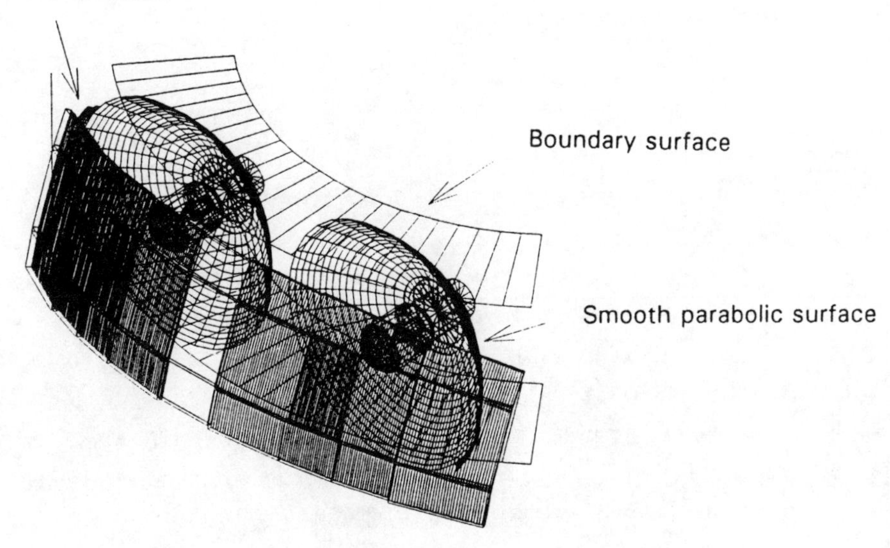

Figure 3: Geometric forms of reflector and coverglass modules built with reference to computer simulation data.

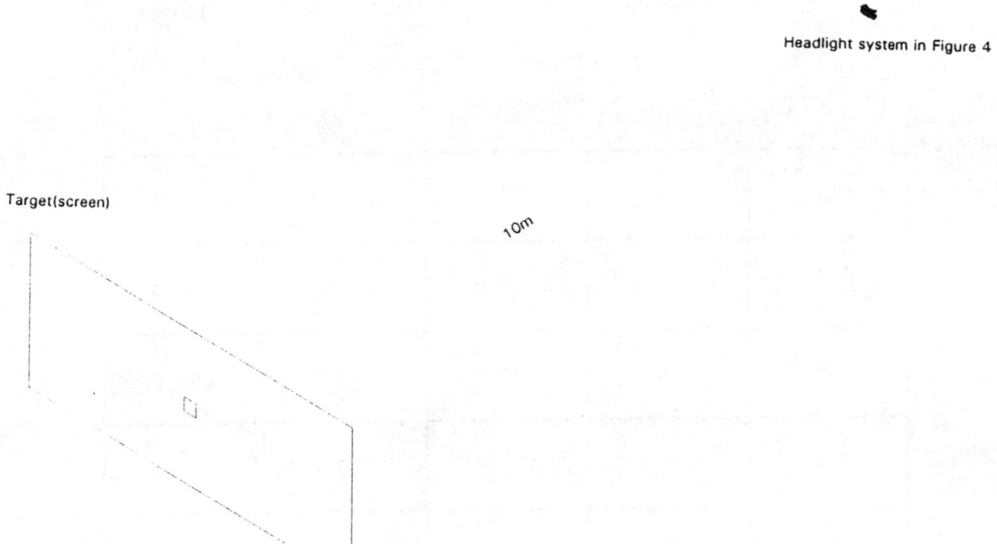

Headlight system in Figure 4

Target(screen)

10m

Figure 4: Measuring system with the headlight systems in Figure 4.

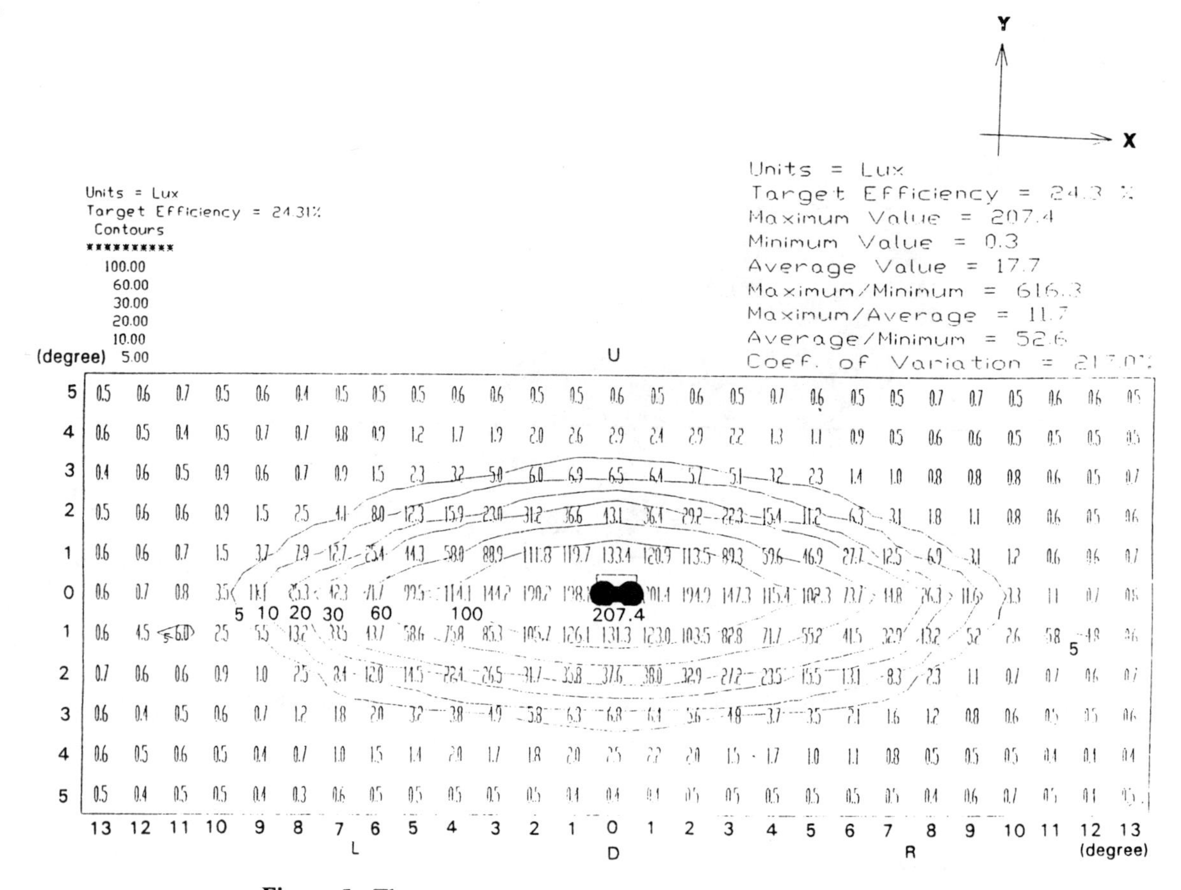

Figure 5: The computer simulation analysis results of the headlight .

unit : lux

(degree)	0.4 (L)	0.2 (L)	0	0.2 (R)	0.4 (R) (degree)
0.4 (U)	215.3	209.6	209.3	209.8	209.5
0.2 (U)	202.8	196.9	193.3	191.6	205.4
0	211.2	194.5	196.2	190.7	214.2
0.2 (D)	194.9	189.2	198.7	192.1	201.4
0.4 (D)	192.7	206.4	210.4	178.1	195.4

Figure 6: The results of an illuminance distribution with the vertical and horizontal view angle set at 0.2 degree.

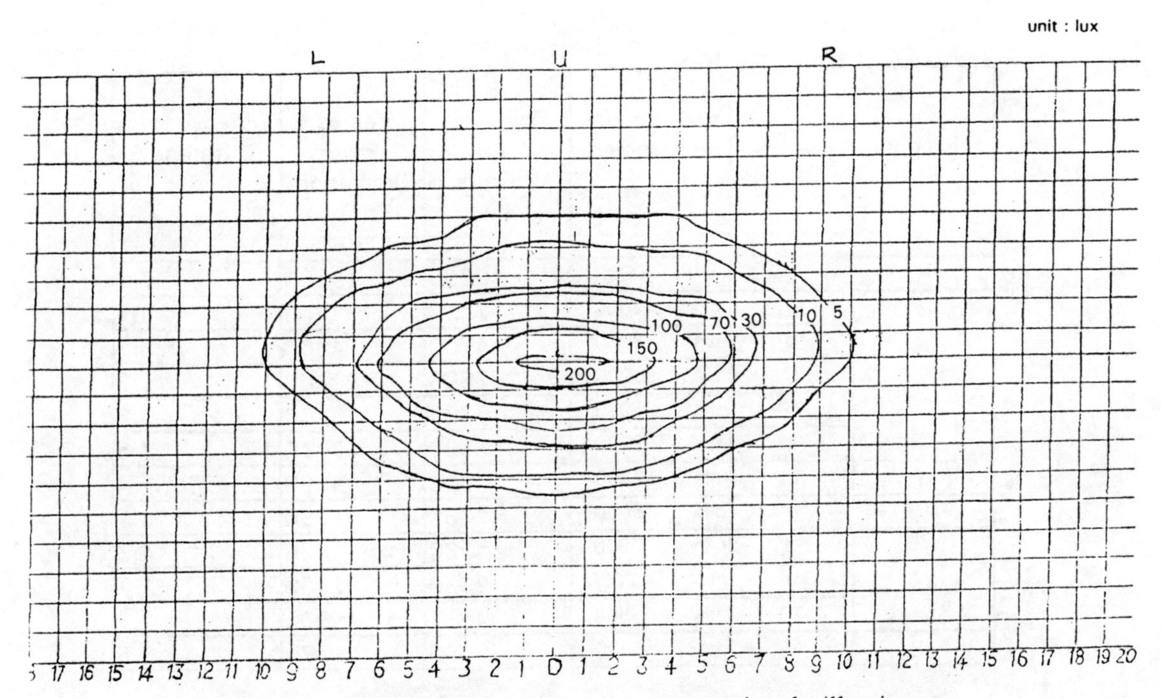

Figure 7: The shape of the contour line representing the illuminance distribution measured by photometer.

LUMEN RATING - 330 LMS.

CANDLEPOWER DATA

ANGLE	CANDLEPOWER	LUMENS
0	0	
5	0	0
10	0	
15	0	0
20	0	
25	0	0
30	3	
35	8	5
40	15	
45	21	16
50	25	
55	27	24
60	29	
65	30	30
70	31	
75	31	33
80	32	
85	31	34
90	31	
95	32	35
100	33	
105	33	35
110	34	
115	34	34
120	33	
125	31	28
130	29	
135	28	22
140	26	
145	24	15
150	23	
155	24	11
160	24	
165	22	6
170	22	
175	20	2
180	18	

Table 1: The candlepower data of incandescent bulbs measured using goniophotometer.

unit : cd

Measuring	points	Requirement	Product Test Results (a)	Design Values (From Simulation Study) (b)	Measured Difference (a-b)/a * 100 (%)
H	V	min 10000	18300	19600	7.1
1/2D	6L	min 2000	7120	7340	3.1
	3L	min 5000	14700	14800	0.7
	V	min 20000	22600	22100	2.2
	3R	min 5000	15800	15200	3.8
	6R	min 2000	6920	7440	7.5
1D	V	min 12000	14600	14800	1.4
2D	V	min 5000	7700	7230	6.1
3D	6L	min 800	930	890	4.3
	V	min 2000	2320	2280	1.7
	6R	min 800	917	901	1.7
4D	V	max 5000	450	475	5.6

Table 2: A Comparative Chart of Optical Design Values Established in Computer Simulation and Actual Tests on Manufactured Products

Distance from filament intensity center to the parabolic focus of the reflector(mm)	Maximum illuminance value(lux)
1	170
0.5	197
0.2	206
0	228
-0.2	209
-0.5	204
-1	176

Note: In comparing the filament intensity center to the parabolic focus of the reflector, positions above the parabolic focus are labeled with positive numbers. Negative numbers are used below the focus point.

Table 3: Filament Placement common tolence Analysis Results

950591

High Intensity Discharge Headlamps (HID) - Experience for More Than 3-1/2 Years of Commercial Application of Litronic Headlamps

Wolfgang Huhn
BMW AG

Guenter Hege
Robert Bosch GmbH

Abstract

A new electronic headlamp system Litronic (light and electronic) was developed by Bosch and was launched in a first series project by BMW in autumn 1991. The 2nd generation Litronic or Xenon headlamps have been available since mid of 1994 in the new BMW 7 series. The unsurpassed light output of the newly developed high intensity discharge lamp for automotive use results in a significantly improved light pattern produced by a specially adapted low beam projector headlamp. Better visibility and visual guidance will essentially enhance driver's safety and comfort without dazzling other traffic. This paper presents the development of the integrated electronic HID headlamp and discusses light pattern, electronics and headlamp design. Furthermore the application of this headlamp to the car and the specific automotive reqirements as well as the experience from more than 3-1/2 years of original equipment are discussed. The high acceptance of this new headlamp technology in the market proves the superior concept in safer and more comfortable night-time driving, in Europe as well as in the USA.

Introduction

First steps in the development of a new high intensity discharge (HID) bulb for use in automotive headlamps were taken early in the 80'ies; 1988 the VEDILIS-group was established in Europe as part of the EUREKA-program. Intensive research and field testing under the guidance of VEDILIS [1] lead to draft regulations now under way for ECE validation and to a first national approval in Germany in march 1991. In september 1991 the world's very first series introduction was accomplished with the BMW 7 series and with the HID headlamp from Bosch [2][3]. Afterwards several other European countries and the USA also granted permission for HID headlamps. Today nearly all major car makers are testing or preparing series production of HID systems, but with the new BMW 7 series model year '95 a next generation HID headlamp as Xenon respectively Litronic headlamp is available in Europe and in the USA.

1. Litronic, the electronic headlamp design

The first series production of a high intensity discharge headlamp introduced in september 1991 has been an application of the 1st generation discharge bulb D1, specially designed projector headlamp and electronic ballast from Bosch [2][3] to an existing headlamp design of the BMW 7 series.

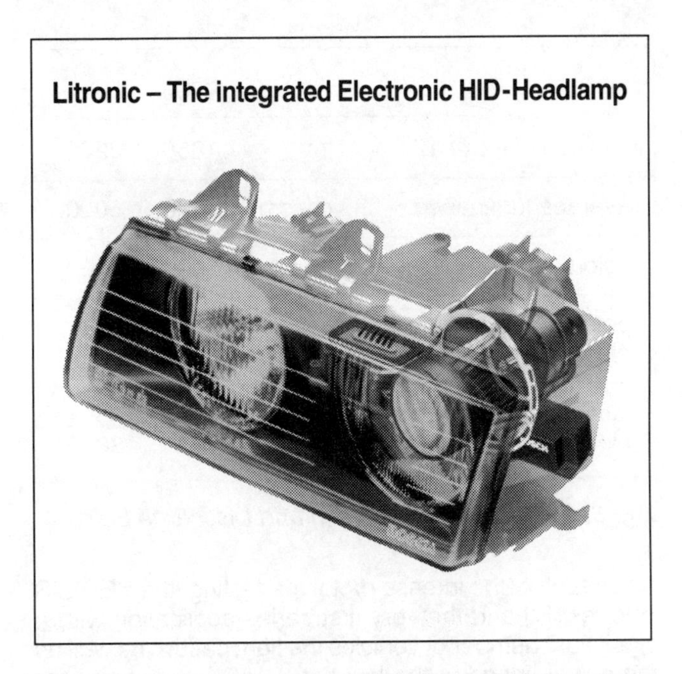

Fig. 1: Litronic, the Integrated Electronic HID Headlamp

The new generation HID headlamp launched with the model year '95 of the new BMW 7 series shown in Figure 1 comprises a 2nd generation high intensity discharge bulb D2S, a 2nd generation projector headlamp, a 2nd generation elecronic ballast for the low beam function [4] and a halogen type reflector with the new H7 bulb for the high beam function all integrated in a common housing. The optional HID headlamp uses the same mechanical and electrical (connector, wire harness, levelling device) interfaces as does the standard all halogen equipped headlamp.

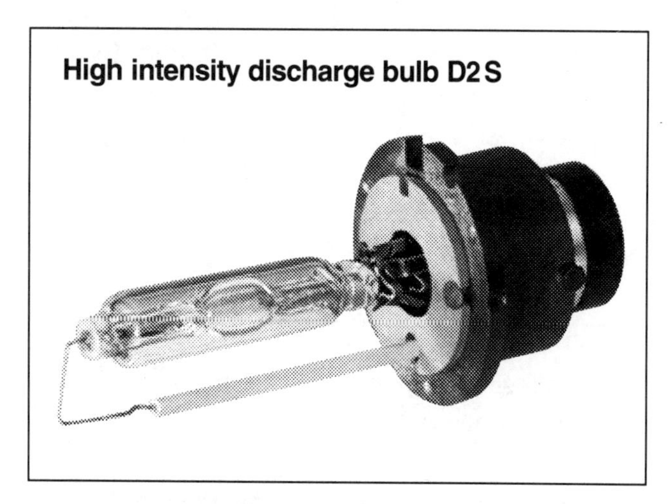

Fig. 2: High Intensity Discharge Bulb D2S

Even though the D2S-bulb has a much improved operational life compared to a conventional halogen bulb it is designed as a replaceable bulb offering a bayonet connector, Figure 2. The intrinsic advantages of the high intensity discharge bulb are depicted against the halogen bulb in Figure 3.

bulb		9006	D2S
luminous flux @ 12.8 V	lm	1050	3200
average luminance	cd/cm^2	1400	6000
colour temperature	K	3000	4200
power consumption incl. electr. ballast @ 12.8 V	W	56	40
luminous efficiency incl. electr. ballast @ 12.8 V	lm/W	19	80

Fig. 3: Comparison of Halogen and Discharge Bulbs

As a result of the intense prototype testing, the VEDILIS field test [1] and the very first series application with a significant number of vehicles the light pattern as well as the signal image of the low beam HID headlamp were improved. The main light output is accomplished by a lens with a diameter of 60 mm. Adding an outer ring reflector will enlarge the illuminated area for a more convenient signal image of the headlamp, Figure 1. The comparison of the halogen and the Litronic light patterns from a bird's eye view is shown in Figure 4, clearly demonstrating the enlarged side illumination and the longer range of the light pattern.

Better overall visibility and orientation in various traffic situations gives rise to a safer night-time driving. Limitation of maximum illumination values together with the later-on described automatic levelling system prevents glare to other traffic.

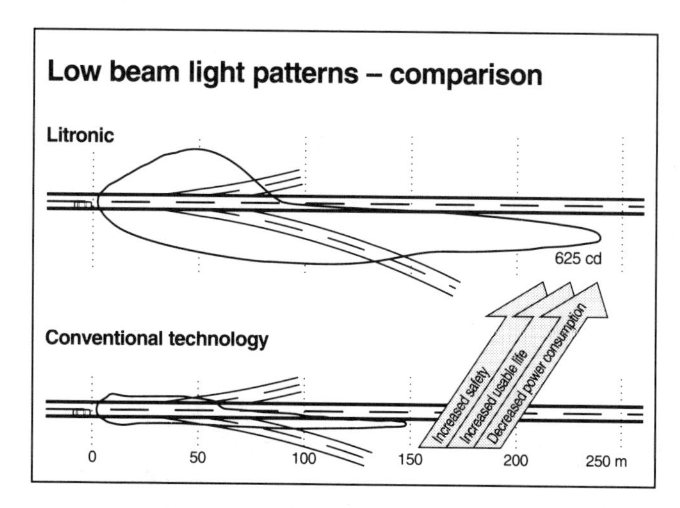

Fig.4: Low Beam Light Patterns - Comparison

The compact and modular design of the 2nd generation electronic ballast allows for a flexible integration into the headlamp design, Figure 5.

The electronic control unit (ECU) incorporates a high frequency DC/DC-converter for conversion of the 12 V battery voltage to the 85 V required by the D2S bulb, a low frequency AC/DC-converter required for the AC-type D2S bulb and a microcontroller for controlling quick run-up as well as regulated operation and safety functions. The ignition module provides for the high voltage ignition pulse of up to 25 kV which may be needed for the reliable ignition of a hot discharge bulb. The socket provides for reliable contact of the replaceable D2S bulb by means of a bayonet interface. To prevent electromagnetic interference (EMI) generated by the AC-type D2S bulb itself, this socket comprises an electrical filter network.

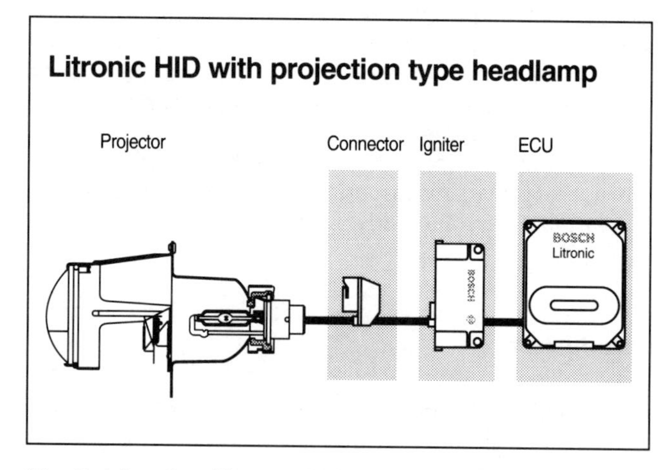

Fig. 5: Litronic - Electronic Ballast

Typical power dissipation of the entire electronic ballast at nominal voltage and temperature is less than 5 W and even including the 35 W of the D2S bulb the resulting 40 W are much less compared to the 56 W of the halogen bulb (Figure 3).

2 Car-specific demands

Traditionally BMW employes a 4-headlamp system (2 low beam headlamps outboard, 2 high beam headlamps inboard). The low beam headlamps only are equipped with the superior HID systems, as in standard traffic situations more than 90 % of the driving takes place with dipped headlamps.

2.1 System characteristics

For the well known halogen headlamps there exist environmental and lighting requirements. With the introduction of the new HID headlamps for the very first time a complex high voltage driven system in the car's front end emerges. Thererfore some new requirements have to be added:
- safety and
- electrical conditions
- system lifetime

The reason for the lower supply voltage specification is the performance under assumed generator damage. The car will run until the voltage breakdown of the battery stops the engine, and accordingly the HID system must perform. In actual car testing the driver of a HID equipped BMW car gets many warnings from the dashboard and from the very slow running car before the engine switches off; almost at the same time the HID headlamp switches off, too.
Redundancy requires two separate electronic ballasts for each of the two low beam headlamps.
Discussion of the rise times of cold run-up and hot restrike is of low practical relevance compared to the performance of existing HID low beam headlamps. Possible colour changes in cold or hot run-ups are normaly not recognized by the driver. Therefore, from a practical point of view all existing draft-regulations are too restrictive in this respect. The operational life of HID bulbs in contrast to halogen bulbs depends much more on the frequency of cold run-ups. In cities with many tunnels and garages a lot of cold-starts occurs preferably in summertime, whereas continuous operation preferably in wintertime causes less wear and tear.

2.2 Crash safety

All our crash-tests (and one real accident during the test phase) showed positive results concerning the safety of the HID headlamps due to the high stability of the ellipsoidal or projector module. Reflector, shade, lens holder and lens itself are forming a closed and rigid cage around the HID bulb. If this cage will be damaged, the bulb will be destroyed also and the HID-system switches off. Not a single case had been found where the unprotected bulb kept burning after a crash.

Safety	
Redundancy	100 % (by separate systems)
Supply voltage start	9 V – 16 V
run	Umin – 16
	Umin: lowest voltage, engine runs without problems
wire cut inside	headlamp switches off
short circuit inside	headlamp switches off
bulb damage	headlamp switches off
Light	
photometric table	National standard, but Emax < 50 lx (Imax < 31250 cd) and increased side values
run up / hot restrike	National standard
Electrical conditions	
EMC	BMW standard
Open connector	No damage
Short cut outside	No damage
BMW standard Load dump	No damage
BMW standard Jump start	No damage
Lifetime	
Switching cycles	8000
Burning time	> 1000 h
Min. light after test standards	according to national
Environmental conditions	
Ambient temperature	– 40 °C to 100 °C
range, shock, vibration, water- and salt-spray, temp. cycling etc.	BMW standard

Fig. 6: HID System Characteristics

2.3 Test runs in real traffic

Colour

The HID light is very distinct by its different colour. This makes it attractive by an innovative appearance. Unless a great number of cars equipped with HID headlamps are on the roads the exceptional appearence could make people very curious. In driving practice the different colour of the new light is well accepted by the drivers. The driver's eyes accept the blueish-white as white and the halogen light of other cars as yellowish-white. Traffic signs and road markings were well recognized in any case.

Driving comfort

In one of our test cars we have the possibility to switch between one HID and two different halogen headlamp sets while driving. We had found no weather (we drove under worst weather conditions) or road conditions where any of the test persons preferred one of the halogen headlamps.

The tested HID systems had very wide illumination angles. This provided good visual guidance and good orientation to the driver especially in fog, snow or rain conditions.

Cut-off

Projector type headlamps provide a high luminuous gradient at the cut-off line resulting in good quality light patterns. In cars with very soft suspension this cut-off could be sometimes disturbing for the driver in case of a rough road or while braking strongly. The oncoming and preceding traffic could be dazzled during strong acceleration or by a heavy loaded trunk (much the same situation with every other headlamp system). A stiffer suspension will reduce such effects.

2.4 Automatic headlamp levelling system

The better solution for the above mentioned problems is an automatic headlamp levelling system, Figure 7. BMW introduced a quasistatic levelling control together with the new 7 series. Due to an addditional front axle sensor it is more precise than a power suspension and the system costs are less expensive.

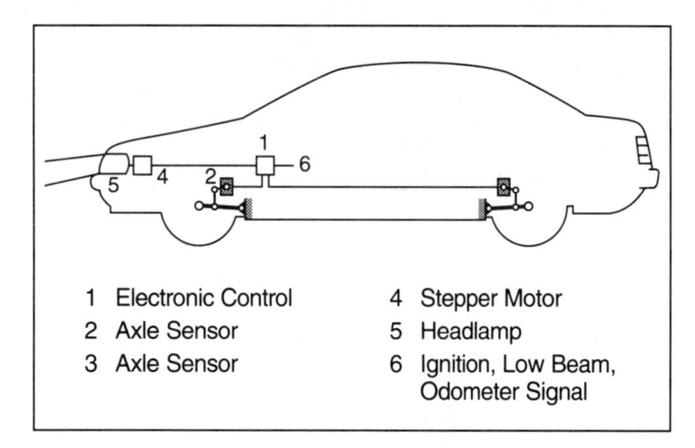

Fig. 7: Automatic Headlamp Levelling Device

The hardware comprises:
– sensor on the rear axle
– sensor on the front axle
– electronic control unit
– stepper motor inside of each headlamp

This system prevents glare and short visibility distances, provided that the headlamp aiming is correct.

3 Practical experience from pilot run and original equipment

The worldwide first series production car with HID headlamps was the BMW 750iL with a national german permit sold September 91. It was an option for 1550 DM and the car was equipped with a standard power suspension. The next steps in 92/93 were swiss [5] and austrian permits and the introduction in the US market based on a positive ETL report.

The 730i, 740i/iL and 750i/iL of model year 94 are available in several european countries (Germany, Austria, Switzerland, Belgium, Spain, Greek, Luxembourg, Portugal) with the optional HID system, in the US it was standard on the 750iL. With the new 7 generation model year 95 the HID system is optonally available in several European countries for 1350 DM and in the USA including an automatic headlamp levelling system.

3.1 Test evaluation from pilot run

Prior to the first series introduction and somewhat in parallel to the VEDILIS field test [1] an intensive test phase conducted by BMW took place. The most important results on a scale of 1 (= excellent) to 5 (= inadequate) and on the basis of 112 drivers representing some 100 000 miles of driving experience with the newly HID headlamps are shown in Figure 8:

Side illumination	1.6
Range	1.7
Visibility on wet roads	2.0
Visibility of traffic signs	2.1

Fig. 8: Driver's Judgement of HID Headlamps

A steadily increasing anticipation over several years of testing has most likely impeded an even better ranking. Regarding the reactions of the other traffic, similar data were found as in the VEDILIS field test [1].

3.2 Evaluation of data from original equipment

A recent questionaire among 38 drivers representing some 400 000 miles of driving experience shows the following results, Figure 9.

improvement in safety	89 %
better side illumination	85 %
better range	71 %
better visibility on wet roads and in fog	47 %
better visual comfort	37 %
better visual guidance	32 %

Fig. 9: Advantages of HID- against Halogen-Headlamps

4. Summary

The presented low beam HID headlamp Litronic or Xenon is the most sophisticated headlamp to date. Modular design of the electronic ballast allows for full integration into the headlamp, the D2S bulb allows for possible replacement even though it has a much longer operational life compared to halogen bulbs. The projector headlamp stands for lighting effiency and well controlled light projection. Superior light output and light pattern in combination with an automatic levelling device provide for best lighting performance and comply with the stringent requirements in light quality.

A very smooth and convenient light distribution raises driving comfort in almost every traffic situation and in combination with the increased illumination in the center, to the sides and in the distance a better visibility and a better visual guidance will largely improve safety during night-time driving.

References

[1] EUREKA project No. 273 VEDILIS
Vehicle Discharge Light System
Final Report, 1992

[2] B. Woerner, R. Neumann:
Motor Vehicle Lighting Systems with High Intensity Discharge Lamps.
SAE Technical Paper Series 900569, 1990

[3] R. Neumann, B. Woerner:
Litronic - new automotive headlamp technology with gas discharge lamp
Automotive Design Engineering 1993,
p. 152-156, 1993

[4] R. Neumann:
Improved Projector Headlamps Using HID (Litronic) and Incandescent Bulbs
SAE Technical Paper Series 940636,1994

[5] C. Schild, H. Schlegel:
Hell wie der lichte Tag? (bright as the daylight? - a headlamp test including BMW/Bosch-HID)
ACS (Automobile Club of Switzerland)
auto 12 / 1991. p. 10 -15

950592

Headlight Beam Pattern Evaluation Customer to Engineer to Customer - A Continuation

Daniel D. Jack, Stephen M. O'Day, and Vivek D. Bhise
Ford Motor Co.

ABSTRACT

The method of communication between the customer and the engineer has been refined to further improve the headlight beam pattern development process. The refinements included: a) reduction of word pairs used for semantic differential scaling and b) use of shortened questionnaire on night-roadway viewing zones. The added benefit of the new questionnaire method allows the engineer to evaluate the customer responses of the beam pattern within specific areas on the road scene. A statistical technique called factor analysis has been used to evaluate and to reduce the large number of semantic differential word pairs used in the previous work by Jack, O'Day and Bhise (1). A comparison of the two questionnaire forms used in the evaluation surveys was completed based on an evaluation of beam patterns in a dynamic drive situation. In addition, weighting factors for night viewing zones on the road surface and for the reduced number of factor analyzed word pairs have been derived to aid in applying relative importance factors for areas of view and word pair topic. The present research suggests a strong relationship between the scores of the reduced word pair/zone questionnaire and the overall positive score of the longer ("short") questionnaire used in this and the previous research.

BACKGROUND

Previous research on the subjective evaluation of headlight beam patterns was presented by Jack et.al. (1) during the 1994 SAE Technical Sessions. The research papers on functional beam pattern requirements are referenced in that paper. Nakata's et.al. (2) "satisfaction factor" (expressions of uniformity, power sensation and balance of light distribution) served as the model for exploration of the present data base. The suggested methodology was utilized as an initial signpost in the analysis conducted on newly acquired data from program beam pattern evaluations.

FACTOR ANALYSIS OF THE "SHORT" QUESTIONNAIRE

The number of semantic differential word pairs in the previous "short" questionnaire was deemed too large for convenient testing if applied to an increased number of viewing zones on the roadway. A statistical technique termed factor analysis was applied to the "Short Questionnaire" data base. A detailed description of factor analysis is provided by Kim (3). A brief explanation of the technique may be of value. Factor analysis has an objective to represent a larger set of variables (word pairs) in terms of a smaller number of hypothetical variables. It examines the inter-relationships among the word pairs - correlation coefficients - by looking at the relationship within some subsets. The process identifies key concepts - locating the higher loading word pairs in each factor - which can provide insight into naming the word pair variables in the reduced set. The results can also point toward flaws in the selection of the initial pairs of words with respect to covering conceptual ideas and show the necessity for judgment calls.

The data obtained on both drives from the previous beam pattern research using the "short" questionnaire was combined and a factor analysis was performed on the word pairs. The factor analysis results (with a varimax rotation in an attempt to simplify and make the results easier to understand) are presented in Figure 1.

The results tended to reveal two strong factors and two weak factors (see columns in Figure 1). The strength of each factor was determined by the number of contributing word pairs and the strength of their relationship to a common conceptual factor. The name of the conceptual factor was suggested by the highest loading word pair. Factor 1 was named SMOOTH (.969). This terminology related in concept with the other word pairs that loaded strongly within Factor 1. Factor 2 was named AIMED (.952). The other word pairs that loaded in Factor 2 also related to where the light was distributed on the road scene. Factor 3 had only one loaded word pair which was not enough of a clue as to it's meaning.

Factor Analysis

One case for each subject for all night drives.

Rotated Loadings	**Factors**			
varimax rotation	Smooth	Aimed		
Word Pairs*	*1*	*2*	*3*	*4*
SMOOTH	**0.969**	-0.046	0.1	0.123
UNIFORMS	**0.956**	0.131	-0.071	-0.235
BLENDED	**0.952**	-0.065	0.007	-0.282
STRAIGHT	**0.948**	0.128	-0.045	0.013
UNIFORM	**0.946**	0.111	-0.085	0.285
UNIFIED	**0.935**	0.163	0.097	0.258
CLEAN	**0.927**	0.013	0.212	-0.041
QUALITY	**0.926**	0.27	0.26	0.001
SOFT	**0.922**	-0.048	0.339	-0.026
EXPRNSIV	**0.916**	-0.015	0.392	-0.022
PLEASANT	**0.885**	0.409	0.185	0.001
BALANCED	**0.942**	0.364	0.212	0.315
CORRECT	**0.824**	0.411	0.135	0.188
SPACIOUS	**0.8**	0.415	0.086	0.399
COMFORTABLE	**0.792**	0.509	0.181	-0.147
FUNCTIONAL	**0.708**	0.574	0.324	0.15
SUFFICIENT	**0.608**	**0.646**	0.236	0.249
SECURE	0.589	**0.783**	0.146	0.112
AIMED	0.16	**0.952**	-0.209	-0.112
CENTERED	0.053	**0.913**	0.163	0.222
STRONG	0.271	**0.827**	0.391	-0.128
LONG	0.131	**0.819**	0.187	-0.33
LEFT	-0.351	**0.782**	-0.427	0.154
ELONGATED	-0.148	**0.658**	0.247	-0.0645
SHARP	-0.226	**-0.232**	**-0.929**	0.054

VARIANCE EXPLAINED BY ROTATED COMPONENTS

1	2	3	4
13.803	6.606	2.098	1.343

PERCENT OF TOTAL VARIANCE EXPLAINED

1	2	3	4
55.211	26.426	8.393	5.373

* = only desired adjective is shown.

Figure 1 - Factor Analysis Results

The first and second factors combined explained about :% of the total variance.

Four word pairs were established for use in evaluation of w beam patterns when the pattern scene was separated into x zonal areas. The six zones chosen for evaluation were the ft / center / right areas in both the foreground and distant eas of the night road scene. The four word pairs are esented in Figure 2. The reason that a word pair was cluded is presented in the second column. The third and urth word pairs were chosen because judgment indicated e need for a "power" factor and also a measure of subjective elings factor.

Word Pair	Reason
Smooth / Choppy	Factor Analysis
Aimed / Misaimed	Factor Analysis
Bright / Dim	Perceived Light Output
Secure / Insecure	Maintain subjective impression of acceptability also Related strongly to <u>both</u> of the first two word pairs shown above

igure 2 - Word Pairs Evaluated Within Zones

PPLICATION AND COMPARISON OF THE QUESTIONNAIRES IN HEADLAMP BEAM PATTERN EVELOPMENT

Program engineers responsible for the development of eadlamps requested that the semantic differential evaluation echnique be applied to a new vehicle program. A first night rive evaluation was conducted with two sets of early rototype headlights, a vehicle with current production lamps nd three competition benchmark vehicles. Program evelopment people utilized the two questionnaires (the old short" form and the new form which required that four word airs be evaluated within six zones on the roadway scene) to valuate the six beam patterns.

Figure 3 presents the beam profile characteristics for each ehicle by evaluated pairs. The 93 Base vehicle beam pattern s presented as a solid bold line. Prototype D is the dashed old line.

Figure 4 presents a summary for the "short" questionnaire of the positive and negative ratings expressed as percentages or each vehicle. The white portion is the percentage of ositive (desirable characteristics listed on the left side of Figure 3) ratings in the "very much" category given by all evaluators. The grey portion is the percentage of "somewhat" ositive responses. The dark grey portion below the zero line s the percentage of "somewhat" negative responses (undesirable characteristics listed on the right side of Figure 3). Finally, the black portion is the percentage of "very much" negative responses. Total positive and total negative response percentages are presented at the top and bottom of each bar.

Figure 5 presents the Average Zone Scores. These are the average of the word pairs (equal weighting) for the six beam patterns in each of the six viewing zones on the roadway. A composite score (an average of the six zone scores) is also presented.

FIRST DRIVE RESULTS - Both the Prototype Y and the Prototype D headlamp beam patterns were evaluated as significantly better than the 93 Base beam pattern. The Prototype Y beam pattern was significantly better than the Prototype D beam pattern. The Prototype D beam pattern needed revision in the center and right foreground areas and also in the center and right distant areas in order to equate with the Prototype Y beam pattern. The optics engineer's task was to provide a beam pattern which got rid of subjective responses such as: Streaky, Dirty, Uneven, Choppy and Spotty.

This information from the first drive evaluation was used by the lamp designers to make adjustments in the beam patterns. Another night drive evaluation was conducted with the new sets of beam patterns. Two vehicles were common to the first drive survey, namely the 93 Base and the Prototype Y. Three new beam patterns: Prototype Da, Prototype Db and Prototype Yb were evaluated. Vehicle T beam pattern was included to bring the vehicle count to six and also because it had never been evaluated previously by our group.

Figure 6 presents the second drive beam profile characteristics for each of the beam patterns. The 93 Base vehicle beam pattern is again presented as a solid bold line. Prototype Da and Db are represented by dashed bold lines with filled and open triangle markers respectively.

Figure 7 presents a summary for the second drive of the "short" questionnaire positive and negative ratings as percentages for each vehicle.

Figure 8 presents the second drive Average Zone Scores for the six beam patterns within each of the viewing areas. A composite zone score for each beam pattern is also presented.

SECOND DRIVE RESULTS - All of the Prototype beam patterns were evaluated as significantly better than the 93 Base beam pattern. The new Prototype Da beam pattern was now equal to the Prototype Y beam pattern from the previous drive survey. The average zone scores for both beam patterns were nearly identical in the three foreground zones. The Prototype Y beam pattern held a slight edge in the distant viewing areas.

Figure 9 presents the Mean Profile Scores on some of the key word comparisons for the two drives.

	First Drive Prototype D	Second Drive Prototype Da
Streaky / Uniform	-.83	+.17
Dirty / Clean	-1.0	+.83
Uneven / Balanced	-.5	+.50
Choppy / Smooth	-.5	0
Spotty / Uniform	-1.5	+.33

Figure 9 - Key Word Comparisons

A substantial change toward more positive scores in the new Prototype Da beam pattern was observed by the drivers in the second drive, when compared to the same word pair scores for the Prototype D beam pattern from the first drive.

However, two beam patterns evaluated produced results worthy of note. First, the revised Prototype Yb beam pattern was better than the Prototype Y carryover beam pattern.

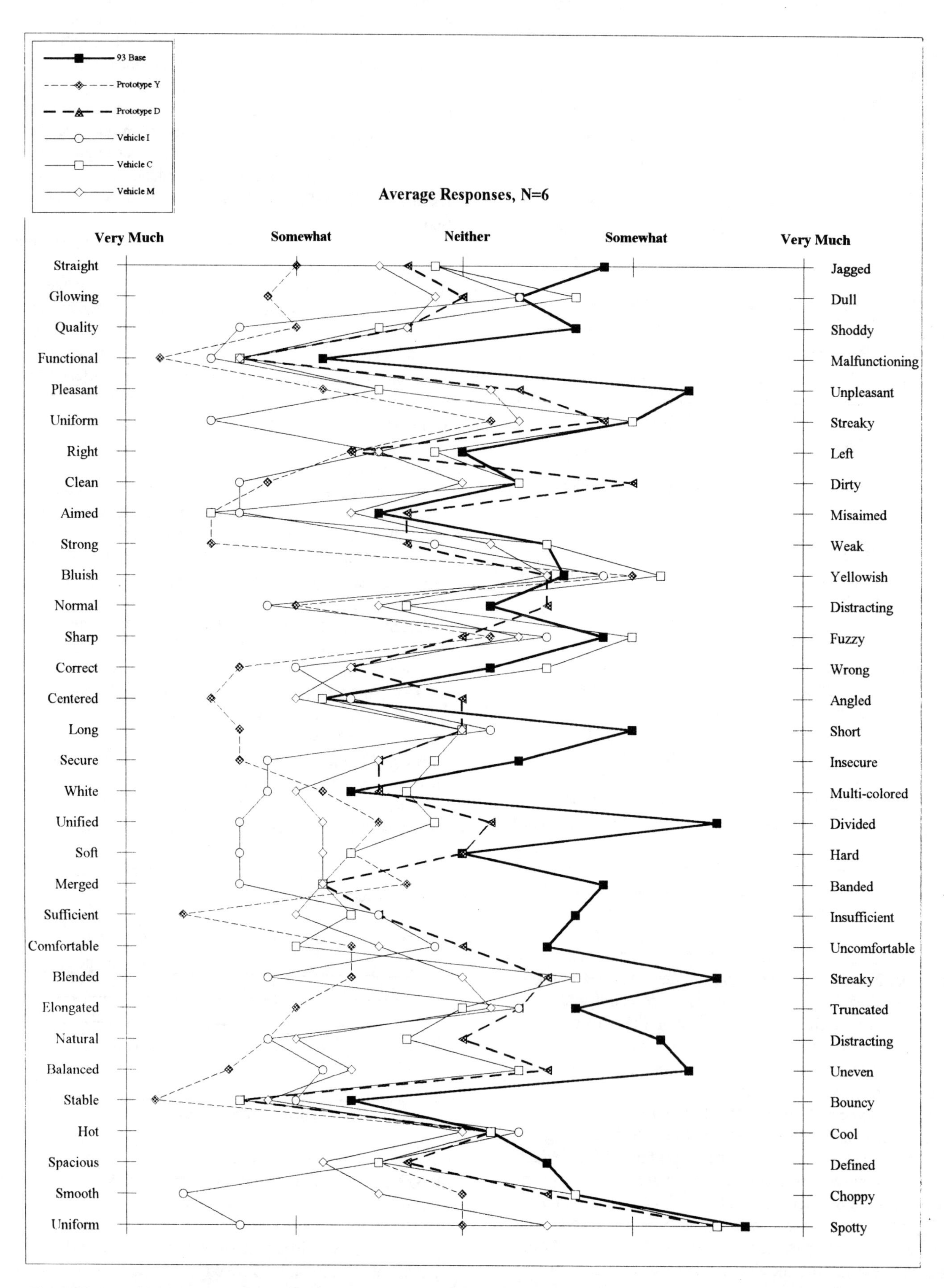

Figure 3 - Evaluation Profile for the First Night Drive

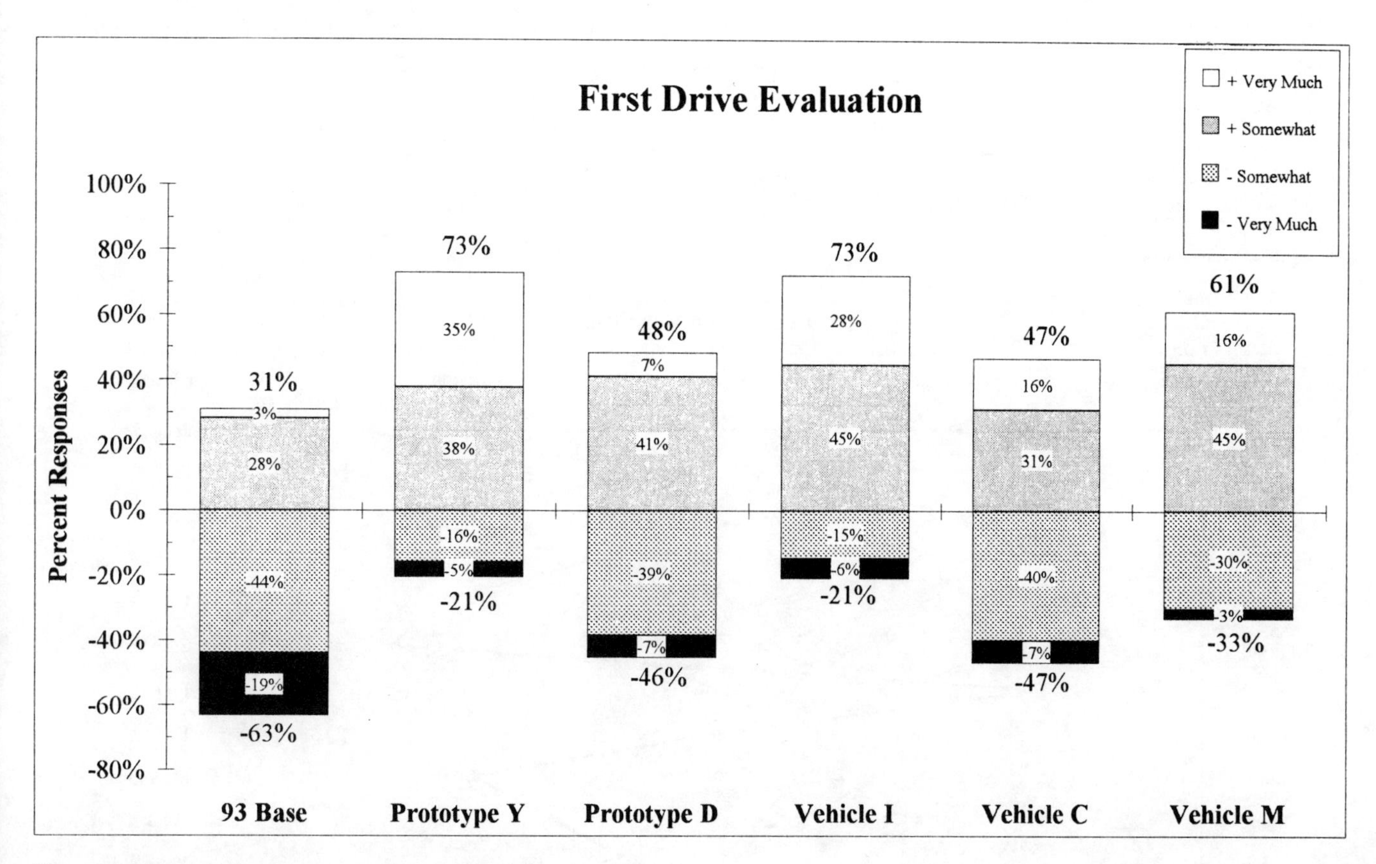

Figure 4 - Percent Positive and Negative Responses

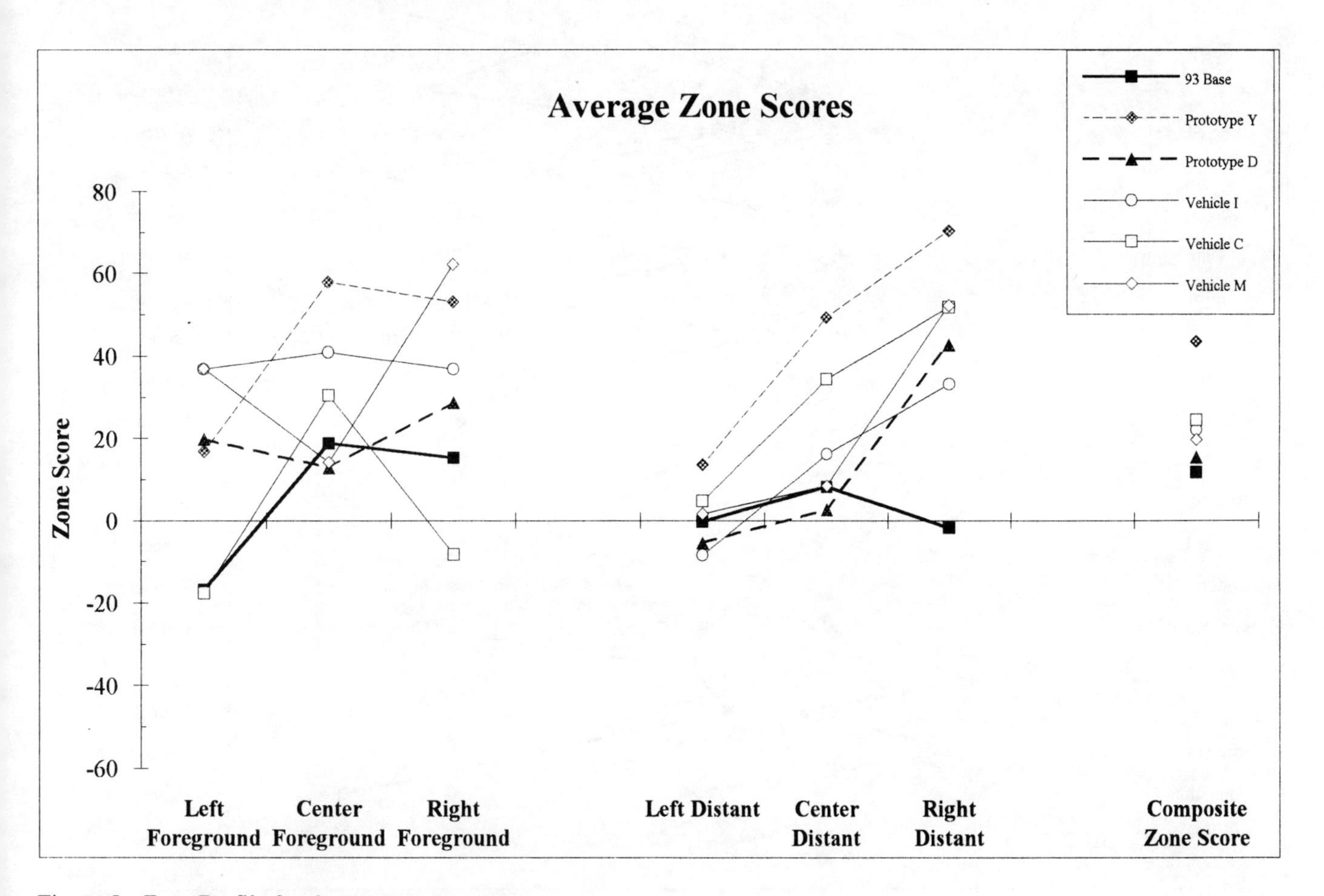

Figure 5 - Zone Profile for the First Night Drive

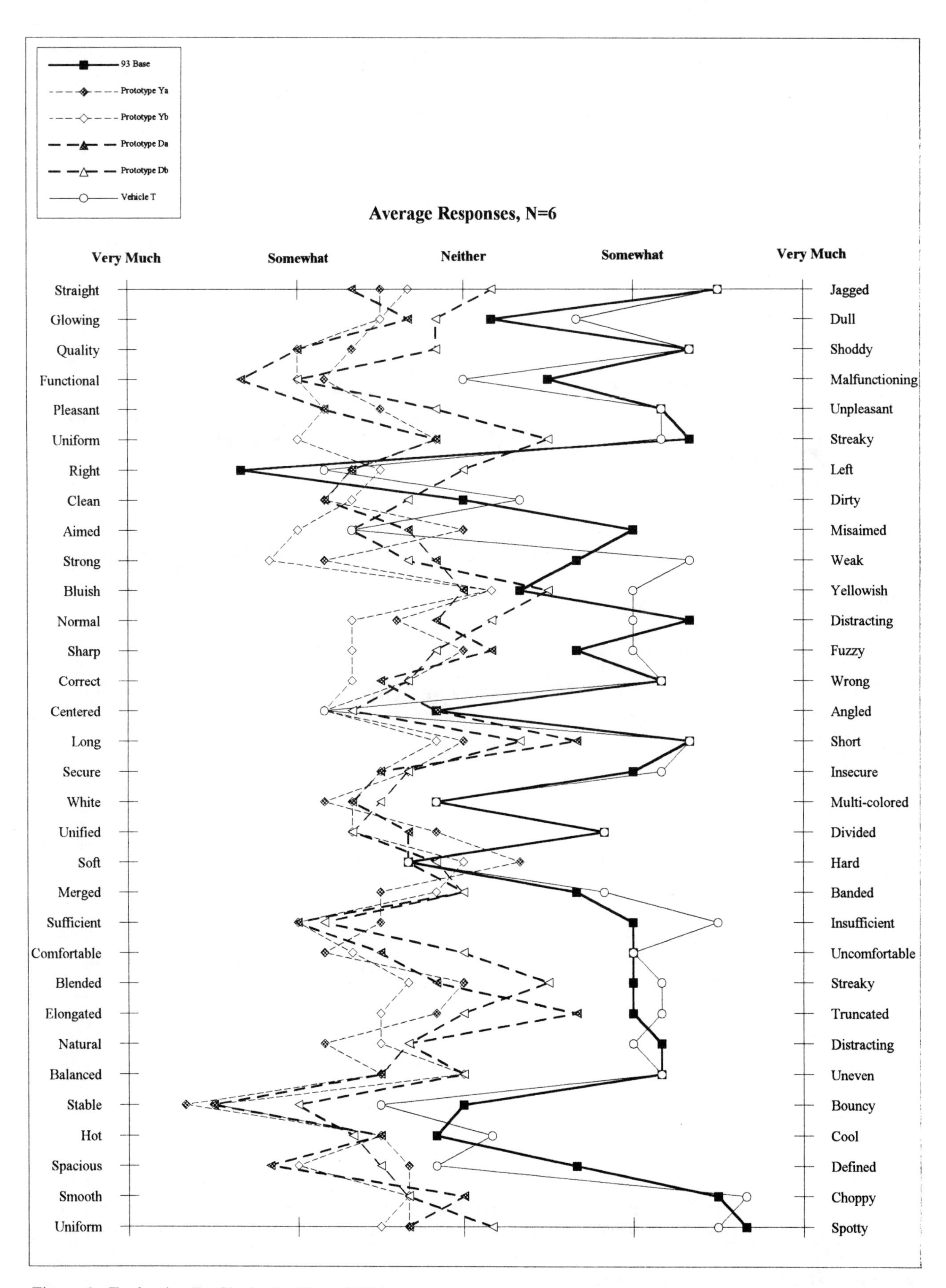

Figure 6 - Evaluation Profile for the Second Night Drive

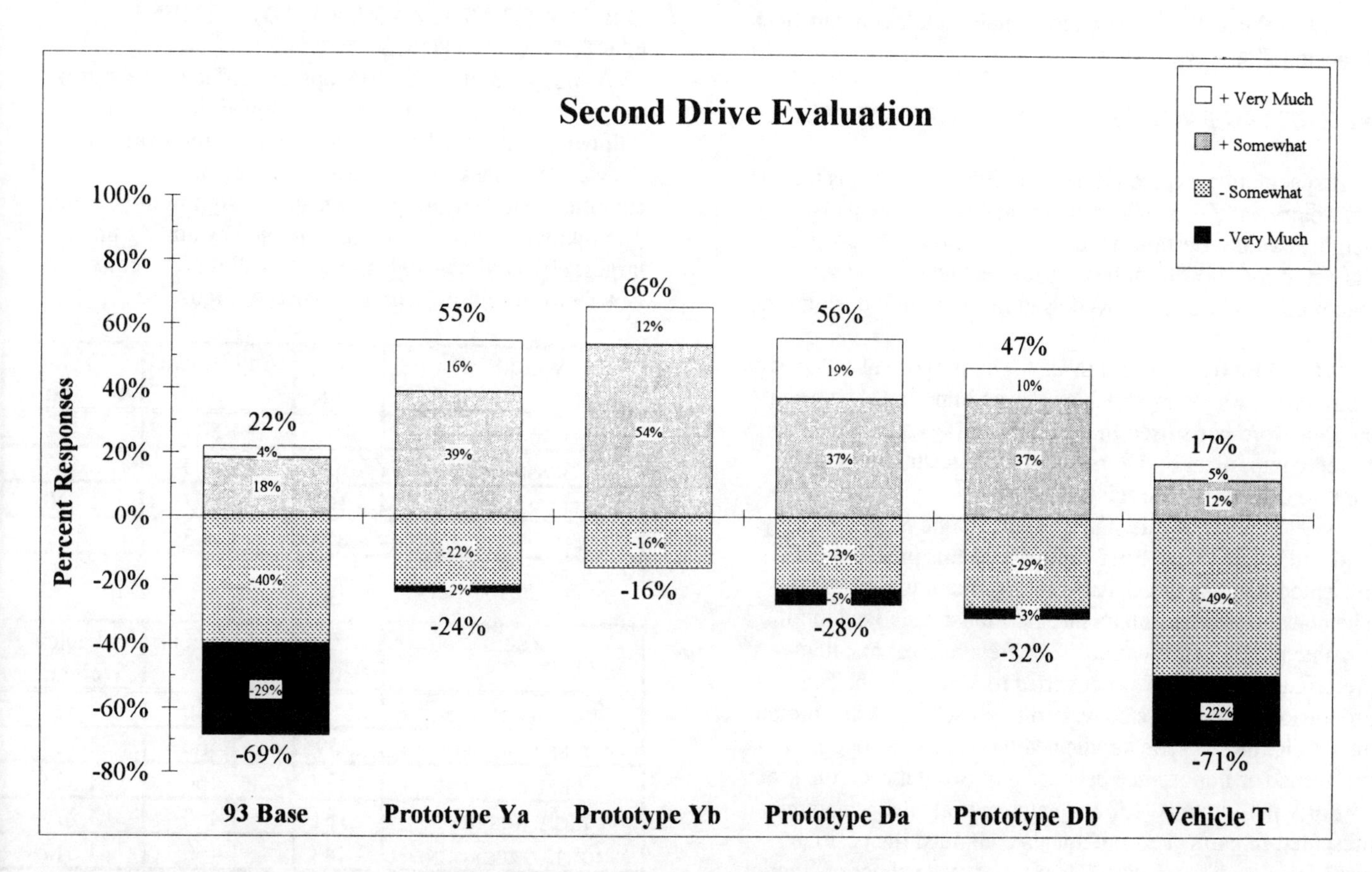

Figure 7 - Percent Positive and Negative Responses for the Second Night Drive

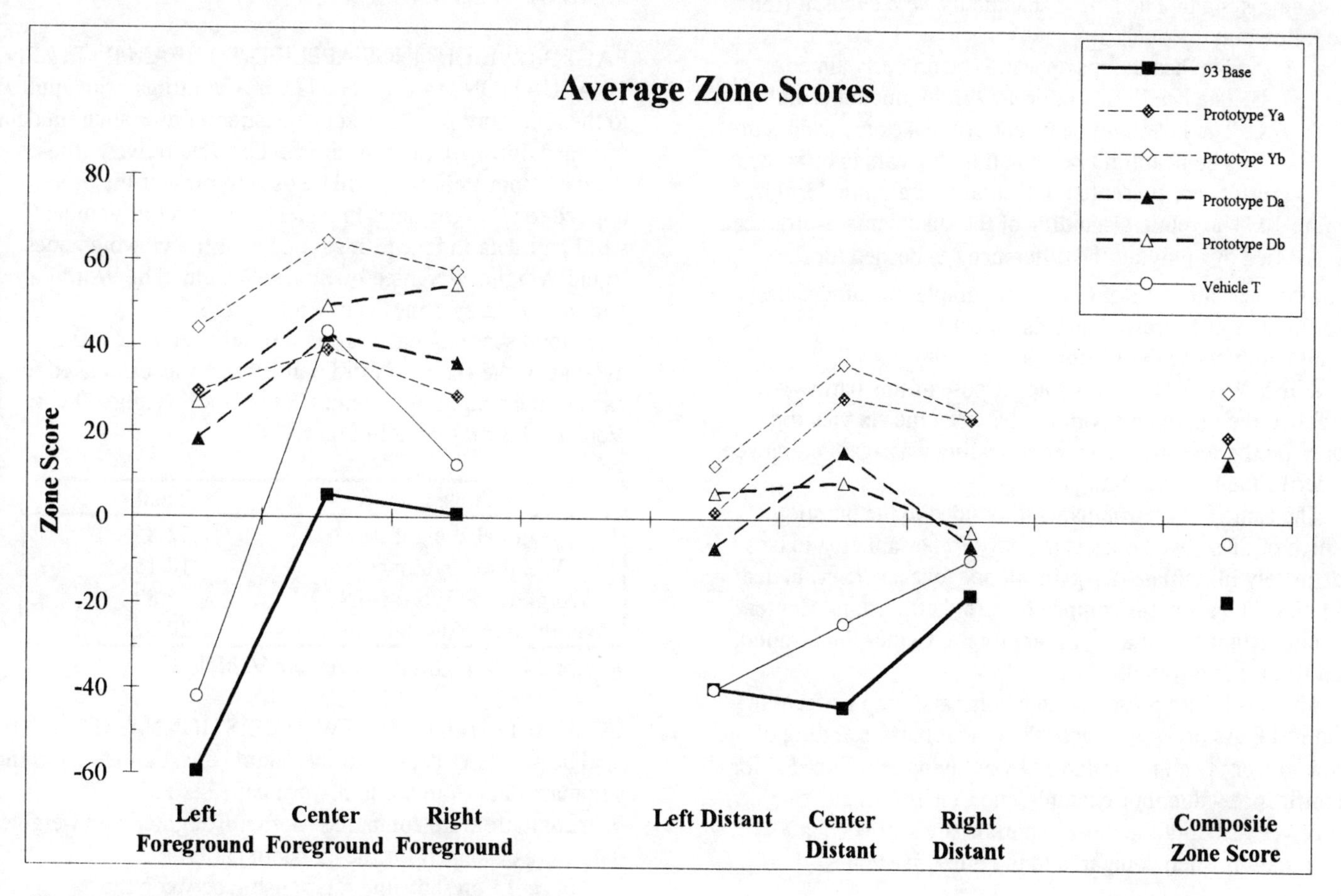

Figure 8 - Zone Profile for the Second Night Drive

Second, Vehicle T became a new benchmark low in our fleet of beam pattern evaluations.

WEIGHTING FACTORS - WORD PAIRS AND ZONES

A speculation expressed in the previous paper was that word pairs and / or viewing zones may not be of equal weighted value. Certain word pairs could be more crucial in characterizing headlight beam patterns and some areas of the road scene could be perceived as more important during night driving.

WORD PAIR WEIGHTINGS - The purpose of this part of the research was to establish a relative importance between the four word pairs used in the zonal ratings questionnaire. A paired comparison or Thurstone scaling technique was employed to derive the relative ratings.

A description of this statistical technique is presented by Guilford (4) or (5). A brief summary of this procedure is presented here for those that want a general understanding. The objects (words, statements, etc.) must be judged in all possible paired combinations. The frequency of one item preferred over another is converted to a proportion. The proportional values are converted to z - scores. When plotted on a scale, the z - scores indicated the relative strength of preference or importance between any two items on the scale.

Forty participants (split by gender) evaluated statements presented in pairs. The statements contained the positive word from each word pair. Their task was to choose which of the two statements was the most important to them in an ideal headlamp system. The paired statements were choosen from the following:

A. The headlamps appear to be properly aimed.
B. The headlights project a bright amount of light.
C. The headlight beam pattern makes me feel secure.
D. The headlight beam pattern appears to be smooth.

The results are presented in tabular and graphic form in Figure 10. The relative standing of the statements (words) can be assessed by applying the difference (Δ) needed for significance value at $P \leq 0.05$ For example: the aim of the headlights was perceived as being most important, significantly more than secure, and smooth.

ZONE WEIGHTINGS - The purpose of this part was to establish the relative importance between the six viewing zones on the roadway. Thurstone scaling was again employed to derive the relative ratings.

The same forty participants responded to the question: Which of these two zones is the most important area to be adequately illuminated? Again, all six zones were evaluated in pairs. The six areas compared were: Left Distant, Center Distant, Right Distant, Left Foreground, Center Foreground, and Right Foreground.

The results are presented in tabular and graphic form in Figure 11. As previously described the relative standing of the viewing zones can be evaluated by applying the Δ needed for significance value. For example: the Center Distant zone was perceived as being the most important area for adequate illumination when compared to the other five zones.

FIRST ATTEMPT: Z - SCORE ADJUSTMENTS - FACTOR WEIGHTINGS

A first attempt at the development of factor weightings for the Word Pairs and Zones was developed that utilized the following scheme. The z - scores for both the word pairs and the viewing zones were adjusted to eliminate the complications of dealing with negative sign multiplication. The relative standings were maintained by adding unity + the largest observed negative z - score to all the z - scores. The new factor weightings are presented in Figure 12

Word Pair	Z Score	Adjustment	Factor Weighting
AIMED / Misaimed	.601	+1.803	2.404
BRIGHT / Dim	.364	+1.803	2.167
SECURE / Insecure	-.163	+1.803	1.640
SMOOTH / Choppy	-.803	+1.803	1.000

Zones	Z Score	Adjustment	Factor Weighting
CENTER DISTANT	.962	+1.460	2.422
CENTER FOREGROUND	.259	+1.460	1.719
LEFT DISTANT	-.137	+1.460	1.323
RIGHT DISTANT	-.181	+1.460	1.279
RIGHT FOREGROUND	-.443	+1.460	1.017
LEFT FOREGROUND	-.460	+1.460	1.000

Figure 12 - Factor Weightings

FACTOR WEIGHTINGS APPLIED TO BEAM PATTERN EVALUATION DATA - The factor weightings were applied to the four word pair / six zone questionnaire evaluation data obtained during both night drives. For illustrative purposes the data from Vehicle C will be used to present the process applied to all of the data. Figure 13 presents the Vehicle C word pair data in four forms for each of the viewing zones: Equal Weight, Weighted by Zones, Weighted by Word Pairs and Weighted by Zones and Word Pairs .

A total score for each of the weighting schemes (the average of the six zone word pair scores) was calculated for each of the weighting schemes. This Total weighted Score for Vehicle C is presented in Figure 14.

Vehicle C	Total
Equal Weight	12.85
Weighted by Zones	18.15
Weighted by Word pairs	15.82
Weighted by Zones & Pairs	20.67

Figure 14 - Weighted Scores for Vehicle C

COMPARISON OF THE TWO QUESTIONNAIRES

The summary data from the "short" questionnaire and the summary data from the four word pairs/six zone questionnaire were compared. Both unweighted and weighted data were evaluated for the 12 beam patterns.

Figure 15 presents the relationship between the unweighted data for both Zone Scores and Large

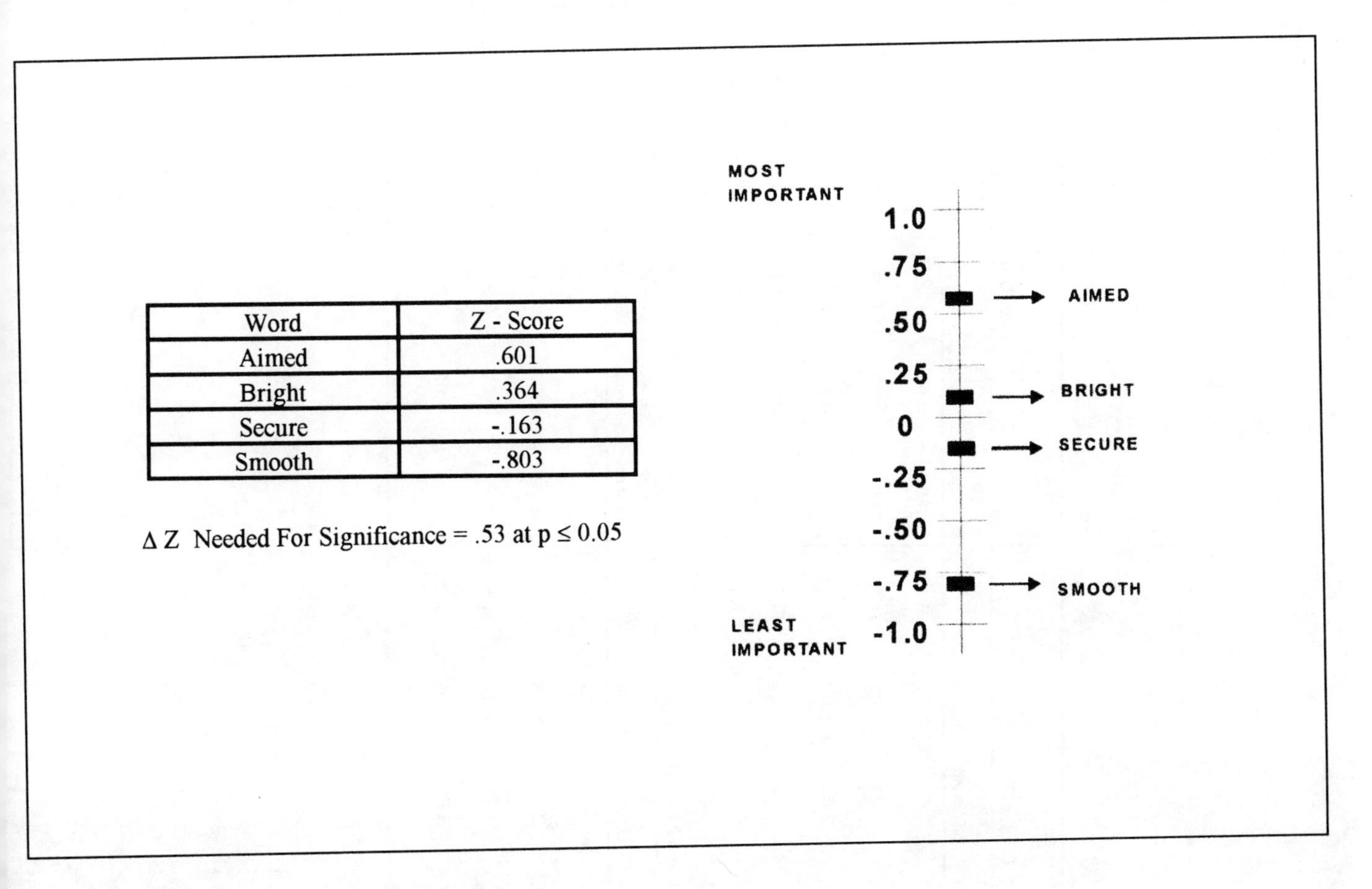

Word	Z - Score
Aimed	.601
Bright	.364
Secure	-.163
Smooth	-.803

Δ Z Needed For Significance = .53 at $p \leq 0.05$

Figure 10 - Relative Word Pair Weightings

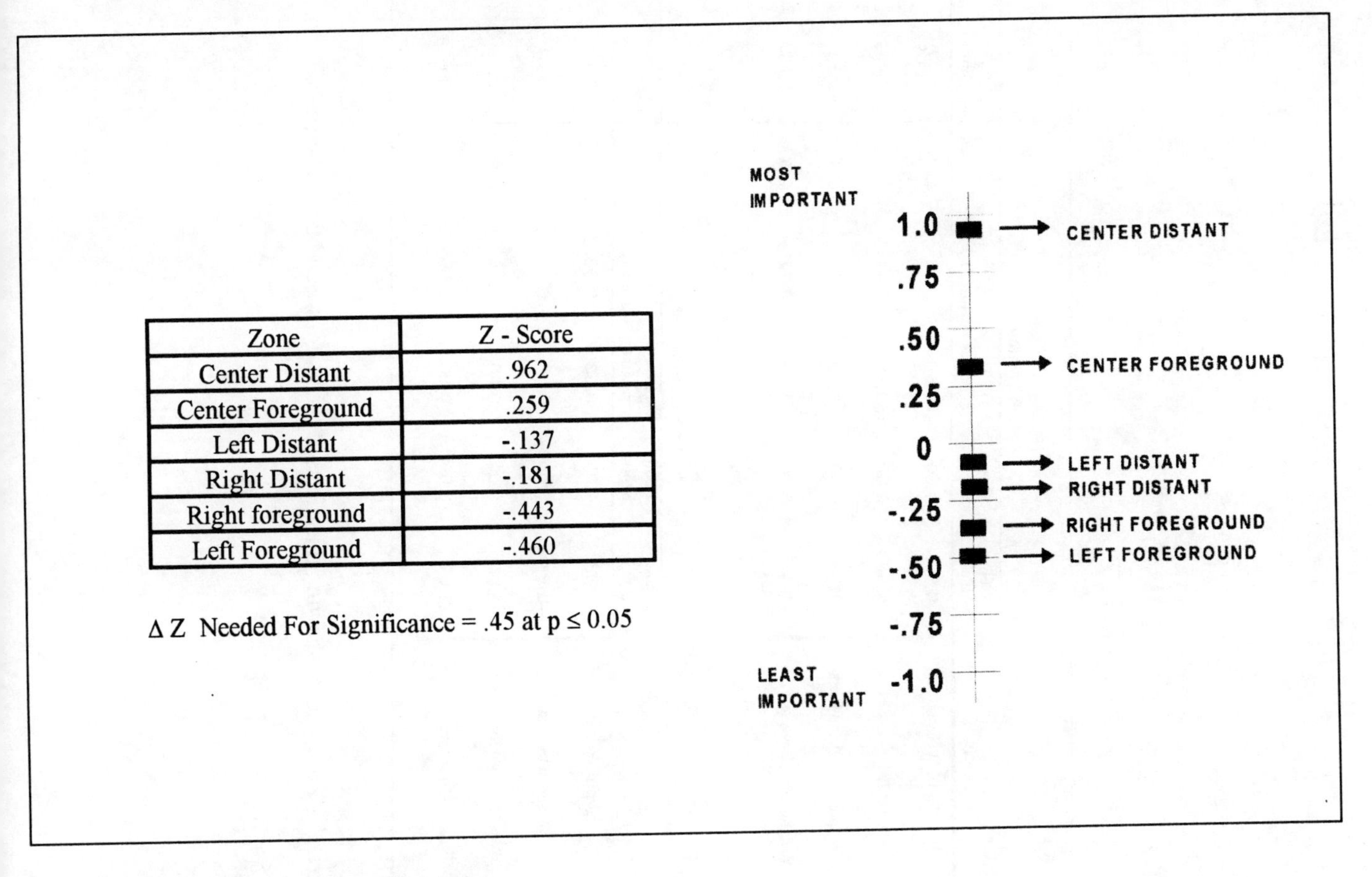

Zone	Z - Score
Center Distant	.962
Center Foreground	.259
Left Distant	-.137
Right Distant	-.181
Right foreground	-.443
Left Foreground	-.460

Δ Z Needed For Significance = .45 at $p \leq 0.05$

Figure 11 - Relative Zone Weightings

Vehicle C	Left Distant					Center Distant					Right Distant				
	Bright Dim	Smooth Choppy	Aimed Misaimed	Secure Insecure	**Score**	Bright Dim	Smooth Choppy	Aimed Misaimed	Secure Insecure	**Score**	Bright Dim	Smooth Choppy	Aimed Misaimed	Secure Insecure	**Score**
Equal Weight	-0.50	0.00	0.50	0.33	**4.17**	0.17	0.67	1.17	0.67	**33.33**	0.83	0.67	1.17	1.33	**50.00**
Weighted by Zones	-0.66	0.00	0.66	0.44	**5.51**	0.40	1.61	2.83	1.61	**80.73**	1.07	0.85	1.49	1.71	**63.95**
Weighted by Word Pairs	-1.08	0.00	1.20	0.55	**4.61**	0.36	0.67	2.80	1.09	**34.15**	1.81	0.67	2.80	2.19	**51.75**
Weighted by Zones & Pairs	-1.43	0.00	1.59	0.72	**6.10**	0.87	1.61	6.79	2.65	**82.72**	2.31	0.85	3.59	2.80	**66.19**

	Left Foreground					Center Foreground					Right Foreground				
	Bright Dim	Smooth Choppy	Aimed Misaimed	Secure Insecure	**Score**	Bright Dim	Smooth Choppy	Aimed Misaimed	Secure Insecure	**Score**	Bright Dim	Smooth Choppy	Aimed Misaimed	Secure Insecure	**Score**
Equal Weight	-0.83	-1.00	0.33	-0.33	**-22.92**	-0.17	0.17	1.33	0.83	**27.08**	-0.33	-1.00	0.33	-0.17	**-14.58**
Weighted by Zones	-0.83	-1.00	0.33	-0.33	**-22.92**	-0.29	0.29	2.29	1.43	**46.56**	-0.34	-1.02	0.34	-0.17	**-14.83**
Weighted by Word Pairs	-1.81	-1.00	0.80	-0.55	**-17.69**	-0.36	0.17	3.21	1.37	**30.35**	-0.72	-1.00	0.80	-0.27	**-8.28**
Weighted by Zones & Pairs	-1.81	-1.00	0.80	-0.55	**-17.69**	-0.62	0.29	5.51	2.35	**52.18**	-0.73	-1.02	0.81	-0.28	**-8.42**

Figure 13 - Zone Scores showing the effects of the various weighting schemes

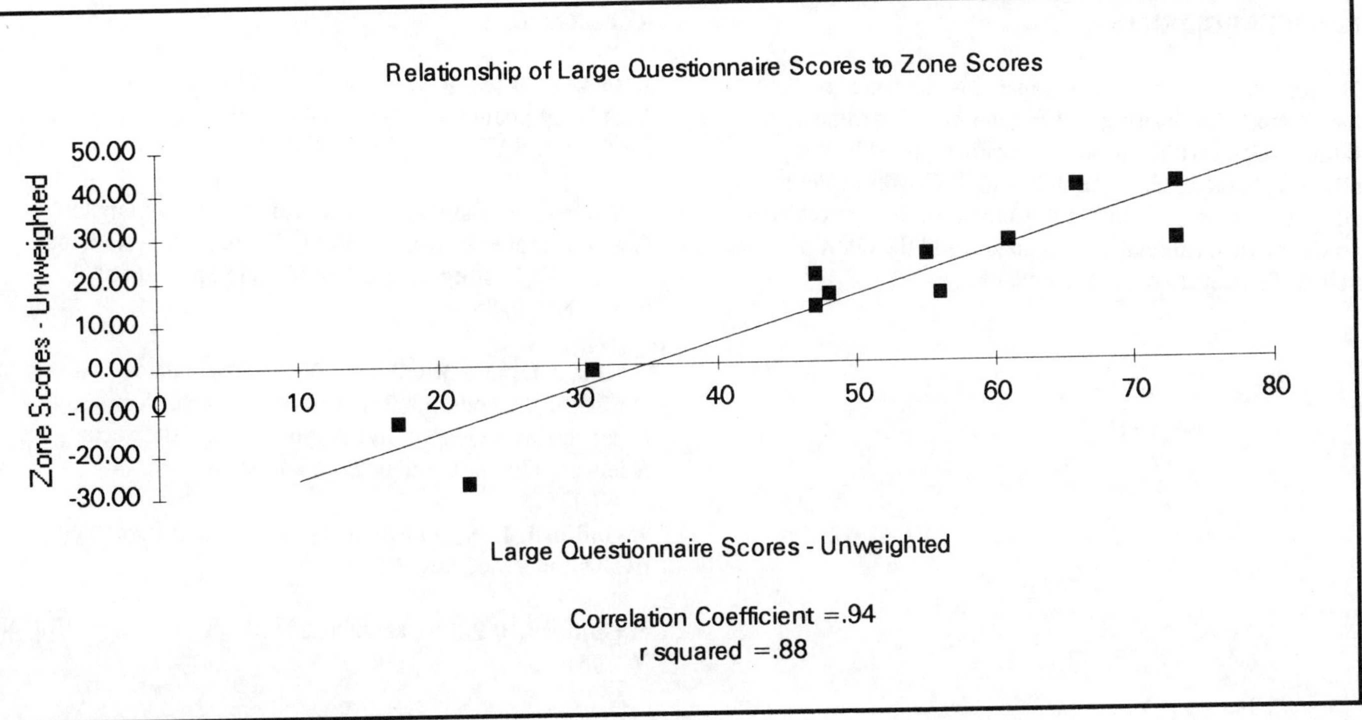

Figure 15 - Relationship of Zone Scores to Large Questionnaire Scores

Questionnaire Scores. The relationship was very strong with a correlation coefficient of .94. This suggests that the two questionnaires could be used interchangeably. Their selection would depend on the specific program objectives. For example, the zone questionnaire could be employed for detailed analysis of the the beam pattern within specific areas on the road.

FUTURE RESEARCH IDEAS

The initial attempts at weighting the word pair / zone data based on their perceived relative importance did not improve the understanding of the data. New weighting schemes and new methodologies for customers to express the relative weightings should be evaluated. A potential methodology would have the customers do the paired comparison tasks related to word pairs and zones while driving at night. Another topic which needs investigation is the zone areas. Where is the perceived separation between the foreground and distant zones? How is this affected by the various environmental factors such as: road type, straight or curved roads, vehicle speed, etc. The zone dimensions and other related matters could be evaluated by using a technique such as eye movement recordings during night driving evaluations.

ENGINEER TO CUSTOMER

As expressed in our previous paper the key part of the longer research project involves the identification of the specific photometric aspects of a beam pattern that make it subjectively pleasing to customers. Once identified, photometric characteristics will be related to the customer's subjective evaluation data using the methodologies already developed.

The relevant identified photometric aspects could then serve as guidelines to specify new beam patterns. These guidelines should supplement the photometric requirements related to SAE and federal standards.

The customer descriptions used during the development of the beam pattern that affected the beam pattern design process would be returned to the customer in the form of a more functional headlamp beam pattern.

ACKNOWLEDGMENTS

The research reported in this paper was conducted by the Human Factors Engineering and Ergonomics Department, Corporate Design with additional support supplied by the Lighting Technology Group and the Small / Medium Vehicle Launch Team, both of Ford Motor Company. We gratefully acknowledge the beam pattern evaluators and the effort of Russell A. Rockett in report construction.

REFERENCES

1. **Jack, D. D.; O'Day, S. M.; and Bhise, V. D.** "Headlight Beam Pattern Evaluation Customer to Engineer to Customer", SAE paper No. 940639 (SP-1033), March, 1994

2. **Nakata, Y.; Ushida, T.; and Takrda, T.** "Computerized Graphics Light Distribution Fuzzy Evaluation System for Automobile Headlighting Using Vehicle Simulation", SAE Report NO. 920816

3. **Kim, J-O; and Mueller, C. W.** (1978) "Factor Analysis Statistical Methods and Practical Issues, Sage University Paper series on Quantitative Applications in the Social Sciences, 07-014. London: Sage Publications.

4. **Guilford, J. P.,** "Fundamental Statistics in Psychology and Evaluation", McGraw-Hill, N.Y. 1950

5. **Guilford, J. P.** "Psychometric Methods, McGraw - Hill, N. Y. 1954

950593

Application of Free Form Reflectors in Modern Headlamp Systems

Henning Hogrefe and Rainer Neumann
Robert Bosch Corp.

Abstract

Car Designers are asking more often for headlamps that adapt well to highly inclined and low vehicle bodies or other exotic shapes e. g. of sports cars. In other cases headlamps are additionally used as stylistic elements and must therefore have a certain appearance e. g. oval shape or a clear, unprofiled cover glass. To realize a headlamp for such applications and at the same time keep it cost-effective with good light quality special headlamp systems have to be developed. Especially the normally used profiled scattering lens in front of the reflector causes complications under these circumstances. Therefore headlamp systems with complex reflector shapes have been developed whose reflectors produce an almost complete light distribution. Some different reflector concepts to realize this idea are presented in this paper and concrete examples for the application of this new technique are discussed.

Introduction

In the automotive field one can recognize overall an increased competition concerning costs, safety features (technology) and styling aspects. The automotive lighting systems and especially the headlights have to balance all three aspects in order to be attractive and competitive. We have just introduced in Europe the plastic lens, which offers a variety of new design solutions for headlamps. We will have headlamps with clear outerlenses, covering an entertainment feature for the customers. And we have introduced on the market the first series production of headlamps with gas discharge lamp (Litronic) in 1991 (Ref. 1). The Litronic system currently represents the most advanced lighting technique worldwide.

To be cost-effective, attractive in a styling point of view and simultaneously presenting an improved light quality in terms of safety for the driver is a big challenge which we try to solve by means of modern headlamp design with free form reflectors. This paper describes our potential to design headlamp light pattern with specially developed computer programs. Taking into account the given boundary conditions (reflector size, inclination of the outerlens, bulb etc.) we will demonstrate that a headlamp can show both: attractive design and high performance in light quality within a reasonable cost frame.

The total light pattern is generated by a summation of filament images or light arc images from all reflector areas (see Fig. 1).

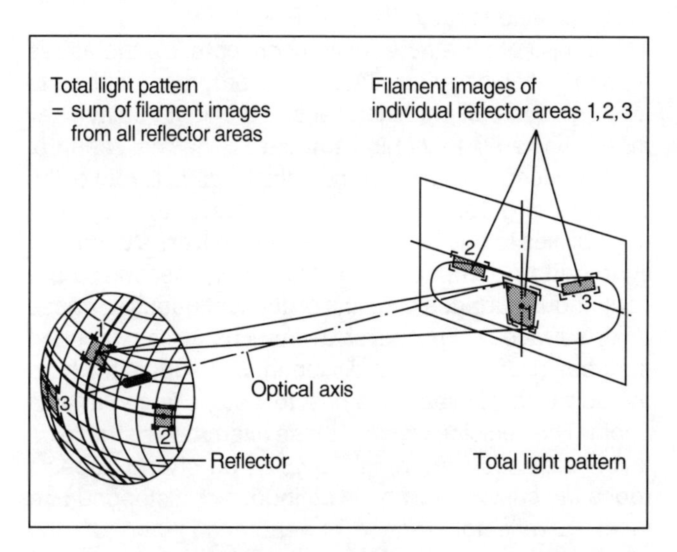

Fig. 1: The light pattern of a reflector can be unterstood as the sum of an infinite number of bulb filament images from the reflector points. A specific light pattern can be achieved by changing the reflector surface in an intelligent way using the fact that the angular orientation of any filament image on the screen is determined by the angular position of the corresponding reflector surface point.

The special arrangement and position of these images is responsible for the quality of the light pattern. The angular position of a reflector point determines in principle the angular orientation of its corresponding filament image: Considering e. g. the most common case of an axially extended filament (e. g. H 7 bulb) one can observe that points which are close to the vertical reflector center section yield a more vertical filament image on the screen, whereas surface points in or closer to the horizontal section deliver more horizontally oriented images. Changing the shape of the reflector surface translates the corresponding filament images while keeping their angular orientation on the screen. The possibility of directing and changing the position of individual filament or arc images and the ability of constructing the desired light pattern point by point is the most important advantage compared to other program systems. Some concrete examples will illustrate our flexibility in computing free from reflectors.

Free Form Reflector Surface Modelling

In order to achieve the outlined performance it is necessary to have the appropriate software tool for reflective surface modelling and lighting evaluation. Our CAL (Computer Aided Lighting) software package described earlier (Ref. 2) is a powerful optics program and it is the ideal platform to perform the optical layout of free form reflectors. With CAL it was already possible since longtime to develop all the known standard and more advanced lighting concepts like H4 paraboloid reflector headlamps, ellipsoidal projector type headlamps (PES) including the lens and even variable focus reflectors (Ref. 3).

To incorporate the new lighting concepts like the above mentioned clear lens reflectors the program just had to undergo some slight extensions. If the light quality shall meet our traditionally high standards, the realization of these concepts requires a fine point by point layout of the optical reflector surface. First, there are several possibilities to build a reflector which delivers a complete light pattern. These range from fully second order continuous surfaces over first order continuous surfaces (i. e. surfaces with one or a few kinks) over facetted surfaces (Ref. 4) to even discontinuous surfaces where the surfaces between one or a few steps is continuous again. The decision which of these alternatives should be taken depends on the light quality, on styling aspects (does the customer want a continuous or a stepped or a facetted reflector surface appearance ?) and on the available mass production methods. In the following we restrict ourselves to the description of second order continuous reflectors since we are convinced that this is the most challenging and interesting technique from the development point of view which in principle contains all other alternatives. Nevertheless, also stepped clear lens reflectors have already been realized by Bosch using the CAL program. Others have developed clear lens low beam reflectors which still have some first order discontinuities and need additional overlayed ripple to achieve a smooth continuously distributed light pattern (Ref. 5).

The needed mathematical description of the continuous surface must allow the variation and optimization of every local area of the reflector. The filament images must be point by point aligned below and as close as possible to the cut-off. For low beam, light from no point of the reflector is allowed to be directed above the cut-off. There must be a pronounced maximum near the center of the light pattern but no maxima or minima besides the central maximum i. e. the light pattern must be homogeneously distributed. Additionally, in most cases, the reflector should collect as much light as possible. These requirements result in a point by point precision of the reflector surfaces in the range of a few hundreds of a millimeter and a point by point angular precision of less than 0.1 degree.

A suitable mathematical description for this application may be given by certain analytical equations, by polynomials, by different spline-methods e. g. Bezier or even simply by points and normals in space.

After extensive tests some of these methods have been incorporated in our CAL system. In Fig. 2 one of the implemented surface descriptions is illustrated graphically. We call our second order continuous reflectors "Homogeneous Numerically Calculated Surfaces" HNS.
At the beginning of the HNS-reflector design the available space (width / height / depth) and type of bulb are needed as input data. With this method, a grid of points is generated on a model-surface in the computer. This grid

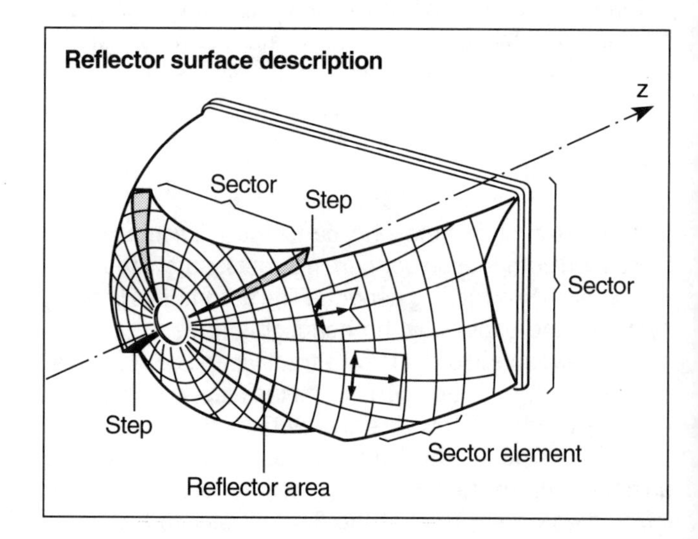

Fig. 2: A mathematical surface description is necessary which allows fine point by point or area by area surface optimization. Arrows indicate radial and tangential surface shifts. Here, surface steps are additionally shown to demonstrate our most general approach.

should be very fine yielding several thousand points distributed over the whole reflector surface. Each point can be thought as the center of a small facette-like surface area of the reflector. Each of these small areas collects a small amount of light from the burning filament of the bulb and images this light onto the measurement screen or onto the street. In an iterative process these facettes can be displaced and oriented to each other in space in order to position the collected light. Some key parameters of the desired light pattern can be entered to control the iteration process: e. g. the shape of the light pattern cut-off, the maximum illuminance or the scattering width etc. (Fig. 3). At the end of the iteration process all surface points are optimized to obtain the desired light pattern and a smooth reflector surface is generated. With this HNS-method already several high performance headlamps have been developed in connection with profiled scattering outerlenses. These headlamps which have been brought to series production deliver excellent light quality.

We now want to construct a reflector which yields only in connection with a proper bulb, for example the H7 bulb, a low beam light pattern that needs no further correction by an outerlens. This can be done using the extended CAL software as described above with a further small adaptation regarding this specific problem.

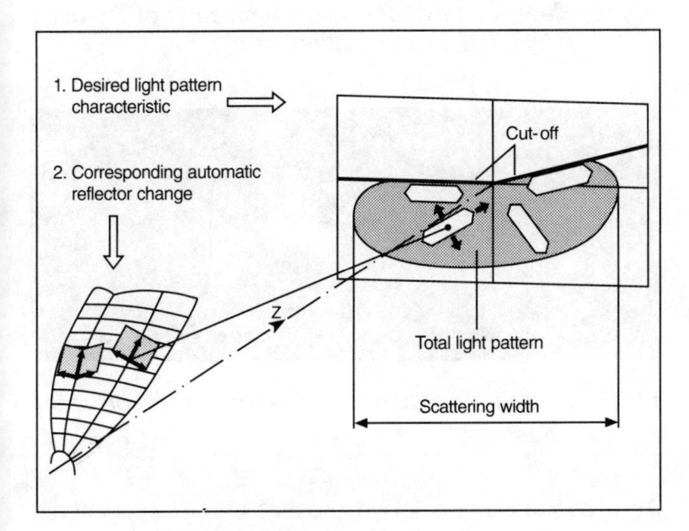

Fig. 3: Some light pattern key parameters can be specified such as the scattering width in order to control the iterative orientation of small reflector areas.

A vehicle light pattern exhibits, depending on the headlamp type (low beam, high beam, fog beam) and the individual properties of the headlamp, a more or less broad horizontal scattering and only little scattering in the vertical direction with the scattering intensity continuously decreasing from the pronounced central maximum to the sides. Also a sharp horizontal cut-off is needed (for low beam and fog beam) to prevent glare light to oncoming traffic. Therefore such reflector must scatter and concentrate the collected light in the horizontal forward direction whereas it only has to concentrate the light in the vertical direction. Although the extension of a real filament guarantees some angular degrees of scattering with every reflector, this task cannot be performed by conventional reflectors with parabolic, ellipsoid or hyperbolic shapes which are usually taken for current headlamp systems using additional active lenses. Fig. 4 shows the horizontal scattering behaviour of some reflector types that deviate from the conventional ones in the desired way. These free form variants are able to scatter the light horizontally and at the same time concentrate a certain amount of the collected light parallel or nearly parallel to the optical axis. Between horizontal and vertical direction there is a transition from free form behaviour as discussed to nearly parabolic behaviour since there must be no additional scattering in the vertical direction. Depending on the required light pattern and the applied bulb, one of the three free form variants of Fig. 4 or even a mixture of them may be the best choice.

To apply these basic strategies to real reflector design some experience has to be collected regarding the precise control of the HNS iteration process. If, on the one hand, too much light is directed to the left and right sides and too little to the center, the maximum in the center of the light pattern will not be satisfactory and also a too high amount of light on the left and on the right will be irritating to the driver. If, on the other hand, too little light is scattered and too much is concentrated, unacceptable inhomogenities in the light pattern will occur and the side illumination will be too sparse. So, a fine balance between these effects is necessary.

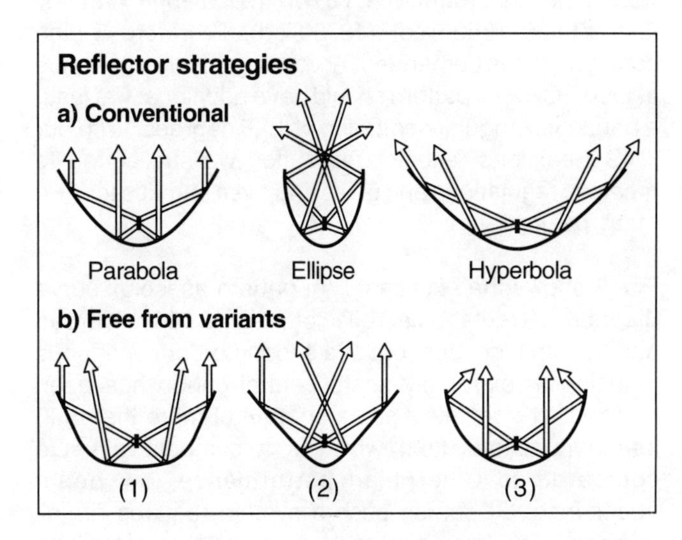

Fig. 4: Imaging strategies of free form headlamp reflectors in comparison to conventional reflectors. A free form reflector must be able to concentrate the light parallel to the optical axis and at the same time distribute it over a large angular width.

Also, from the introductory remarks about the principal angular orientation of the filament images and regarding the desired light concentration below and close to the cut-off, it is clear that horizontally positioned reflector points (i. e. points within ± 30-50 degrees from left and right horizontal axis) are of higher value for the generation of the light pattern than vertically positioned points. Therefore, during reflector layout, increased emphasis should be given to these reflector areas and the available reflector width should not be too small.

Clear lens headlamp examples

We used the described reflector modelling capabilities for concrete headlamp realization. The most demanding application is the European (ECE) low beam light pattern. In this case the regulation requires a very sharp cut-off allowing only for 0.7 lux (437.5 cd) glare light directly above the cut-off just 0.57 degree vertically separated from points below the cut-off where 12 lux minimum illuminance (7500 cd) are required. In current headlamps with good light performance we reach approximately 25 - 30 lux (15625 cd - 18750 cd). Therefore a high contrast ratio is demanded. According to our experience a good ECE light pattern should have additionally at least a horizontal angular scattering of ±25 degrees. With our HNS clear lens reflector realization we succeeded to meet the regulations and goals and even surpass them in many points.

Fig. 5 shows the excellent light pattern as isolux curve diagram of a rectangular reflector which is 160 mm broad and 90 mm high. Just out of a second order continuous reflector an extremely sharp central cut-off has been realized. The reflector has an output of more than 500 lumen (luminous flux) which is a competitive value compared to other high performance low beam headlamps. Of course such a system requires a well designed, low tolerance light source with no extra light coming from areas outside the burning filament (e. g. bulb glass reflections). Therefore, for our sample the H7 bulb was used.

When designing free form reflectors it is very helpful to see the light pattern of different variants in direct comparison in a realistic road scene instead of only having isolux curve diagrams.

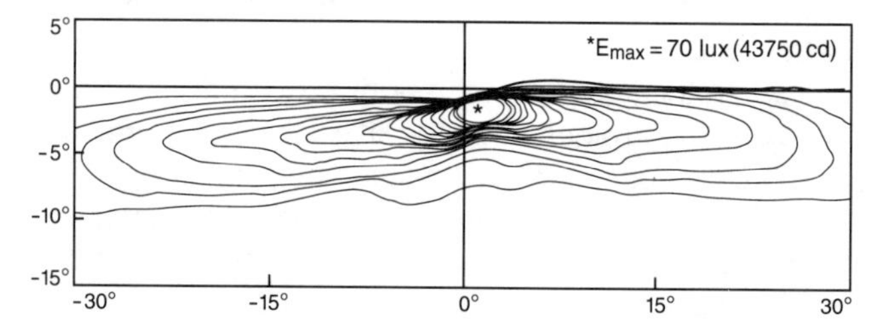

Fig. 5: Light pattern of a clear lens low beam reflector. Isolux lines on a measuring screen (25 m).

With our so-called visualization simulation system (Ref. 6) we can simulate the appearance of any calculated or measured light pattern in realistic scenes which is shown for the light pattern of Fig. 5 in Fig. 6. Again, here we see that our clear lens reflector gives a fully satisfactory illumination. This has of course been verified by night time driving tests. Fig. 7 shows, again generated by the visualization program a simulated screen projection of this light pattern as it could be seen on the screen of a lighting laboratory.

Fig. 6: Real scene visualization simulation of the clear lens reflector light pattern of Fig. 5.

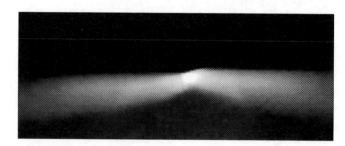

Fig. 7: Screen projection visualization simulation of the light pattern of Fig. 5.

As a second example we present an HNS fog beam headlamp. Using a concept similar to that shown in Fig. 4b (2) we call our free form fog lamps CD fog lamps (Converging Diverging). Our sample reflector with the isolux curves and visualization simulation shown in Figs. 8 and 9 is 100 mm broad and 70 high, has a luminous flux of about 500 lumen and has an impressive scattering width of more than ±40 degrees. With this technique even reflectors as small as 60 x 40 mm^2 yield a good light quality.

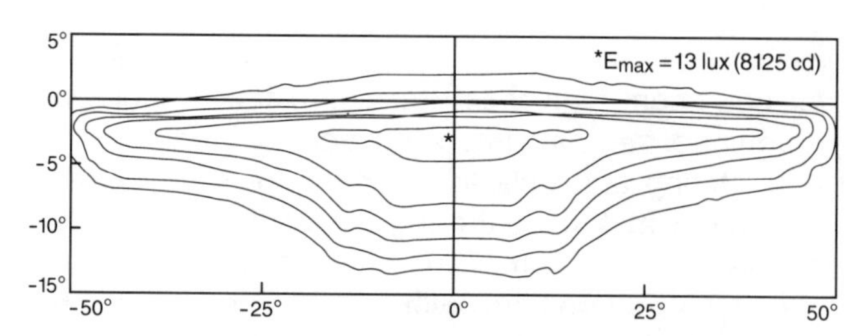

Fig. 8: Light pattern of a clear lens CD fog lamp. Isolux lines on a measuring screen (25 m).

The availability of clear lens fog beams is especially important since the geometric design conditions (i. e. the available space) for real fog lamps are often extremely limited and also strongly inclined scattering outerlenses must often be used which degrade the light quality of conventional systems.

Fig. 9: Real scene visualization simulation of the CD fog lamp light pattern of Fig. 8.

Summary

With these examples we have shown that the optical layout and realization of clear lens reflectors is possible using adequate software tools. Other applications are possible e. g. clear lens high beam reflectors or "semi clear lens reflectors" where the reflector makes some prescattering and a simplyfied outerlens completes the light pattern. Our CAL has the flexibility to realize any of these advanced lighting concepts. A headlamp covering all light functions (low beam, high beam, fog lamp and turn signal lamp) with a clear outer lens can be designed for ECE as well as for SAE requirements. By means of the visualization program the results can be seen on realistic road scenes in a very early stage of development. CAL offers the possibility to combine styling aspects when designing a headlamp with high performance in light quality, thus contributing to improved safety in night time traffic.

References

(1) B. Wörner, R. Neumann:
Motor Vehicle Lighting Systems with
High Intensity Discharge Lamps,
SAE Technical Paper Series 900569, 1990

(2) R. Neumann, H. Hogrefe:
Computer Simulation of Light Distributions
for Headlamp Systems,
SAE Technical Paper Series 910827, 1991

(3) R. Neumann:
High Efficiency Headlamp Systems with
Variable Focus,
SAE Technical Paper Series 890687, 1989

(4) T. Fujita, T. Ichihara, H. Oyama:
Development of MR (Multi Reflector) Headlamp
(Headlamp with Slant Angle of 60 Degrees,
Contributable to Future Vehicle Body Styling),
SAE Technical Paper Series 870064, 1987

Y. Nakata:
Multi B-Spline Surface Reflector optimized
with Neural Network,
SAE Technical Paper Series 940638, 1994

(5) N. Nino, H. Ishida:
Development of Controlled Surface Reflector
for Headlamps,
SAE Technical Paper Series 920813, 1992

(6) A. Bonvin, Ch. Liétar:
Visualization of Headlamps Light Distribution –
The Road into the Show-Room,
Automobiltechnische Zeitschrift (ATZ) 91, No. 9,
p. 524, 1989, in German

950594

New Optical Simulation Systems Revolutionise Headlamp Development

John F. Monk
Magneti Marelli

ABSTRACT

A new generation of optical simulation software has been developed and is now in use to optimise the design and performance of vehicle headlamps that are destined to set new standards of lighting performance.

The new system differs from previously available products in that it has greater prescriptive powers in generating the new reflector form to satisfy the input requirements. Furthermore, the structure of the software and its use make it possible to streamline the headlamp design and development process, resulting in time and cost savings.

This paper discusses the principles, structure and performance features of the new software, and goes on to examine the changes possible in methodology used to organise the headlamp development and optimisation process. Finally a case study is given to illustrate the system performance.

INTRODUCTION

It is widely acknowledged and appreciated that many significant advances have taken place in the product and manufacturing technologies used for vehicle headlamps in recent years, resulting in startling improvements in optical performance.

Most notable among these advances have been the development of new materials and manufacturing processes to enable greater freedom in the design of the headlamp, which in turn provided the stimulus for optical designers to create new and more efficient optical forms. Details of these developments are described elsewhere (1,2,3).

The move away from traditional round or rectangular shaped reflectors, with simple parabolic optical forms, to increasingly complex reflector geometries has been enabled by the increasing use made of computers, first to generate the new reflector forms, and then to calculate the optical characteristics and performance of the complete headlamp system.

This evolution in product performance and supporting technologies has been rapid and of fundamental importance. However it has not been without problems. In the opinion of the author some of the most serious of these problems relate to the fact that the new approaches to the computer generation of the reflector form and the calculation of optical characteristics have not yet been exploited in a way that optimises the headlamp development process. In particular there appears to be further scope for improvement in the aspects of (a) tighter headlamp performance specifications to be agreed between customer and supplier at the outset of the project, (b) more ample and demonstrable headlamp performance detail during product development phases, and (c) elimination of un-necessary design stages with associated time and cost savings.

The Lighting Division of Magneti Marelli has already played a leading role in the development of the new materials, manufacturing processes and advanced optical solutions for vehicle headlamps. In addition it has seen the need to develop a new generation of optical simulation software which is now being used to optimise the design and performance of new products. The calculated performances predicted by the system are also used to construct a vivid photorealistic representation of a road scene illuminated by the designed lamps. The quality and accuracy of this representation, and the facility with which the road scene can be manipulated at will, in real time, are unrivalled at this moment.

As a result, this tool provides a revolutionary approach to headlamp development, which enables headlamp performance to be optimised and faithfully demonstrated in software. Its application is being used to fundamentally change the traditional design and development methodology for vehicle headlamps. Significant benefits are now coming from the greater certainty of product performance from the outset of the project, reduced time to manufacture and cost, reduced prototyping activity and cost and greater standardisation potential.

EXISTING METHODOLOGIES FOR COMPUTER ASSISTED DESIGN AND OPTIMISATION OF HEADLAMPS

The principal steps involved in the process of headlamp design and development up to the stage of off-tool samples are shown in flow chart format in Figure 1. As may be expected, the car manufacture provides the imputs defining the available installed volume and the general styling guidelines. At the same stage it is also necessary to establish the optical performance targets for the product, this task requiring the close collaboration of supplier and

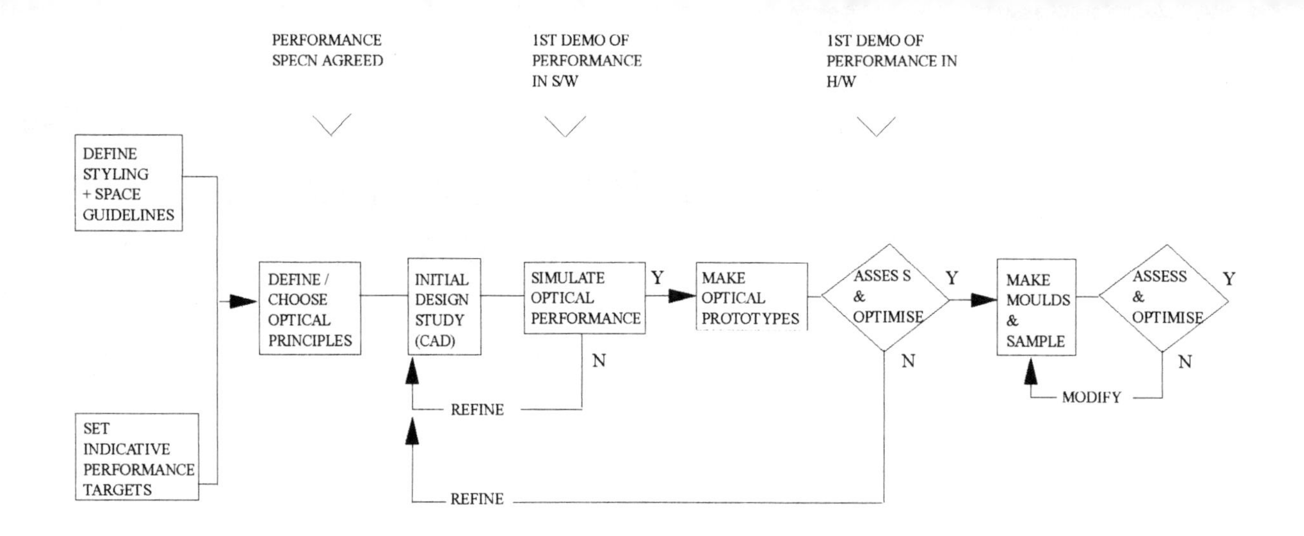

FIG.1. Existing Methodology for Design and Optimisation of Headlamp Optical System

customer. Usually the targets are oriented towards the optical performance of an existing product which can serve as a benchmark for comparison. From this phase it is necessary to confirm and agree the optical principles to be employed, which strongly influence the performance, dimensions and aesthetic characteristics of the final product. For a project involving a new, rather than existing reflector geometry, it is necessary to create an initial reflector design in CAD media before a computer simulation of optical performance can be attempted. The commonly used simulation software packages for this purpose are based on an integration of repetitive ray tracing calculations with the result presented in the form of isolux curves of equal illuminance as if projected on a flat screen placed at 25 metres distance in an optical dark room (Fig.2). An optimisation loop can be established at this stage, before an optical prototype is produced, lensing is developed and further system optimisation carried out prior to a design freeze and manufacture of definitive (production) tooling.

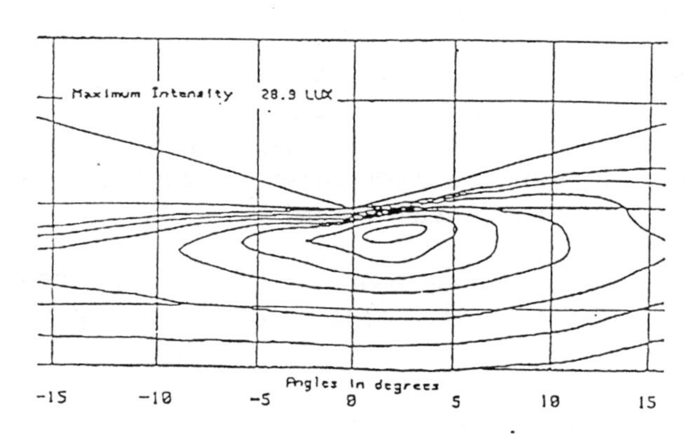

Fig.2 Beam Pattern from a European Complex Surface Reflector in Isolux Format.

The use of optical simulation software obviously plays an important role in the described process. However the whole development process cannot be considered as optimised for the following reasons:

(a) in the first phase of setting indicative performance targets for the headlamp, the only benchmarks available are existing products. This means that in the case that the newly requested product is required to differ significantly from previous products, for example in optical principles, basic geometry or relationship of geometry to required performance - as is becoming increasing common, then the benchmark that may be agreed for starting the project may differ significantly in performance from the ultimately developed solution.

Certainly, with the information likely to be available at this point, it is uncertain that a genuine product performance specification could be written with confidence..

(b) Since, in the case of a new reflector geometry, there is no CAD definition of reflector form available to start with, then it follows that there is no possibility of using optical simulation software to provide objective guidance during the critical first project phases in which the optical performance specification is usually agreed.

(c) As can be seen from Figure 1, in this sequence of events, the first simulation of the optical performance of the new headlamp is likely to be available several weeks into the project.

(d) Because the design evolution from initial concept to the simulation output following the design study may be substantial, it is highly likely that optical prototype hardware is always necessary to give a high level of confidence to all parties at this point in the project.

THE NEW GENERATION OPTICAL SIMULATION SOFTWARE

When the Lighting Division of Magneti Marelli took the decision to specify and develop a new generation of optical simulation software, it approached the project with the objective of achieving a major breakthrough in the functionality and usefulness of system, and not just an incremental improvement of the existing solutions.

To this end, three principle targets were set

(i) To develop a software that would be strongly autonomous and prescriptive in its functional capacity.

That is to say that, unlike previous systems which calculate and visualise headlamp performance based on precise data inputs (eg designs, dimensions, shapes, component choices) which have to be decided in a previous step prior to the simulation run, the new software would be expected to search and discriminate between many different potential solutions, starting only from an input of the required headlamp function performance (in terms of desired isolux distribution) and the principle product dimensions.

(ii) The visualisation of the simulated headlamp performance would be more realistic and reactive than previously available system,

and importantly (iii) the characteristics and performance of the system would enable simulation to take place at the earliest possible moment in the design sequence, and permit a seamless dialogue to be maintained with the CAD design process.

The system development has been successful in reaching these goals, and is already confirming the benefits that come from the possibility of changing the headlamp design process.

A brief description of the main features of the system is now given with reference to Figures 3 - 5.

As mentioned previously optical simulation systems are already in commercial use, amongst which the most advanced allow simulation of the full optical system (reflector, bulb and lens) with output in isolux, grey scale or limited road scene format (4). The architecture of the Magneti Marelli system (Fig.3) is significantly different to previous products.

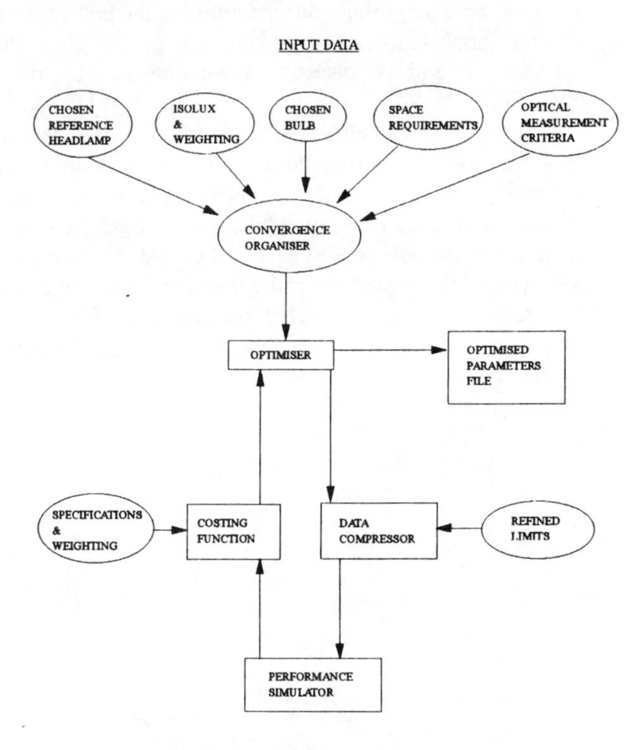

FIG.3. Functional Diagram of the Simulation Software Architecture

The input requirements for the study consist of the lamp dimensions (coming from the available installation volume), a selected reference headlamp, the optical measurement criteria, the selected bulb and the requested isolux illuminance distribution with suitable weighting factors given for the distribution. The required isolux distribution for the new product is created directly on the terminal screen and then entered into the study files. A simple example is shown in Figure 4.

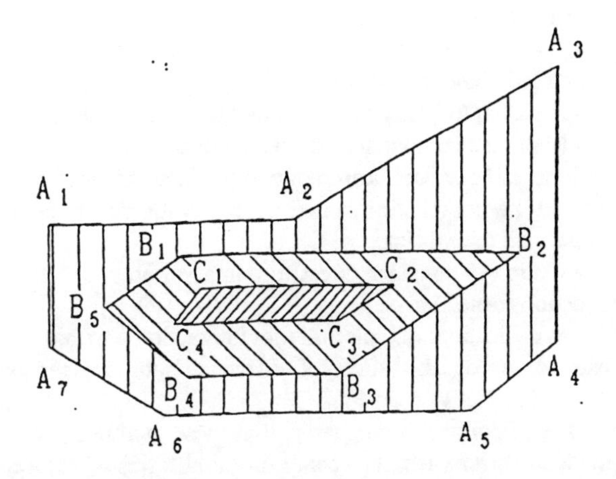

FIG.4. Schematic Version of Isolux Input to Simulation Study

It should be noted that the study does not require the provision of the mathematical description of the required reflector form. On the contrary, the study is organised to provide proposals of the required form as an output of the study.

The system then carries out a preliminary structuring of the input data in the convergence organiser, prior to the first phase of optimisation of the input data and parameters. At this point there is still a large body of data and potential options to be analysed. It is therefore necessary to carry out a second data compression process with the application of more refined limits than were used in the previous phase. At this point we have the required data and formatting to carry out the detailed optical performance calculations,.

The final control of the output is made in the costing function which compares the best achieved optical results with the required (weighted) specification. In this context low cost is perceived to be linked to the closeness that the simulated result approaches the original requirement.

The optical calculation processes carried out in the performance simulator do not make use of traditional ray tracing methods, but result from the use of newly developed calculation algorithms derived from the physical laws relating to energy transfer. The bulb is defined as a photometric solid and the filament as a surface with the possibility of subdivision for greater accuracy. The complete energetic simulation can also take account of multi-reflections into the reflector from the bulb shield or other reflective elements.

Since the optimised performance simulation has been reached through an iterative process of refinement involving changing and retesting the reflector geometry then it follows that the principle output of the process are reflector form options defined directly by the software.

The visualisation facilities offered by the system offer new standards in terms of realism and capability. Obviously the conventional displayed outputs include the isolux and grey scale formatted results. it is however in the capability to display the headlamp performance in simulated photorealistic road scene conditions that the system demonstrates its outstanding capabilities. Most importance among these are:

- Freedom to create complex road scenes with precalibrated real elements without significant penalty in terms of calculation time.
- Freedom to rotate and / or change the point of view of the road scene in real time.
- Freedom to "walk into" the road scene in real time.
- Multitask comparisons displayed in split screen format (eg head-to-head comparison of two products: isolux format results v photorealistic road scene etc), with manipulation maintained in real time.
- Option to consider direct illumination only, or global reflection scene analyses.
- Verification of colorimetric interactions within the road scene (eg the effect of different light sources on coloured reflective road signs).

It is important to emphasise that these facilities can be provided without resort to compromise in terms of reduced calculation accuracy.

IMPACT OF THE NEW SIMULATION SOFTWARE ON THE PRODUCT DESIGN AND OPTIMISATION PROCESS

From the description of use given in the previous section and the flow chart of a proposed methodology for the new design process (Fig.5), it can be seen that the simulation process can be started at a much early stage in the design process than previously possible, since the system does not require a reflector design to function. This benefit is of great importance because it allows the car manufacturer and lighting supplier to agree the optical principles and performance specification in a more precise and objective manner than previously possible at the start of the project. It also means that the confirmation of the headlamp performance, based on the detailed optical simulation study, is also available earlier than previously possible in the design process.

The net result of the streamlining of the design process plus the fact that customer and supplier are working to a more clearly defined and demonstrable target means that the whole development process can be shortened with related cost savings. It also offers the prospect of reducing the amount and cost of downstream re-assessment, revision and modifications.

CASE STUDY

The case study chosen to illustrate the use of the simulation software involved the development of a completely new reflector for the dip beam function for a customer who was looking for high total luminous output from a restricted surface area asymmetric reflector. Initial discussion and optical calculation suggested that the optical system necessary could not be based on parabolic or derivatives of the parabolic optical principle. Instead the optical system would need to make use of all the reflector surface to give a high output from the limited area.

Therefore the problem at the start was that there was not an available relevant existing product to demonstrate to the customer and use as a valid benchmark for the specification and design of the new product.

Given the required lamp dimensions it was therefore necessary to derive the allowable dimensions for the dip beam function and then create the required isolux distribution for the new product directly as the input into the simulation program, this being done in an interactive manner involving the appropriate specialists of the customer and supplier (Fig.6). With these basic inputs, given suitable weighting, plus definition of the bulb and the selected optical measurement criteria to be used the optimisation and performance simulation run could be carried out.

The iterative calculation process was then allowed to progress with monitoring until a satisfactory output was achieved. At this point the output of the study was critically evaluated in isolux, photorealistic road scene and other useful formats (see Figs. 7 to 11). At this point the customer was provided with a clear and objective demonstration of the expected performance of the new headlamp, and authorised the hard tooling development phase.

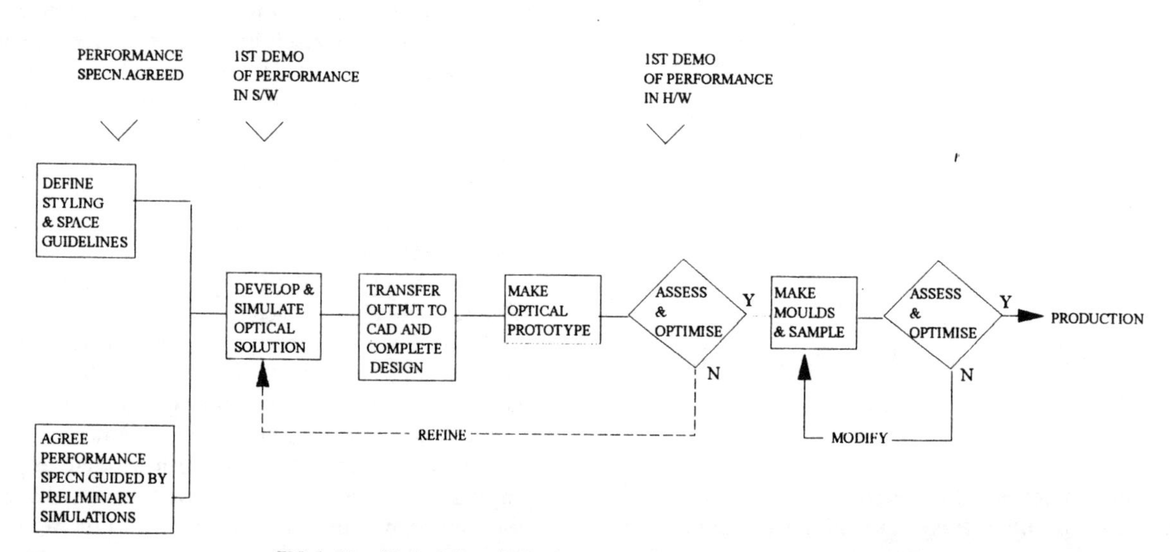

FIG.5. New Methodology for Design and Optimisation of Headlamp Optical System

The development of the headlamp through prototyping and pre-production sample development phases proceeded very smoothly, with little need for optical fine tuning, and certainly no major problems or revisions. The product reached production comfortably within the planned deadline in a total development tune some 15% less than would have previously been possible using traditional software tools and design methodology.

SUMMARY

The use of optical simulation software programmes to assist in the design of vehicle headlamps is already known and practised.

The newly developed Magneti Marelli system differs from other approaches in that it not only offers new standards in optimisation and demonstration of headlamp performance in software, but it can also be a powerful tool in streamlining the development process and achieving closer and more effective working partnership between customer and supplier.

Many important benefits can stem from this new methodology, including greater certainty of product performance from the outset, reduced time to manufacture and cost, reduced prototyping activity and cost and greater standardisation potential.

ACKNOWLEDGEMENTS

The author would like to express his thanks to Mr S Zattoni who has been responsible for the conception and successful management of this project for his help in the preparation of this paper and also the team of software specialists who have worked to realise the finished product.

REFERENCES

1. Monk John F, and Morgan Leonard "The Manufacture of a New Generation of Headlamp Reflectors". Proceedings of Institute of Mechanical Engineers Vol.199 No.B3, 1985.

2. Spencer, Charles W and Manunta, Giorgio, "High Performance Low Profile Headlamps", SAE Technical Paper #870060, SAE International Congress and Exposition, 1987.

3. Curtis Anthony, "The Age of Enlightenment" Car Design and Technology Magazine Vol.1 Issue 2, 1991.

4. Neuman Rainer, and Hogrefe Henning, "Computer Simulation and Light Distributions for Headlamp Systems", SAE Technical Paper #910827, SAE International Congress and Exposition, 1991.

ABOUT THE AUTHOR

John Monk is currently the Innovation and Design Services Manager of the Lighting Division of Magneti Marelli based at Venaria in Italy.

He is a Chartered Engineer, Fellow of the Institute of Materials and Member of the Institute of Electrical Engineers whose working life has been principally involved in research and development within the Motor Industry.

In particular, he is an acknowledged expert in the field of polymer technology and is most known for the innovation of materials and processes for the manufacture of plastics headlamp reflectors that have revolutionised the design and performance of this family of products worldwide.

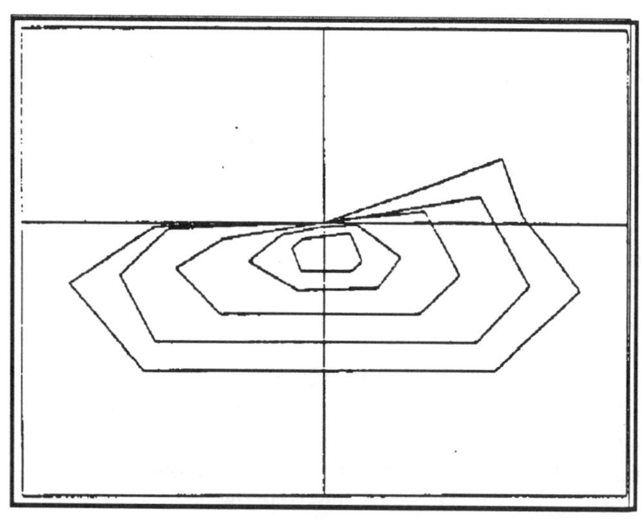

Fig. 6 Required Isolux distribution for the dip beam function

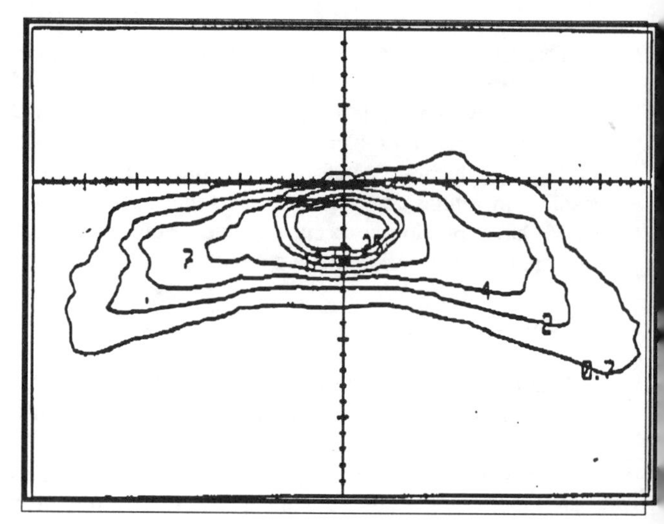

Fig. 7 Intermediate study output of dip beam isolux distribution

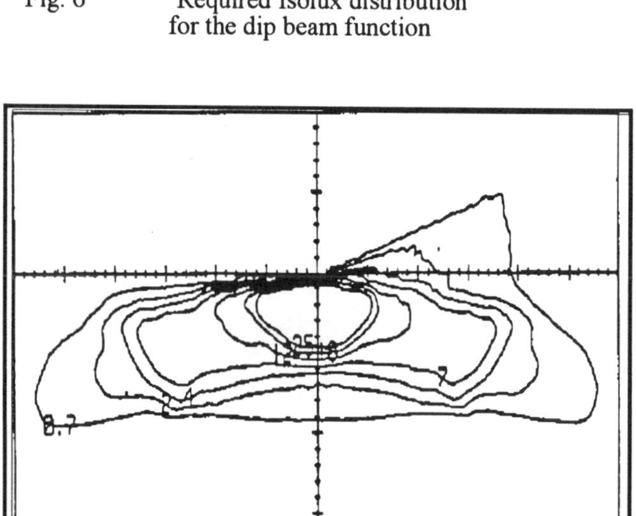

Fig. 8 Final study output of optimised dip beam distribution

Fig. 9 Photorealistic road scene simulation of optimised dip beam

Fig. 10 An example of a "walk-in" examination of the fig.9 scene

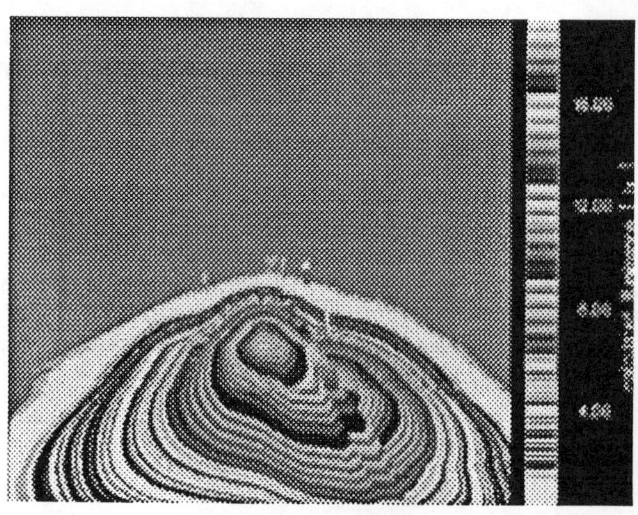

Fig. 11 Stepped isolux representation of the fig.9 road scene

950595

Development of High-Speed Photometric Measurement System

M. Sasaki
Koito Manufacturing Co., Ltd.

ABSTRACT

Due to the increased performance and reduced prices of workstations and personal computers, highly advanced, computerized analysis has become a common procedure, including simulations such as the generation of virtual road surface images on CRT displays - images generated by irradiating the road with headlamps.

Analysis of this type is carried out based on many thousands of illuminance value points scanned over a wide angular range by a goniophotometer. The accuracy of the analysis is dependent on the angle resolution of the scan data, the number of data, and to carry out the scan at the desired resolution, it is not uncommon to require several hours or more for a single lamp. Therefore, in order to achieve a realistic scan time, there is a tendency in the field to settle for a reduction in the scan points and a corresponding decrease in the accuracy of the analysis.

In response to this situation, we have developed a system which scans tens of thousands of illuminance values in hundredth or less of the time previously required, in order to fully exhibit the power and efficiency of a high-performance analysis system. While previous systems required five hours to scan 27,000 points at 0.2-degree resolution, the new system completes the same scan in two minutes without any practical loss of accuracy.

This new system utilizes the same geometry as the previous one, in that it employs a goniometer and one photometer set at a position of 10m, 25m, or 60 feet, and it retains complete compatibility in terms of angle and accuracy of illuminance values.

BEAM DISTRIBUTION SCANNING AND EVALUATION

With respect to legally established regulations regarding the beam distribution performance of headlamps, the distribution of luminous intensity (in the U.S., Japan, etc.) of the lamp proper, and the illuminance of flat screens (Europe), are stipulated at anywhere from about five to twenty points.

This method is thought to be sufficient from the perspective of the quality control requirements of mass production, but it is insufficient for the purposes of evaluation and research of visibility under actual driving conditions.

By definition, the beam distribution performance of headlamps should be evaluated by the magnitude of illuminance of each objects; for example, the various parts of the road surface, traffic signs, oncoming vehicle, pedestrians and the like. However, since even in the case of identical headlamps, illuminance will vary greatly depending upon the position at which they are mounted on the vehicle, and also depending upon the direction of aim, in the method of measuring illuminance directly on the road surface, the number of scans is necessarily determined by the number of mounting conditions. Under these circumstances, the accepted practice is to scan the distribution of luminous intensity that emits from the lamp in every direction. Using this method, if each lamps takes only one scan, it is possible to calculate the illuminance of objects on the road even in cases where lamp conditions such as the mounting position and the direction of aim are altered. (Fig. 1)

THE NEED FOR GREATER RESOLUTION AND SPEED

In the case of conventional headlamps, we made use of luminous intensity distribution data measured vertically and horizontally at intervals of one degree within a range from 30° left to 30° right, and 6° up to 12° down. (Fig.2a) The isocandela curve in the proximity of the cut-off line for this data appears unnaturally terraced, but it is clear that in reality, the luminous intensity distribution for the actual lamps can hardly be so uneven. The most economical way to faithfully display this data, which is such an obvious departure from actual conditions; is to apply smoothing by utilizing such methods as the Spline function, and indeed, in lamps with a comparatively broad beam distribution pattern, a sufficient degree of accuracy is achieved. However, in the case of low beams which possess a clear cut-off line, these interpolations do not provide sufficient accuracy, and not only does this have no significance in the realm of physics, but it even has the danger of leading to faulty analytic results. (Fig.3)

Fig. 2b is the result of scanning at 0.2° intervals, and while smoothing has not been applied, the curves are significantly smoother in comparison to Fig.2a. This degree of detail in data is the minimum required for use in visibility analysis, and furthermore, in order to evaluation of such factors as irregularities in the beam distribution pattern, resolution of 0.1° is sometimes necessary. In response to the conditions of the objects to be measured and the required accuracy levels, we scanned in a range of between 0.1° and 1° (intervals).

However, with the current measurement systems, the 1° intervals in Fig.2a resulted in 1159 measurement points, and the time required was 13 minutes, while at 0.2° intervals, with 27,391 points, 5 hours were required, so a substantial increase in speed was clearly desirable.

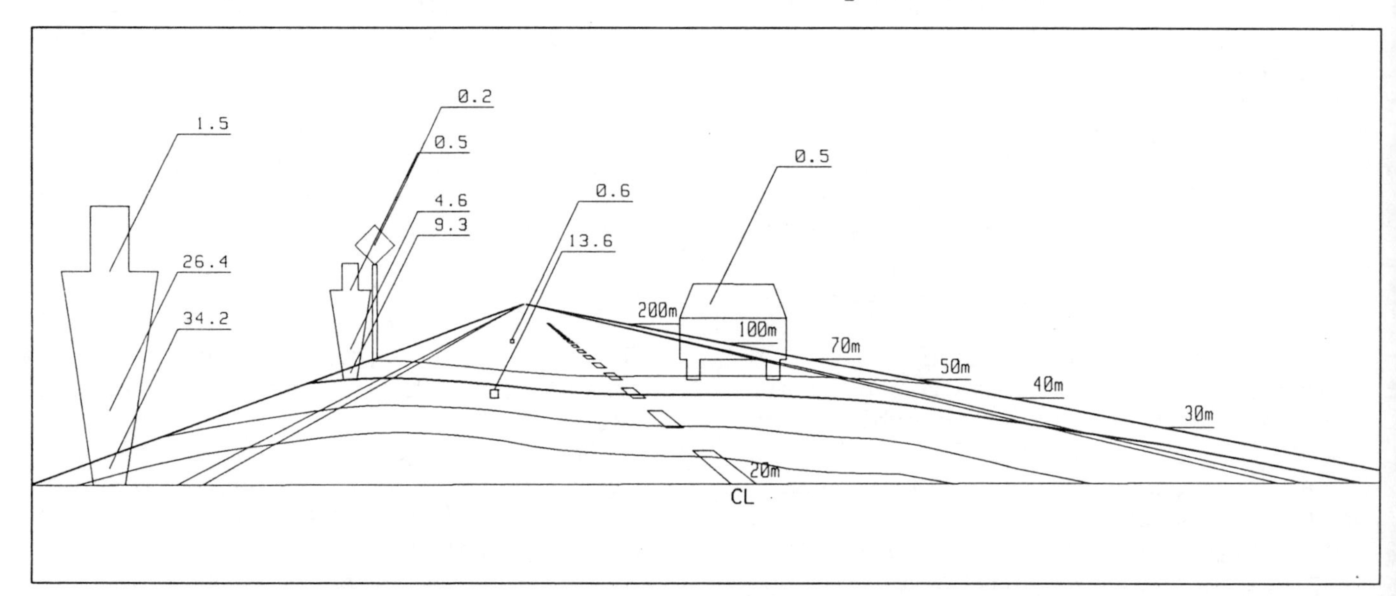

Fig. 1 - Road Surface Illuminance Distribution Data Calculated from Luminous Intensity Distribution Data

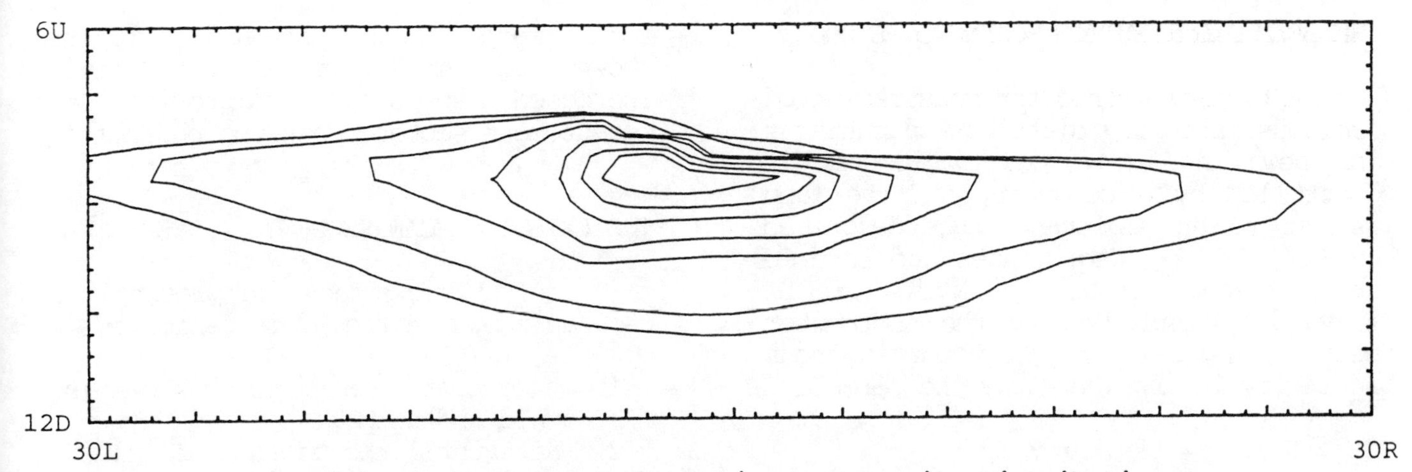

Fig. 2a - An Example of the Luminous Intensity Distribution of Low Beam Measured at 1° Intervals

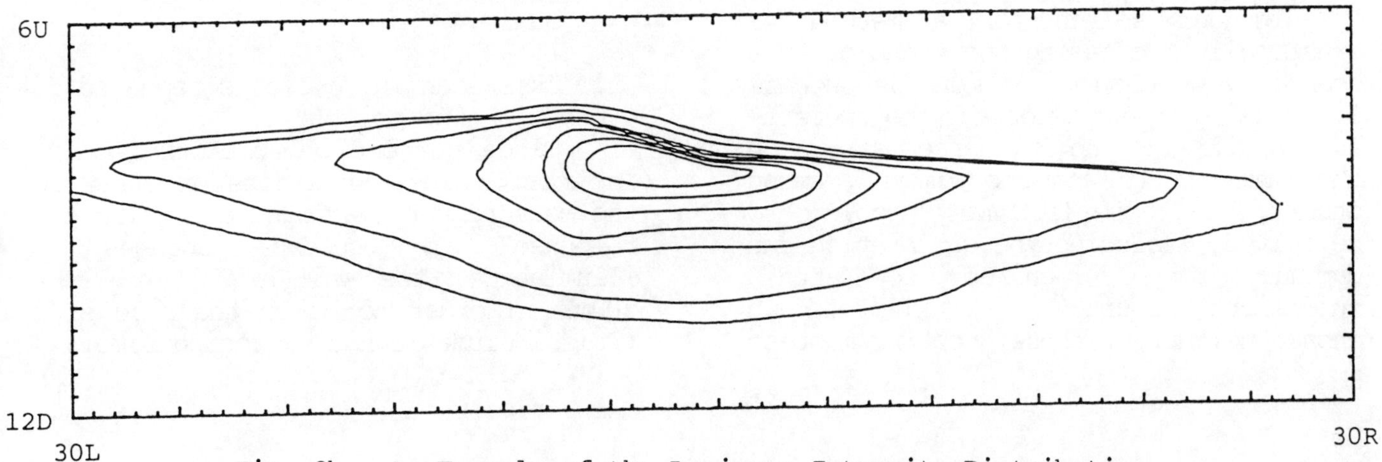

Fig. 2b - An Example of the Luminous Intensity Distribution of Low Beam Measured at 0.2° Intervals

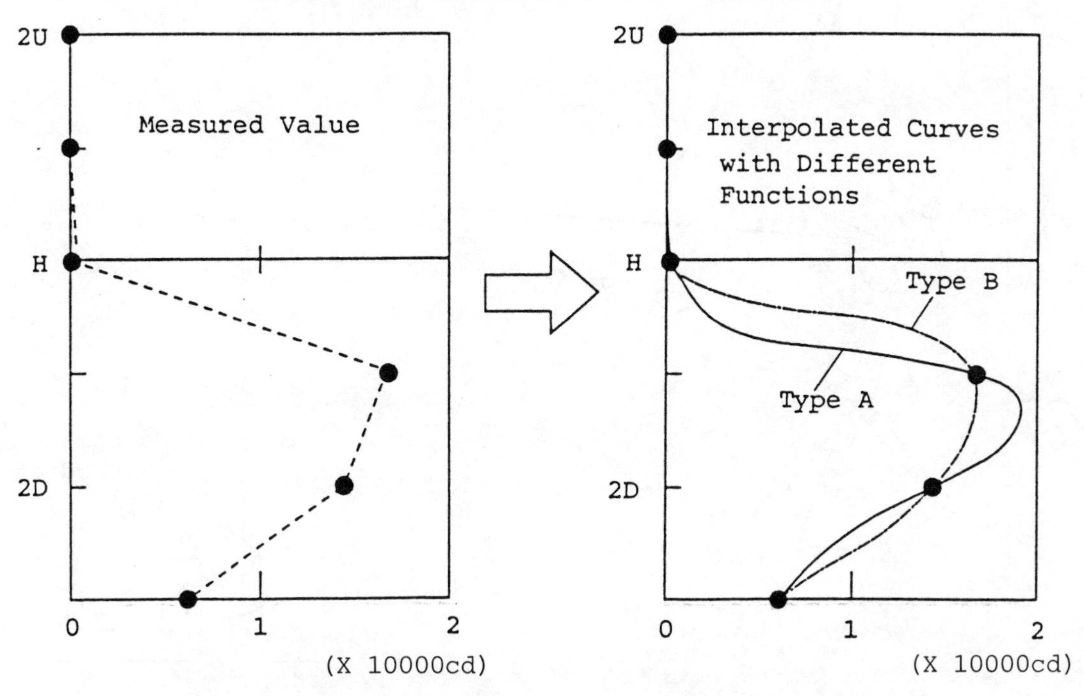

Fig. 3 - Which is Correct? We have no way to know that in this data.

THE BEAM DISTRIBUTION SCANNING METHOD

A typical method for measurement of luminous intensity distribution patterns is shown in Fig.4. The goniometer is controlled by a computer, and the lamp is set to an arbitrary angle described by "u" and "v" coordinates and secured in position. Herein, illuminance "E" (u,v) is measured using the photometer set at distance "r", and the luminous intensity is arrived at by the formula;

$$I(u,v) = E(u,v) \cdot r^2$$

This is repeated over the entire range of the necessary u,v.

In our search for a method of drastically increasing the speed of this scanning, we thought of methods such as "increasing the number of photometers" and "digitizing the luminance pattern on the screen with a video camera", among others, but these left many problems to be solved, including the calibration and dynamic range required to attain sufficient accuracy. In the end we chose neither of these, opting instead for a method using the same structure shown in Fig.4 unaltered, achieving increased speed by improving the measurement sequence and the photometer.

THE EXISTING SCAN SEQUENCE - STEP SCAN

The scanning sequence presently in use, as shown in Fig.5, is as follows;

1. Goniometer rotates to the scanning angle and stops.
2. Waits until the photometer value stabilizes.
3. Reads photometer.
4. Rotates to the next measurement angle.

This scanning cycle is repeated for every scanning point.

Since current photometers require, in worst cases, in excess of 200mS for the measurement readings to stabilize in response to sudden changes in illuminance, the wait state is set to 300mS in order to be on the safe side. (This includes range switching time.)

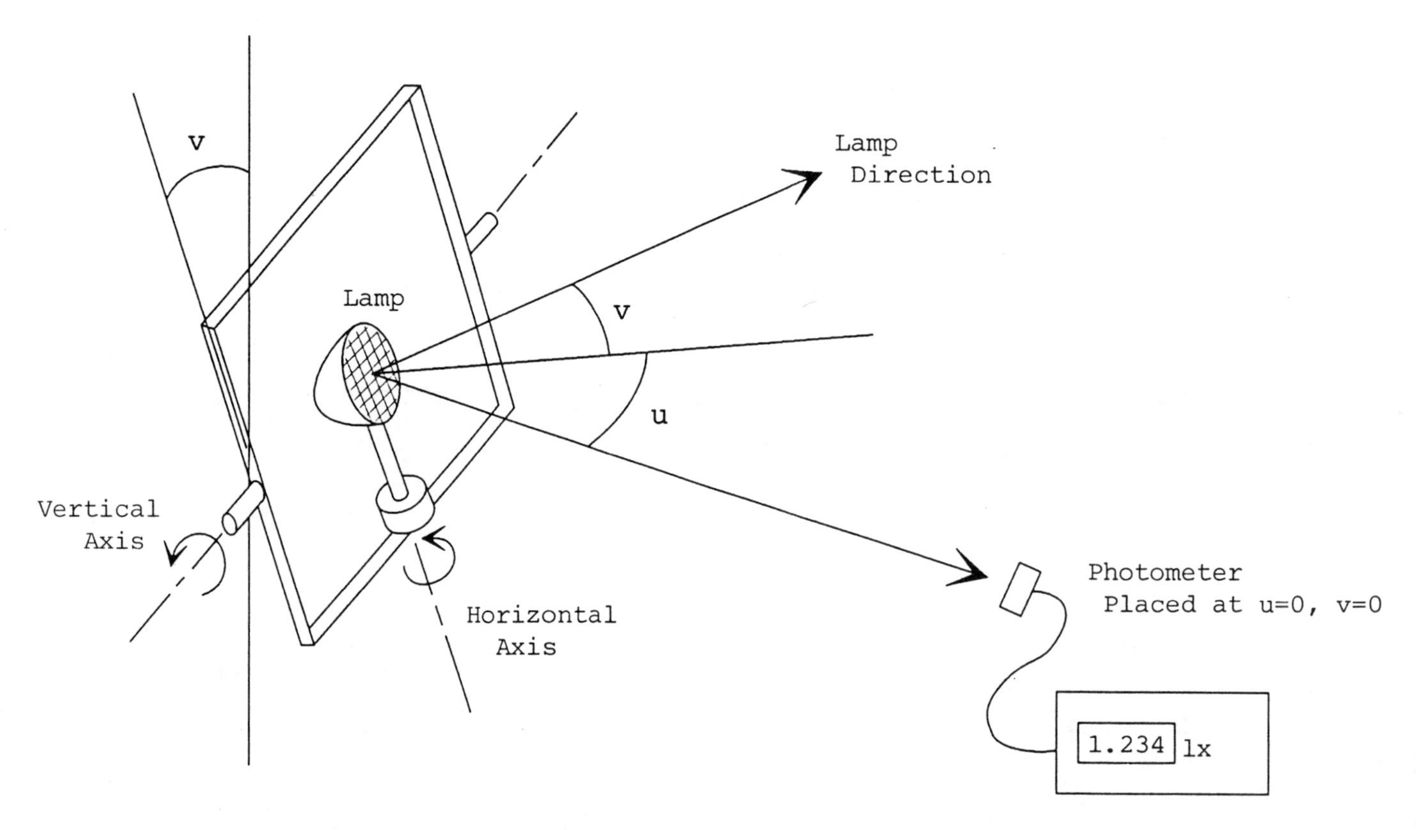

Fig. 4 - Scanning of Luminous Intensity Distribution Using Goniophotometer

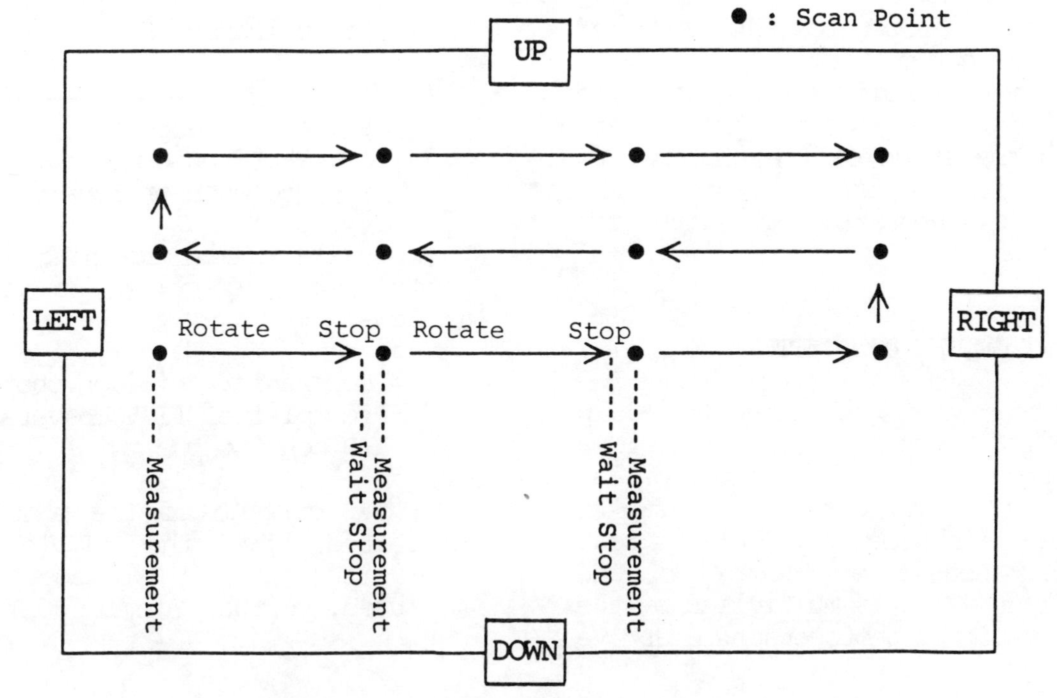

Fig. 5 - Present Step Scan Sequence

However, even if the speed of the photometer is increased, the greatly increased strain and wear on the motor and gears due to repeated acceleration and deceleration of the goniometer each time a point is measured leads to the conclusion that the current speed has virtually reached its technological ceiling, and no dramatic increase can be expected.

NEW SCAN SEQUENCE - CONTINUOUS SCAN

In order to radically reduce the scan time, it is absolutely essential for the goniometer to take scans while spinning continuously, rather than stopping for each scanning point.

More specifically, the goniometer scans as it moves from the left extreme to the right extreme while fixed at a constant vertical angle. (This will be called Line 1). Next, the vertical angle is moved one step, and the next line is continuously scanned from the right edge to the left edge. (Fig.6).

The advantage of this method is that, while reducing the scanning time, since there are no sudden accelerations or decelerations at each scanning point, the life of the motor and the gears is

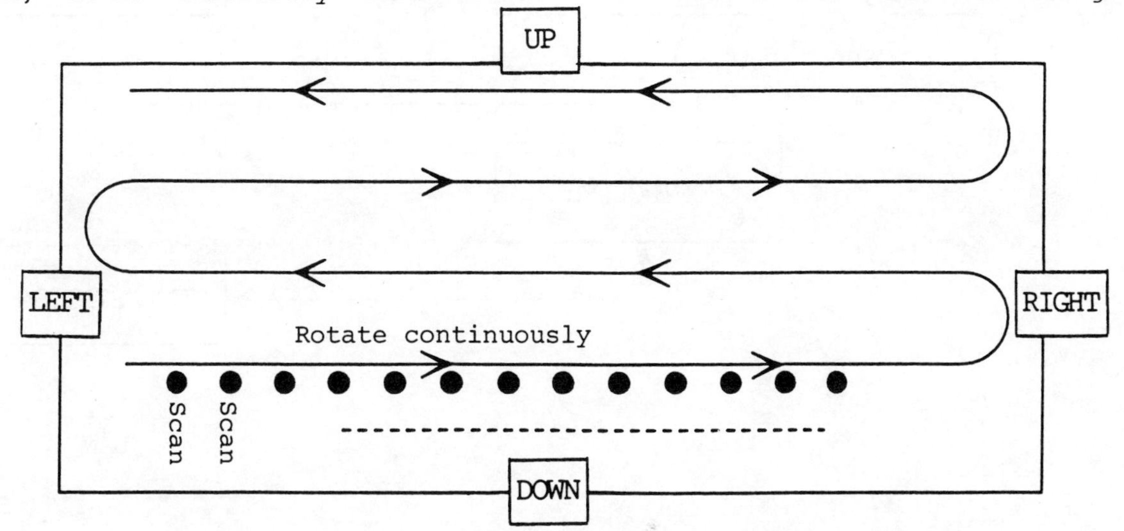

Fig. 6 - Newly Developed Continuous Scan Sequence

extended in comparison to that evidenced using the current method.

The engineering elements required in order to realize this method are increasing the speed of the photometer, extending its dynamic range, and accurately synchronizing it with the drive section.

IMPROVEMENT OF THE PHOTOMETER

The beam radiated from the lamp undergoes photoelectric conversion in a silicon photodiode, and is retrieved as an amperometric value proportionate to the incident light amount. This is read by a highly sensitive amperemeter, and the resultant value is multiplied by the sensitivity of the photosensor to arrive at the illuminance value.

$$E = A / k \quad \text{-------------------} \quad (1)$$

E : illuminance [lx]
A : Photosensor Output Current [A]
k : Photosensor Sensitivity [A/lx]

In the actual scanning system, the photometer is composed, as in Fig.7, as follows;

1. Photosensor (silicon photodiode)
2. Preamplifier (I/V conversion)
3. A/D converter

and the current of the photosensor is converted by the I/V conversion amplifier into a voltage value and measured, so that Formula (1) is

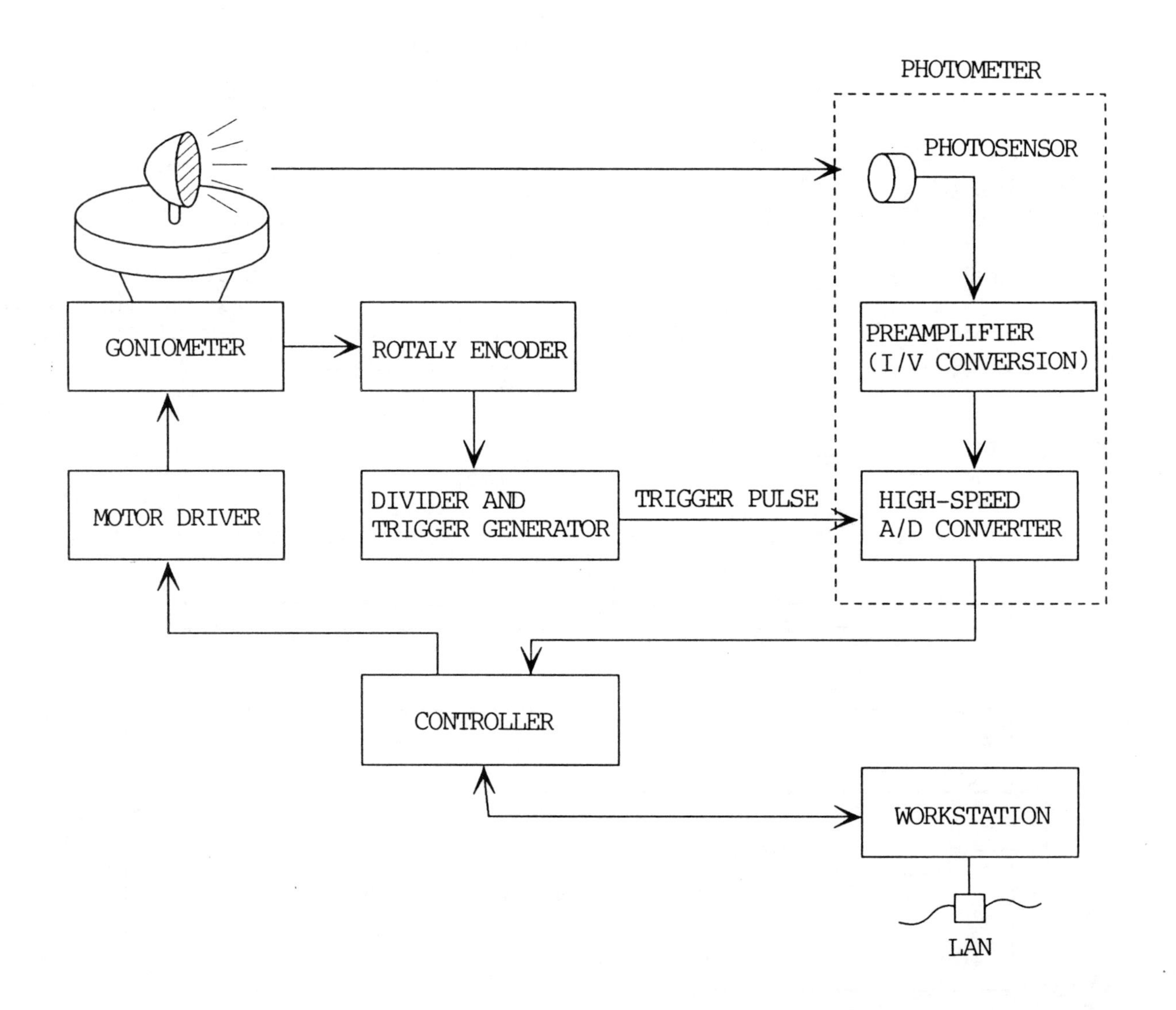

Fig. 7 - Scan System Diagram

$$E = V / (k.a) \text{ ------------------ } (2)$$

V : voltmeter reading [V]
a : amp gain [V/a]

The sensitivity of the previously-used photodiode was 3×10^{-9} [A/lx], so in order to measure a standard headlamp without range switching, the amp gain was set to 10^6 [V/A], and an A/D converter with 21-bit resolution and a read rate of over 1,000 times per second.

Next, we shall consider the effect of photometer response delay upon time variation in incident light amount.

This photometer response delay is determined by the following three factors;

1. photosensor response
2. pre-amp response
3. A/D converter head amp response

and among these, 1. and 3. are of sufficiently high speed, so that 2., the pre-amp response, determines the response of the entire system. (Integration time of the A/D converter will be considered later).

Considering the scan of lamps with beam distribution such as that in Fig.8, if the goniometer is rotated from left to right at a constant speed, the response speed of the photo-sensor is sufficient, so the input current waveform of the pre-amp can be considered to faithfully correspond to the true beam distribution values as shown in Fig.9, put the pre-amp output varies during the delay time (t1, t2) shown in the same diagram. This delay time is called settling time, and if the goniometer speed is "s" [degrees per second], then the waveform (=beam distribution pattern) is observed as scanned when it has moved only from the true position "X",

$$\Delta X = st \text{ [degrees] ---------------- } (3)$$

then the waveform is also deformed.

When the speed of the goniometer increases, this ΔX cannot be ignored, and even in the same beam distribution, a $2\Delta X$ discrepancy occurs between the scan from left to right and the scan from right to left. (Fig.10)

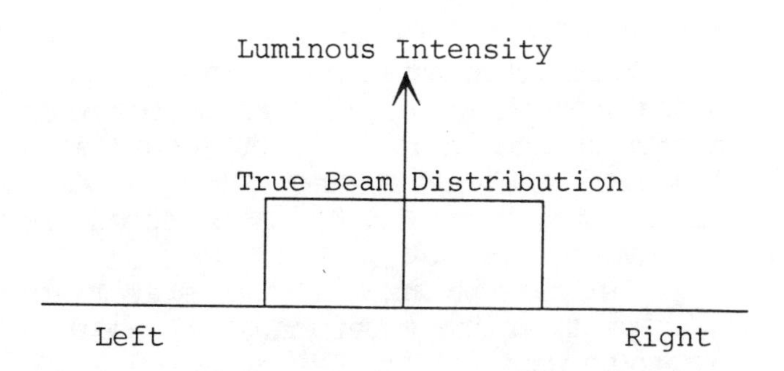

Fig. 8 - Considering This Type of Beam Distribution Lamp

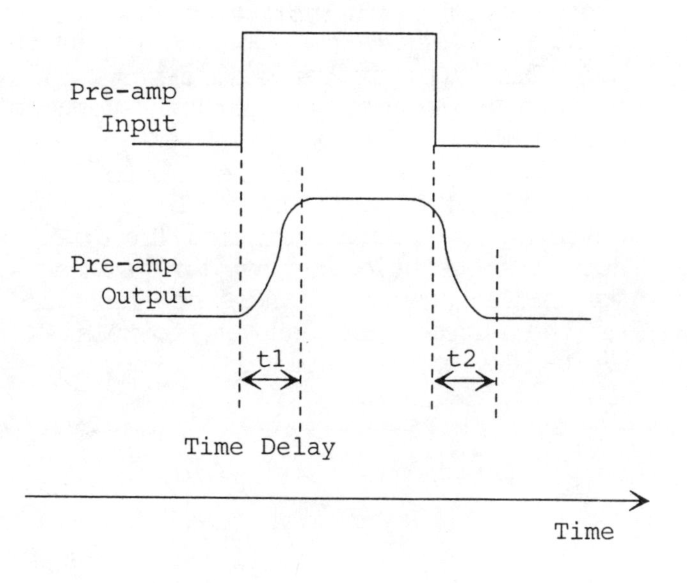

Fig. 9 - Pre-amp Input/Output Waveform

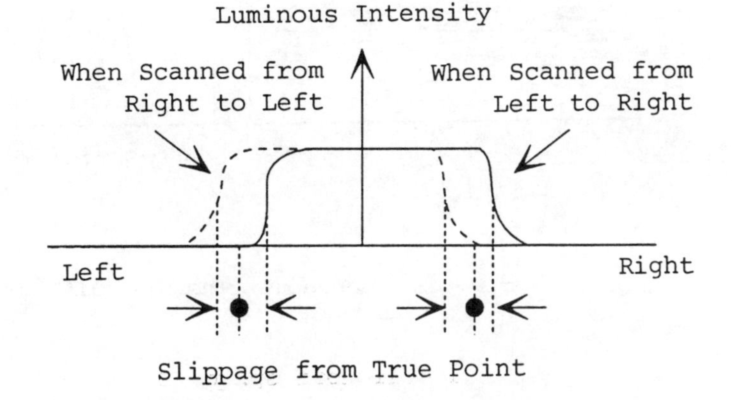

Fig. 10 - Scanning Result where Position Slippage has Occured

Next, we will consider the typical beam distribution pattern f(x), which is the function of the angle "x".

When the moving at a set speed along x from x0 to x1, the photosensor expresses the incident light amount as a function "f" of time "t". (For simplicity, when x=x0, t=0, speed is abbreviated as 1.)

At this time, if the system is linear, then the waveform y(t) received as the scan result is expressed as a convolution of the input signal f(t) and impulse response h(t) of the measurement system.

$$y(t) = \int_0^t f(\tau)h(t-\tau)d\tau \quad \text{----------(4)}$$

This y(t) is the beam distribution pattern y(x) received as the scanning result.

It is possible to revert to f(t), that is, the true beam distribution pattern f(x), by calculating from y(t), but this operation is complex, so it is more advantageous to increase the speed of the photometer, and use formula (4) as closely approximated to

$$y(t) \fallingdotseq f(t) \quad \text{--------------------(5)}$$

as possible.

In the actual scan system, it is possible to confirm, by means of a simple method, whether or not an approximation of formula(5) can be achieved.

When the scan sequence is carried out while alternating directions as in Fig.6, scanning one line at a time from left to right and then right to left, if waveform distortion occurs due to response delays in the scanning sequence, the vertically-oriented line of the isocandela curve drifts left or right alternately, resulting in a jagged line, and if the delay can be ignored, this becomes a straight line, These zigzags become more pronounced the greater the extremity of variation in illuminance, so by measuring the light source causing the sudden fluctuation in luminous intensity horizontally, and if zigzags do not occur, then this can be judged to be sufficient for practical purposes.

In our experiments, with a rotation speed of 33° per second, we obtained sufficient accuracy when the settling time of the amp was 0.5mS or less. (However, we made angle corrections to be mentioned below.)

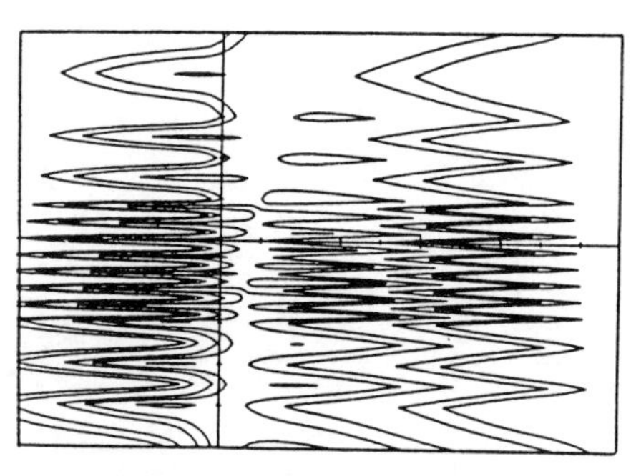
Settling Time 200mS

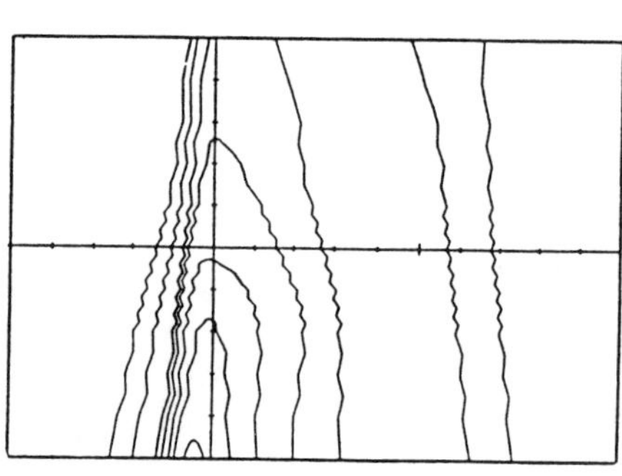
Settling Time 5mS

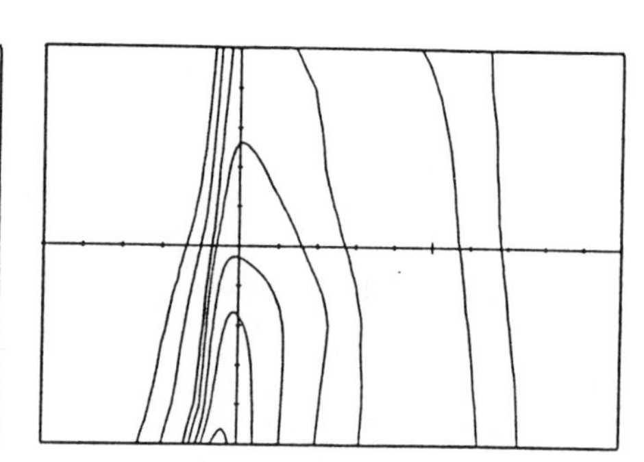
Settling Time 0.5mS

Fig. 11 - Pre-amp Setting Time and an Example of Position Slippage

PHOTOMETER NOISE

There are two chief types of noise that interfere with measurement;

1. Low frequency noise caused by induction from power line (100V/200V).
2. Particularly high frequency spectral white noise.

When we observed the amp output of our present photometer with an oscilloscope, we observed a signal of induction noise of 60Hz (power line frequency). However, since the integration time of the A/D converter was a product obtained by multiplying power line cycle time (16.7 mS for 60Hz) by integers, this was cancelled and had little effect on the measurements. On the other hand, it is impossible to set the integration time to one cycle of the power line frequency, since high speed measuring requires high speed sampling in a range between several mS and 100μS. Therefore, the induction noise from the power line must be restricted to the lower possible level, but reducing the noise by using low pass filters increases the settling time, eliminating this method as a solution.

Central to our approach are mounting methods for the devices such as ground point, shield and so on. By adjusting these as we observed the waveform using an oscilloscope, we were able to reduce the noise level to a mere ±0.01[lx].

The higher frequency spectral white noise is averaged out by the integration time of the A/D converter, thus having little impact, but in order to enhance sensitivity, it is desirable to use an amp with the lowest possible noise level.

ACCURACY OF GONIOMETER ANGLE DETECTION AND A/D CONVERTER ACTION TIMING

When a stepping motor is used to drive the goniometer, it is a simple task to control it by synchronizing the angle and time, as long as the motor is driven at a constant speed. (However, this does not apply to acceleration /deceleration intervals in trapezoid drives.)

However, in cases where servomotors are used, it is impossible to guarantee that the speed will be constant, so the actual angle must continuously be monitored by an encoder affixed to the goniometer axle, synchronizing it and carrying out the scan.

Here, if the encoder output pulse number is n-pulse/degree, there will be a ±1/n detection discrepancy in addition to the gear backlash. This is true not only for continuous scans, but also for current scanning methods, and since with our turntable n=100, a discrepancy of "backlash ±0.01°" is an integral factor in the system.

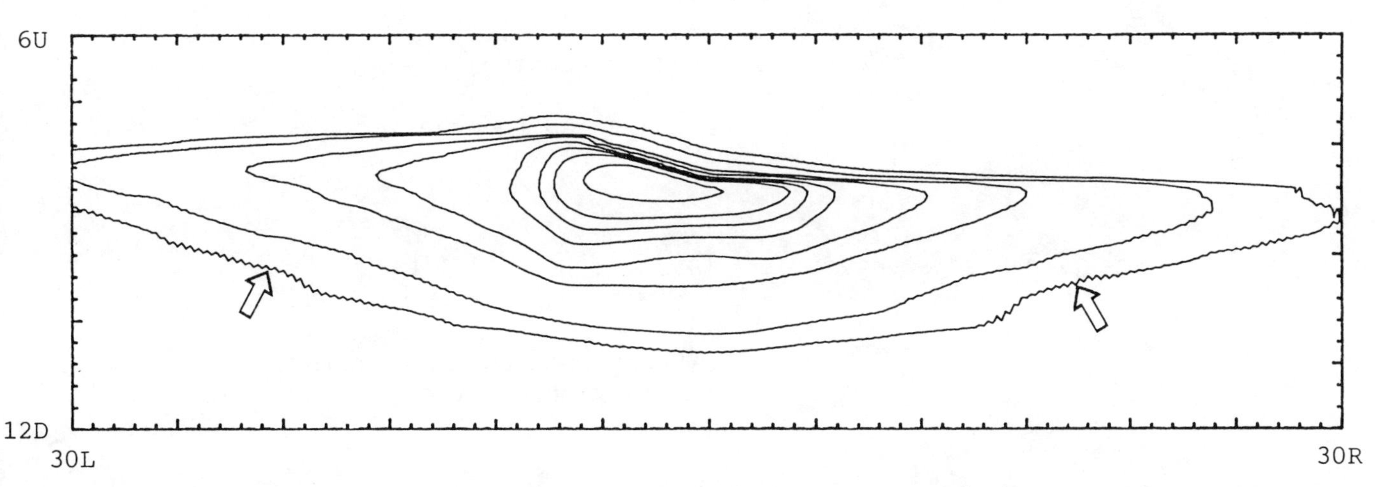

Fig. 12 - Example of 60Hz Noise for ±0.2°[lx]

In addition to this, as shown in Fig.13, in the case of the continuous scan, delays occur due to the trigger delay "t1" and integration time "ts" of the A/D converter.

The relationship between the trigger generator dividing ratio "1/N" and the scan interval "P" is;

$$P = N/n \quad (6)$$

At this time, the time delay in the divider is sufficiently negligible to be ignored, but when the scan begins, if the count value for the previous scan remains in the divider (this is the counter of N progress), then a discrepancy of no more than P-1/n occurs, so it is necessary to clear the count value prior to beginning the scan for each line. (For example, if the value 7 remains in a divider of N=10, then when the next scan begins, the trigger will generate with three encoder pulses, and the entire scanning position for this line will be shifted by 0.07°.)

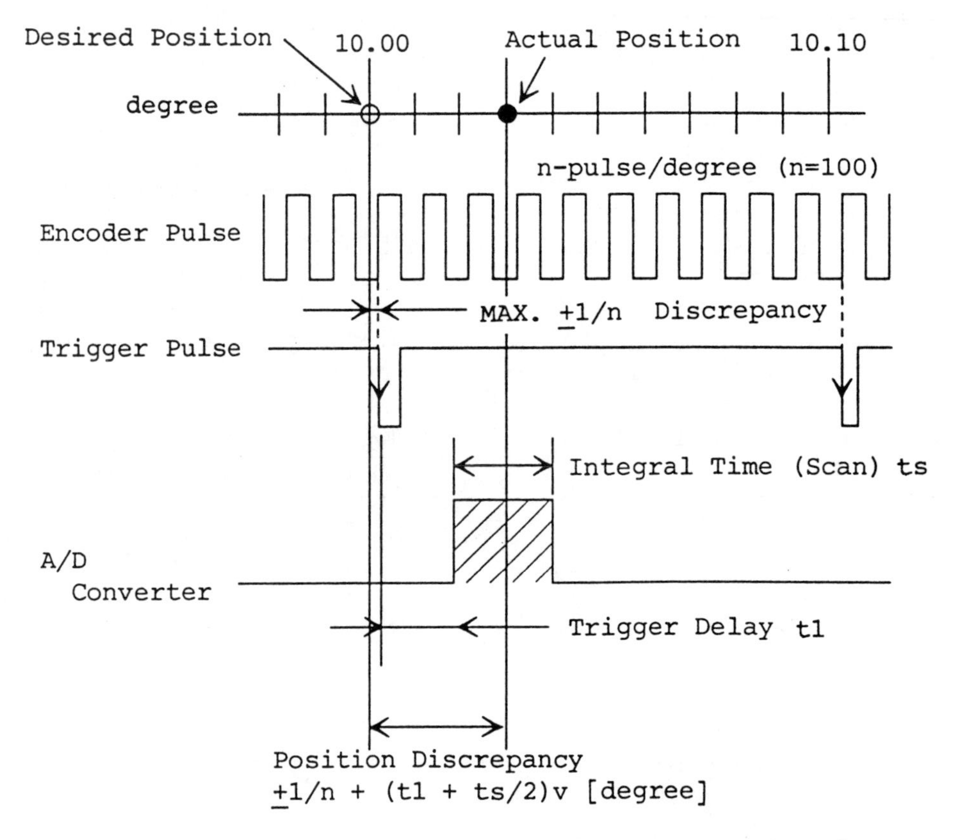

Position Discrepancy : $\Delta X = (\pm 1/n) + B + (t1 + ts/2)v$
B : Position Error due to Backlash

By compensating this value, position accuracy approaches $\pm 1/n$.

Fig. 13 - Angle Slippage Caused by the Encoder and the A/D Converter

Therefore, angle slippage caused by the encoder and the A/D converter is;

$$\Delta X = (\underline{+}1/n) + B + (tl + ts/2)v \text{ ----(7)}$$

B : position discrepancy caused by backlash

v : rotation speed [degrees per second]

Herein, in the case of the continuous scan, the "B" of one scan line is thought to be a constant, so the second and third terms in the right side of formula (7) can be compensated for as constants. In fact, it is possible to approach a position accuracy of $\underline{+}1/n$ without any mathematical operations to adjust the scanning results, simply by making a prior adjustment of $B+(tl+ts/2)v$ to the scan initiation position.

SETTING THE SCANNING INTERVAL

The aperture angle of the photosensor as viewed from the lamp position is;

$$Ad = Tan^{-1}(D/L) \text{ ------------------(8)}$$

D : photosensor surface diameter

L : distance to photosensor

Thus, if the photosensor used is D=20mm, then Ad is 0.11° at 10m and 0.06° at 60 feet, and so this is the resolution limit, determined by the size of the photosensor.

On the other hand, viewed from the perspective of the scanning time and memory capacity, it is necessary to avoid adding scanning points which have no significance. Insignificant scanning points refers to those points whose position it is possible to surmise with sufficient accuracy through interpolation, without scanning data on their position.

Stated in terms of beam distribution data, near the cut-off line the slightest vertical deviation can lead to marked changes in luminous intensity, so it is impossible to accurately ascertain the beam distribution unless the scan is made with 0.2° vertical detail, but in areas where changes in luminous intensity are minimal (places where brightness is flat), even scanning at 2° intervals allows sufficient accuracy through interpolation.

In other words, in areas of the beam distribution pattern which have high spatial frequency the sampling cycle should be small, while those with low frequencies have a large cycle, there is no alternative but to apply the sampling theorem;

$$2f\,max \leq 1/Ts \text{ -------------------(9)}$$

f max : maximum frequency

Ts : sampling cycle

and by varying Ts, obtaining sufficient results even with a small sampling number.

In scanning of beam distribution patterns for headlamps (particularly low beams), the actual reason why a nearly horizontal scan requires vertical resolution of 0.2° is not that "if the lamp height is at 60cm, 0.6° down is 57m and 0.4° is 86m, so a 0.2° shift results in a change of 29m", but that since an area just below the horizon called the "cut-off line" exists where the luminous intensity changes radically (=area where spatial frequency is high), so if the sampling cycle for this area is not kept small the data will be lost, and it will become impossible to reproduce the data using interpolation. Therefore, in lamps where there is no cut-off line, and the beam distribution is extremely loose both vertically and horizontally, scanning at the usual 1° intervals enables sufficient long-range visibility data to be obtained through interpolation.

In this way, the required scanning interval changes according to the lamp beam distribution pattern, so if the automatic scanning system itself could intelligently determine and scan at the most appropriate interval for the beam distribution pattern, it would be possible to obtain optimum data in a minimum of time for lamps of any beam distribution.

In fact, the methods for developing this type of intelligent scanning system are relatively simple, so we plan to implement it in the future, but at the present point in time, we set several types of intervals as standards based on usual headlamp beam distributions, and the person responsible for the scanning selects them as required.

Example of Scan Interval Settings
Vertical Scanning Range :
from 12° down (12D) to 6° up (6D)
Scanning Interval :
0.2° (2D-1U)
0.5° (4D-2D, 1U-3U)
1° (12D-4D, 3U-6U) total 35 lines
Horizontal Scanning Range:
from 30°left (30L) to 30°right (30R)
Scanning Interval:
0.1° total 601 points
Total Number of Points : 21,035

Here, if we proceed based on usual headlamp beam distributions as explained above, the 0.1° horizontal interval is excessive, but in this system, the reasons for using the 0.1° interval are;

1. When scanning, there are limitations on the rotation speed of the goniometer, so the scanning time does not change irrespective of whether the horizontal is 0.1°, 0.2° or 1°.
2. When evaluating irregularities in the beam distribution, a resolution of 0.2°, or in some cases even 0.1°, is required.
3. If the technology for high speed scanning at 0.1° can be achieved, the resolution could be reduced merely by resetting the scanning program.

On the other hand, if the vertical scan interval is reduced, the number of scanning lines increases, and scanning time increases proportionately. Therefore, we have set the interval to the minimum necessary.

COMPATIBILITY WITH PREVIOUS SYSTEMS

We checked to confirm that the scan values of new system and previous automatic scan values and manual scan values did not vary by more than an average of 1% , or at very most, 2%.

CONCLUSION

The vast majority of previous scan samples have been limited to 1° intervals due to scanning time limitations in previous systems. By implementing this system, it is possible to carry out all beam distribution scans at high resolution, and moreover to do so in shorter scanning times than the previous systems could at 1° intervals, so a larger number of samples can be obtained. This increased data is stored on magneto-optical disk, and represents a valuable tool in the quest to develop ever better headlamps.

950597

Evaluation of the SAE J1735 Draft Proposal for a Harmonized Low-Beam Headlighting Pattern

Michael Sivak and Michael J. Flannagan
The University of Michigan Transportation Research Institute

ABSTRACT

This study evaluated the SAE J1735 Draft Proposal for a low-beam headlighting pattern in relation to the current standards in the U.S., Europe, and Japan. The approach consisted of the following steps: (1) identifying a set of 15 important visual performance functions (including seeing and glare) for low-beam headlamps; (2) defining the relevant geometry relative to the visual performance functions; (3) setting criterion values of illumination for each of the visual performance functions based on the available empirical data; and (4) evaluating the standards relative to the criterion values by using the worst-allowed-case approach (evaluating the *minima* specified by the standards for seeing functions, and the *maxima* for glare functions).

The results indicate that the SAE J1735 Draft Proposal tended to require better performance than the current U.S., European, and Japanese standards.

INTRODUCTION

In a previous study (Sivak, Helmers, Owens, and Flannagan, 1992) we evaluated several recent proposals for the low-beam headlighting pattern, along with the then current U.S., European, and Japanese standards. A later version of one of these proposals—the SAE J1735 Draft Proposal (a follow-up to the SAE Proposal 7A)—is currently being considered by international lighting experts for adoption as a harmonized beam pattern for use throughout the world. Consequently, the present study was designed to evaluate the SAE J1735 Draft Proposal in comparison with the current U.S., European, and Japanese standards. [The U.S. and European standards underwent some modifications since our previous study. Furthermore, Sivak et al. (1992) evaluated the proposed standards against the Japanese Industrial Standard (JIS, 1984); the present study used the current Japanese governmental standard (JASIC, 1993).]

A major difference between the SAE J1735 Draft Proposal and its precursor—Proposal 7A (SAE, 1991)—is that the SAE J1735 Draft Proposal contains the four test points that were recommended by GTB (Groupe de Travail "Bruxelles 1952") for adoption as common test points worldwide. (These four test points, in turn, are based on the four test points recommended by Sivak and Flannagan, 1993.)

APPROACH

We used the same approach as in our previous study that evaluated several proposed and existing standards (Sivak et al., 1992). The approach consists of the following steps:

(1) Identifying a set of 15 important visual performance functions (including seeing and glare) for low-beam headlamps.
(2) Defining the relevant geometry relative to the visual performance functions.
(3) Setting criterion values of illumination for each of the visual performance functions based on the available empirical data.
(4) Evaluating the standards relative to the criterion values by using the worst-allowed-case approach (evaluating the *minima* specified by the standards for seeing functions, and the *maxima* for glare functions).

As we have argued previously (Sivak et al., 1992), an evaluation of the standards using computer models such as CHESS (Bhise et. al., 1976; Bhise et al., 1977) is not feasible. Such models require input of a detailed candela matrix to define the beam to be evaluated. On the other hand, standards contain only a limited number of test points, lines, or zones. The great majority of the beam is not explicitly controlled, and only *limits* (minimum and/or maximum), which fall short of specifying actual beam characteristics, are given for the points or regions of concern. Thus, they do not constrain the beam to any particular candela matrix.

The present approach focuses on what is allowed by standards (and not on the performance of lamps that were manufactured to meet different standards). We believe that the present approach is the correct way to evaluate standards, particularly with respect to their harmonization potential, because harmonization raises the possibility that standards will be applied to technologies that have not traditionally been used in certain jurisdictions.

THE PROPOSED AND EXISTING STANDARDS

We evaluated the SAE J1735 draft proposal in relation to the current standards in the U.S., Europe, and Japan. These four standards are summarized in Tables 1 through 4.

Table 1. The current U.S. standard (FMVSS, 1993) for a 2-lamp system.

Test point or region	Minimum (cd)	Maximum (cd)
10U to 90U		125
4U, 8L	64	
4U, 8R	64	
2U, 4L	135	
1.5U, 1R to R		1,400
1.5U, 1R to 3R	200	
1U, 1.5L to L		700
0.5U, 1.5L to L		1,000
0.5U, 1R to 3R	500	2,700
H, 8L	64	
H, 4L	135	
0.5D, 1.5L to L		3,000
0.5D, 1.5R	10,000	20,000
1D, 6L	1,000	
1.5D, 9L	1,000	
1.5D, 2R	15,000	
1.5D, 9R	1,000	
2D, 15L	850	
2D, 15R	850	
4D, 4R		12,500

Table 2. The current European standard (ECE, 1992). The original specifications of the test-point locations were converted from cm on a vertical surface at 25 m to angles. Similarly, the original specifications of illuminances at 25 m were converted to luminous intensities.

Test point or region	Minimum (cd)	Maximum (cd)
4U, 8L	62	438
4U, V	62	438
4U, 8R	62	438
2U, 4L	125	438
2U, V	125	438
2U, 4R	125	438
H, 8L	62	438
H, 4L	125	438
0.6U, 3.4L (B50L)		250
0.6D, 3.4L (75L)		7,500
0.6D, 1.1R (75R)	7,500	
0.9D, 3.4L (50L)		9,375
0.9D, V (50V)	3,750	
0.9D, 1.7R (50R)	7,500	
1.7D, 9L (25L)	1,250	
1.7D, 9R (25R)	1,250	
Zone I (1.7D to D)		200% of the actual value at 0.9D, 1.7R
Zone III (above line H, 9L; H, V; 2.4U, 9R, or above line H, 9L; H, V; 0.6U, 0.6R; 0.6U, 9R)		438
Zone IV (corners: 0.9D, 5.1L; 0.9D, 5.1R; 1.7D, 5.1R; and 1.7D, 5.1L)	1,875	

Table 3. The current Japanese standard for a 4-lamp system (JASIC, 1993), converted to right-hand traffic.

Test point or region	Minimum (cd)	Maximum (cd)
10U to 90U		125
1.5U, 1R to R		1,000
1U, 1L to L		700
0.5U, 1L to L		1,000
0.5U, 1R to 3R		2,700
0.5D, 1L to L		2,500
0.5D, 2R	6,000	20,000
1D, 6L	1,000	
1.5D, 9L	1,000	
1.5D, 2R	15,000	
1.5D, 9R	1,000	
2D, 15L	700	
2D, 15R	700	
4D, 4R		12,500

Table 4. The SAE J1735 Draft Proposal (SAE, 1994).

Test point or region	Minimum (cd)	Maximum (cd)
1.5U, V to 3R	200	1,000
0.5U, 1.5L	125	650
0.5U, 1R to 2R	500	2,400
0.5D, 4R	5,000	
0.6D, 1.3R	10,000	
0.9D, 3.5L	1,800	12,000
0.9D, V	4,500	
2D, 15L	1,000	
2D, 9L	1,250	
2D, 9R	1,250	
2D, 15R	1,000	
4D, 20L	300	
4D, 4R		50% of the maximum in Zone I, but ≤ 12,500
4D, 20R	300	
Zone I (corners: 0.5D, 0.5R; 0.5D, 2.5R; 2D, 2.5R; 2D, 0.5R); tested at the maximum in the zone	15,000	
Zone II (corners: 1D, 5L; 1D, 5R; 2D, 5R; 2D, 5L, except Zone I); tested at the four corners	1,875	
Zone III (corners: 0.5U, 8L; 0.5U, 3L; 2U, V; 2U, 8R; 4U, 8R; and 4U, 8L); tested at the four extreme corners, and at the lower boundary line from 5L to 5R	80	650
Zone IV (4U to 10U, 15L to 15R); tested at 4U from 15L to 15R, and at V from 4U to 10U		525
Zone V (10U to 90U, 45L to 45R); tested at 45U from 45L to 45R, and at V from 10U to 90U		125 (438 within a 2° conical angle)

VISUAL PERFORMANCE FUNCTIONS OF LOW BEAMS

The visual performance functions that we considered important for low-beam headlamps in our previous evaluation of standards (Sivak et al., 1992) are listed below. They are grouped into functions dealing with illumination above horizontal, distant field, foreground, and overall considerations.

Above horizontal

- illumination of a traffic sign on the right shoulder
- illumination of an overhead traffic sign
- glare illumination towards oncoming traffic
- glare illumination towards traffic ahead via rearview mirrors
- illumination prone to scatter in adverse atmospheric conditions (fog, rain, and snow)

Distant field—between 1.75D and horizontal (more than 25 m in front of the vehicle)

- illumination directed towards targets on the right side of the road at intermediate distances
- illumination on hills
- illumination on sags

Foreground—below 1.75D (up to 25 m in front of the vehicle)

- illumination directed towards targets on the right side of the road at near distances
- homogeneity of the beam (aesthetic and comfort considerations)
- illumination prone to glare reflection from wet pavement

Overall

- lateral spread (lane keeping, and aesthetic and comfort considerations)
- relation between seeing illumination and glare illumination
- reliability of visual aiming
- effects of misaim

The following sections will (1) define the relevant geometry for the visual performance functions, (2) set criterion illuminance values for the visual performance functions based on available empirical data, and (3) evaluate the standards in relation to the performance criteria by considering the worst allowed case.

Several caveats are in order. First, to the extent that substantial variations exist in many relevant factors (such as road geometry, headlamp mounting height, and seated eye height), the proposed geometries cannot capture the full range of the properties of a given headlamp.

Second, the selected performance criteria of illumination are, obviously, subject to revision. Their place here is to provide some reasonable benchmarks for quantifying headlamp performance in terms of likely consequences for vehicle operators. Thus, these criteria may be improved through further research on human factors, and they may need revision to accommodate future changes in the driver-vehicle-highway environment.

Third, this research considered only automobiles. Other vehicles, such as trucks and buses were not considered. There are several visibility- and glare-related differences in the design of automobiles on one hand, and trucks and buses on the other hand (Sivak and Ensing, 1989). Of primary importance are differences in seated eye position of the driver and headlamp mounting height (Cobb, 1990), affecting the visibility of retroreflective traffic signs (Sivak, Flannagan, and Gellatly, 1991), visibility of other targets, and glare.

Fourth, the headlamps are treated as being both mounted in the same physical location—in the center of the vehicle. This approximation disregards the differential contribution of light from two mounted headlamps towards a given point in space (cf. Burgett, Matteson, Ulman, and Van Iderstine, 1989). Any errors introduced by this assumption decrease as the relevant distance increases.

Fifth, the test points to be evaluated were not always addressed by every standard. In such instances, we had to rely on the nearest controlled test point. In most cases, the information that we used came from a controlled test point within ±0.75° of the desired test point. This approach is justified given the current variability of aims for on-the-road vehicles, with the standard deviations of 0.8° for horizontal aims and 0.9° for vertical aims (Olson, 1985). Furthermore, where there was an option of two approximately equidistant controlled points, we used the one nearer to the desired point vertically, because, in general, gradients are steeper vertically than horizontally. (The specific controlled test point that we relied on is always identified in the comparisons that follow.)

Sixth, many real-world conditions that lead to decrements in visual performance were not included in the present analysis. These conditions include dirty or scratched headlamps (Cox, 1968), dirt on retroreflective targets (Anderson and Carlson, 1966), atmospheric attenuation, voltage drop, as well as changes in vision of older drivers (Sivak, Olson, and Pastalan, 1981).

Seventh, the actual illumination directed towards a given point in space is not prescribed by the examined standards, which present only minima or maxima, and therefore an actual beam pattern cannot be described. (In the few cases of simultaneous minima and maxima for a given test point, the ranges are still quite substantial.) Thus, our analysis necessarily evaluated the worst case allowed by a given standard. Consequently, we used the specified *minima* in our visibility evaluations, and the *maxima* in glare evaluations. (More specifically, the present analysis evaluated the worst case allowed by the standards, but under relatively optimal driver and environmental conditions—see the preceding point.)

Illumination of a traffic sign on the right shoulder (0.5U, 2.25R)

Geometry. The illumination directed towards a shoulder-mounted traffic sign was evaluated by assuming the following geometry on a two-lane roadway:

- Longitudinal separation between the sign and the headlamps: 150 m. This value was selected because it represents a reasonable sign-legibility distance (Sivak, Flannagan, and Gellatly, 1991).
- Lateral separation between the sign and the headlamps: 6.15 m. This is based on a lane width of 3.7 m, and a lateral separation of the sign from the edge of the roadway of 4.3 m (Woltman and Szczech, 1989).
- Vertical separation between the sign and the headlamps: 1.5 m. This is based on a headlamp

mounting height of 0.6 m (Cobb, 1990), and a sign mounting height of 2.1 m (Woltman and Szczech, 1989).

The angle corresponding to the preceding geometry is 0.5U, 2.25R.

Criterion illuminance. The criterion traffic-sign-illuminance value was set at 0.02 lx based on the following considerations:

- The observation angle for the given geometry is 0.31°.
- The coefficient of retroreflection of the sign material is 150 cd/lx/m^2 at an observation angle of 0.2°. A typical value for a white, encapsulated, sign material is 300 cd/lx/m^2; a realistic in-use value is 50% of the new value (Alferdinck, 1984).
- Assuming the relative reflectance of sign material at 0.2° is set equal to 1, then the relative reflectance at 0.31° is 0.777 (interpolated from the data in Sivak, Flannagan, and Gellatly, 1991 on the effect of the observation angle on the relative reflectance).
- The computed coefficient of retroreflection of the sign material for an observation angle of 0.31° is 116 cd/lx/m^2 (150 x 0.777).
- The desired minimum luminance of the sign material is 2.4 cd/m^2. This value was recommended by Sivak and Olson (1985) and Jenkins and Gennaoui (1992) as a *minimum (replacement)* value of sign materials, based on a literature review of available studies on the effects of sign luminance on their legibility. (In comparison, the *optimal* luminance was found by Sivak and Olson to be 75 cd/m^2.)
- To obtain sign luminance of 2.4 cd/m^2 using a sign material with a coefficient of retroreflection of 116 cd/lx/m^2, the illuminance must be 0.02 lx (2.4/116).

Findings. Table 5 ranks all standards in terms of the decreasing combined luminous intensity from both lamps directed towards 0.5U, 2.25R. This table also lists the nearest controlled test point for each standard (column 4), the difference between the logarithms of the intensity in question and the highest intensity in this direction from all standards (column 5), the resultant illuminance at the sign (column 6), and the difference between the logarithms of the illuminance in question and the criterion illuminance of 0.02 lx (column 7). Two standards exceeded this performance criterion.

Illumination of an overhead traffic sign (2U, V)

Geometry. The illumination directed towards an overhead traffic sign was evaluated by considering the following geometry on a two-lane roadway:

- Longitudinal separation between the sign and the headlamps: 150 m.
- Lateral separation between the sign and the headlamps: 0 m.
- Vertical separation between the sign and the headlamps: 5.5 m. This is based on a headlamp mounting height of 0.6 m (Cobb, 1990), and a sign mounting height of 6.1 m (Woltman and Szczech, 1989).

The angle corresponding to the preceding geometry is 2U, V.

Criterion illuminance. The criterion traffic-sign illuminance was set at 0.02 lx based on the same considerations as in the preceding section dealing with traffic signs on the right shoulder. (The observation angles are very similar, 0.33° for the overhead sign and 0.31° for the shoulder sign.)

Findings. Table 6 ranks all standards in terms of the decreasing combined luminous intensity from both lamps directed towards 2U, V. This table also lists the nearest controlled test point (column 4), the difference between the logarithm of the intensity in question and the highest intensity in this direction from all standards (column 5), the resultant illuminance at the sign (column 6), and the difference between the logarithm of the illuminance in question and the target illuminance of 0.02 lx (column 7). None of the standards met this performance criterion.

Table 5. Illumination of a traffic sign on the right shoulder. The relevant angle (0.5U, 2.25R) corresponds to the following assumed separations between the headlamps and the sign: lateral 6.15 m, vertical 1.5 m, and longitudinal 150 m. The criterion illuminance value was set at 0.02 lx.

1	2	3	4	5	6	7
Rank	Standard	cd (both lamps)	Nearest controlled point	Δlog cd from the best	lx @ 150 m	Δlog lx from 0.02
1.5	SAE	1,000	0.5U, 2R	0	.044	+.34
1.5	U.S.	1,000	0.5U, 2.25R	0	.044	+.34
3.5	Europe	?	no nearby minimum	N.A.	N.A.	N.A.
3.5	Japan	?	no nearby minimum	N.A.	N.A.	N.A.

Table 6. Illumination of an overhead traffic sign. The relevant angle (2U, V) corresponds to the following assumed separations between the headlamps and the sign: lateral 0 m, vertical 5.5 m, and longitudinal 150 m. The criterion illuminance value was set at 0.02 lx.

1	2	3	4	5	6	7
Rank	Standard	cd (both lamps)	Nearest controlled point	Δlog cd from the best	lx @ 150 m	Δlog lx from 0.02
1.5	SAE	400	1.5U, V	0	.018	-.05
1.5	U.S.	400	1.5U, 1R	0	.018	-.05
3	Europe	250	2U, V	-.2	.011	-.26
4	Japan	?	no nearby minimum	N.A.	N.A.	N.A.

Glare illumination towards oncoming traffic (0.5U, 3.5L)

Geometry. The glare illumination directed towards an oncoming driver was evaluated by assuming the following geometry on a two-lane roadway:

- Longitudinal separation between the oncoming driver and the headlamps: 50 m.
- Lateral separation between the oncoming driver and the headlamps: 3 m.
- Vertical separation between the eyes of the oncoming driver and the headlamps: 0.5 m. This is based on a headlamp mounting height of 0.6 m (Cobb, 1990), and a driver eye height of 1.1 m (Cobb, 1990).

The angle corresponding to the preceding geometry is 0.5U, 3.5L.

Criterion illuminance. The criterion illuminance value was set at 0.7 lx based on the following considerations:

- In a typical nighttime situation, discomfort glare reaches the value 4 on the de Boer scale (de Boer, 1967) at approximately 0.56 lx (-0.25 log lx) at the eye (Schmidt-Clausen and Bindels, 1974; Olson and Sivak, 1984a). (The de Boer scale is a nine-point scale with adjectives for odd points only. "Disturbing" corresponds to 3, and "just acceptable" corresponds to 5.)
- The transmittance of the windshield is assumed to be 0.85, which is typical of untinted glass at rake angle of about 45° (Owens et al., 1992).
- To achieve the illuminance at the driver's eyes of 0.56 lx after the light passes through the windshield, the illuminance at the surface of the windshield needs to be 0.7 lx (0.56/0.85).

Findings. Table 7 ranks all standards in terms of the increasing combined luminous intensity from both lamps directed towards 0.5U, 3.5L. This table also lists the nearest controlled test point (column 4), the difference between the logarithm of the intensity in question and the lowest intensity (in this direction) from all standards (column 5), the resultant illuminance at 150 m (column 6), and the difference between the logarithm of the illuminance in question and the criterion illuminance of 0.7 lx (column 7). Two standards exceeded this criterion.

Glare illumination towards traffic ahead via exterior rearview mirrors (1.25U, 8.25R)

Geometry. The glare illumination directed towards a driver ahead via the left exterior rearview mirror was evaluated by assuming the following geometry on a four-lane roadway:

- Longitudinal separation between the mirror and the headlamps of the glare car: 15 m.
- Lateral separation between the mirror and the glare headlamps in the left adjacent lane: 2.2 m.
- Vertical separation between the mirror and the glare headlamps: 0.3 m.

The angle corresponding to the preceding geometry is 1.25U, 8.25R.

Table 7. Glare illumination towards an oncoming driver. The relevant angle (0.5U, 3.5L) corresponds to the following assumed separations between the eyes of an oncoming driver and the headlamps: lateral 3 m, vertical 0.5 m, and longitudinal 50 m. The criterion illuminance value was set at 0.7 lx.

1	2	3	4	5	6	7
Rank	Standard	cd (both lamps)	Nearest controlled point	Δlog cd from the best	lx @ 50 m	Δlog lx from 0.7*
1	Europe	500	0.6U, 3.4L	0	.20	-.54
2	SAE	1,300	0.5U, 3L	+.41	.52	-.13
3.5	Japan	2,000	0.5U, 3.5L	+.60	.80	+.06
3.5	U.S.	2,000	0.5U, 3.5L	+.60	.80	+.06

*Negative values are desirable, indicating values lower than the maximum criterion illuminance at the eye.

Table 8. Glare illumination towards traffic ahead via exterior left rearview mirror. The relevant angle (1.25U, 8.25R) corresponds to the following assumed separations between the mirror and the glare headlamps: lateral 2.2 m, vertical 0.3 m, and longitudinal 15 m. The criterion illuminance value was set at 11 lx.

1	2	3	4	5	6	7
Rank	Standard	cd (both lamps)	Nearest controlled point	Δlog cd from the best	lx @ 15 m	Δlog lx from 11*
1	Europe**	876	1.25U, 8.25R	0	3.89	-.45
2	SAE	1,300	2U, 8R	+.17	5.78	-.28
3	Japan	2,000	1.5U, 8.25R	+.36	8.89	-.09
4	U.S.	2,800	1.5U, 8.25R	+.50	12.44	+.05

*Negative values are desirable, indicating values lower than the maximum criterion illuminance at the eye.

**The European specifications allow two different types of the cutoff to the right of vertical (see Table 2). The point under discussion (1.25U, 8.25R) is controlled for the horizontal cutoff option. For the 15° inclining cutoff option, this point is in an uncontrolled zone, and thus the European specification would be ranked 4.

Criterion illuminance. The criterion illuminance was set at 11 lx based on the data of Olson and Sivak (1984b), which showed that, *given the geometry of interest*, a value of 4 on the de Boer discomfort scale is reached at illuminance of approximately 7.5 lx. When corrections for mirror reflectivity (.80) and windshield transmittance (.85) are applied, the target illuminance becomes 11 lx (7.5/(.85 x .80)). (The target illuminance here is substantially greater than in the oncoming-glare situation discussed in the preceding section. The primary factor responsible for this difference is the increased glare angle in the present situation; the secondary factor is the loss of light due to the reflectivity of the mirror.)

Findings. Table 8 ranks all standards in terms of the increasing combined luminous intensity from both lamps directed towards 1.25U, 8.25R. This table also lists the nearest controlled test point (column 4), the difference between the logarithm of the intensity in question and the lowest intensity in this direction from all standards (column 5), the resultant illuminance at 15 m (column 6), and the difference between the logarithm of the illuminance in question and the criterion illuminance of 11 lx (column 7). Three standards exceeded this performance criterion.

Illumination prone to scatter in adverse atmospheric conditions (fog, rain, and snow) (10U, V)

Geometry. One aspect of the performance under adverse atmospheric conditions was evaluated by considering the amount of illumination directed towards 10U, V. The logic here is that a good beam pattern minimizes the amount of light scatter due to adverse atmospheric conditions (such as fog, rain, and snow) by minimizing the illumination directed toward areas where no targets or signs are likely. The selected vertical angle (10U) corresponds to an overhead sign (6.1 m above the roadway) at 34 m, too short a distance to be of likely importance for sign detection or legibility.

Criterion illuminance. There is insufficient empirical data to set a criterion value.

Findings. Table 9 ranks all standards in terms of the increasing combined luminous intensity from both lamps directed towards 10U, V. This table also lists the nearest controlled test point (column 4), and the difference between the logarithm of the intensity in question and the highest intensity (in this direction) from all standards (column 5). (The lamps currently manufactured to meet the European standard produce less light above horizontal than do lamps manufactured to meet the U.S. standard [Sivak, Flannagan, and Sato, 1993]. However, as indicated in the introduction, the present study evaluates what is *allowed* under a given standard. In that respect, the European standard allows more light at 10U and above than does the U.S. standard.)

Table 9. Illumination prone to scatter in adverse atmospheric conditions (10U, V).

1	2	3	4	5
Rank	Standard	cd (both lamps)	Nearest controlled point	Δlog cd from the best
2	Japan	250	10U, V	0
2	SAE	250	10U, V	0
2	U.S.	250	10U, V	0
4	Europe	876	10U, V	+.54

Illumination directed towards targets on the right side of the road at intermediate distances (0.5D, 1.25R)

Geometry. The illumination provided for detecting targets on the right side of the lane of travel at intermediate distances was evaluated by assuming the following geometry:

- Longitudinal separation between the target and the headlamps: 75 m.
- Lateral separation between the target and the headlamps: 1.6 m.
- Vertical separation between the target and the headlamps (i.e., headlamp mounting height): 0.6 m.

The angle that corresponds to the preceding geometry is 0.5D, 1.25R.

Criterion illuminance. The criterion illuminance was set at 33 lx to permit visual performance that is midway between capabilities in daylight and moonlight. This illuminance is equivalent to the mid-point of log ambient illumination during civil twilight, which occurs when the sun is less than 6° below the horizon and covers levels ranging from 330 to 3 lx (Leibowitz, 1987). Over this range, visual recognition performance falls from near-optimal levels in daylight to near-minimal levels in moonlight. Assuming the criterion illumination and a reflectance of 10%, object luminance is 1 cd/m^2. At this level, visual acuity is about 50% and peak contrast sensitivity is about 33% of photopic values (Owens, Francis, and Leibowitz, 1989). Historically, the dark boundary of civil twilight (3 lx) has been used widely as a benchmark for setting the limit of useful visual recognition. The 3 lx criterion may be a useful value for activities that are not visually challenging, such as farming or sailing, but is inappropriately low for visual demanding tasks, such as driving (Leibowitz and Owens, 1991). The criterion of 33 lx is not out of line with other current estimates of necessary illumination for perceiving unexpected low-contrast targets. For example, Kosmatka's (1992a) calculations for a 7% reflectance target indicate that the illuminance needs to be 32 lx (341,000 cd at 104 m), while Fisher's (1970) analysis (also for a 7% reflectance target), leads to 91 lx (1,200,000 cd at 115 m). Padmos and Alferdinck (1988) accept Fisher's intensity requirement of 1,200,000 cd, but use a distance of 110 m, for target illuminance of 99 lx.

Findings. Table 10 ranks all standards in terms of the decreasing combined luminous intensity from both lamps directed towards 0.5D, 1.25R. This table also lists the nearest controlled test point (column 4), the difference between the logarithm of the intensity in question and the highest intensity (in this direction) from all standards (column 5), the resultant illuminance at 75 m (column 6), and the difference between the logarithm of the illuminance in question and the criterion illuminance of 33 lx (column 7). None of the standards met this performance criterion.

Illumination on hills (1.25D, 2R)

Geometry. Driving on hills was evaluated by considering the illumination directed towards right side delineation using the following geometry:

- Longitudinal separation between the delineation and the headlamps: 50 m.
- Lateral separation between the delineation and the headlamps: 1.85 m.
- Radius of curvature: 3,000 m.
- Headlamp mounting height: 0.6 m.

The angle corresponding to the preceding geometry is 1.25D, 2R.

Criterion illuminance. The criterion illuminance was set at 6.4 lx based of the following considerations:

- Specific luminance of the road delineation: 0.1 cd/lx/m^2.
- Road delineation with specific luminance of 0.1 cd/lx/m^2 was found by Helmers and Lundquist (1991) to be visible at about 50 m.
- The headlamps used by Helmers and Lundquist are similar to the low-beam headlamp documented in Helmers and Rumar (1975). Using the isocandela diagram in Helmers and Rumar (1975), we estimated that each lamp directed approximately 8,000 cd towards the delineation at 50 m, for the resulting illuminance of 6.4 lx (16,000/50^2).

Findings. Table 11 ranks all standards in terms of the decreasing combined luminous intensity from both lamps directed towards 1.25D, 2R. This table also lists the nearest controlled test point (column 4), the difference between the logarithms of the intensity in question and the highest intensity in this direction from all standards (column 5), the resultant illuminance at 50 m (column 6), and the difference between the logarithms of the illuminance in question and the criterion illuminance of 6.4 lx (column 7). Three standards met or exceeded this performance criterion.

Table 10. Illumination directed towards targets on the right side of the road at 75 m (0.5D, 1.25R). The criterion illuminance was set at 33 lx.

1	2	3	4	5	6	7
Rank	Standard	cd (both lamps)	Nearest controlled point	Δlog cd from the best	lx @ 75 m	Δlog lx from 33
1.5	SAE	20,000	0.6D, 1.3R	0	3.56	-.97
1.5	U.S.	20,000	0.5D, 1.5R	0	3.56	-.97
3	Europe	15,000	0.6D, 1.1R	-.12	2.67	-1.09
4	Japan	12,000	0.5D, 2R	-.22	2.13	-1.19

Table 11. Illumination directed towards delineation at the right road edge at 50 m on a hill (1.25D, 2R). The criterion illuminance was set at 6.4 lx.

1	2	3	4	5	6	7
Rank	Standard	cd (both lamps)	Nearest controlled point	Δlog cd from the best	lx @ 50 m	Δlog lx from 6.4
2	Japan	30,000	1.5D, 2R	0	12.0	+.27
2	SAE	30,000	*	0	12.0	+.27
2	U.S.	30,000	1.5D, 2R	0	12.0	+.27
4	Europe	15,000	0.9D, 1.7R	-.30	6.0	-.03

*The minimum of 15,000 cd per lamp has to be met at least at one point in Zone 1 (corners: 0.5D, 0.5R; 0.5D, 2.5R; 2D, 2.5R; 2D, 0.5R).

Illumination on sags (0.25D, 2R)

Geometry. Driving on sags was evaluated by considering the illumination directed towards right side delineation using the following geometry:

- Longitudinal separation between the delineation and the headlamps: 50 m.
- Lateral separation between the delineation and the headlamps: 1.85 m.
- Radius of curvature: 3,000 m.
- Headlamp mounting height: 0.6 m.

The angle that corresponds to the preceding geometry is 0.25D, 2R. (For a level road, this angle corresponds to a longitudinal separation of 138 m between the delineation and the headlamps.)

Criterion illuminance. The criterion illuminance was set at the same level (6.4 lx) as in the above analysis for delineation on a hill.

Findings. Table 12 ranks all standards in terms of the decreasing combined luminous intensity from both lamps directed towards 0.25D, 2R. This table also lists the nearest controlled test point (column 4), the difference between the logarithms of the intensity in question and the highest intensity in this direction from all standards (column 5), the resultant illuminance at 50 m (column 6), and the difference between the logarithms of the illuminance in question and the criterion illuminance of 6.4 lx (column 7). Two standards met or exceeded this performance criterion.

Illumination directed towards targets on the right side of the road at near distances (1.25D, 3.75R)

Geometry. The illumination directed towards targets on the right side of the lane of travel at near distances was evaluated by considering the following geometry:

- Longitudinal separation between the target and the headlamps: 25 m.
- Lateral separation between the target and the headlamps: 1.6 m.
- Vertical separation between the target and the headlamps (i.e., headlamp mounting height): 0.6 m.

The angle that corresponds to the preceding geometry is 1.25D, 3.75R.

Criterion illuminance. The criterion illuminance was set at the same level (33 lx) as in the above analysis for a target at 75 m.

Findings. Table 13 ranks all standards in terms of the decreasing combined luminous intensity from both lamps directed towards 1.25D, 3.75R. This table also lists the nearest controlled test point (column 4), the difference between the logarithm of the intensity in question and the highest intensity (in this direction) from all standards (column 5), the resultant illuminance at 25 m (column 6), and the difference between the logarithm of the illuminance in question and the target illuminance of 33 lx (column 7). None of the standards met this performance criterion level.

Table 12. Illumination directed towards delineation at the right road edge at 50 m on a sag (0.25D, 2R). The criterion illuminance was set at 6.4 lx.

1	2	3	4	5	6	7
Rank	Standard	cd (both lamps)	Nearest controlled point	Δlog cd from the best	lx @ 50 m	Δlog lx from 6.4
1.5	SAE	20,000	0.6D, 1.3R	0	8.0	+.10
1.5	U.S.	20,000	0.5D, 1.5R	0	8.0	+.10
3	Europe	15,000	0.6D, 1.1R	-.12	6.0	-.03
4	Japan	12,000	0.5D, 2R	-.22	4.8	-.12

Table 13. Illumination directed towards targets on the right side of the road at 25 m (1.25D, 3.75R). The criterion illuminance was set at 33 lx.

1	2	3	4	5	6	7
Rank	Standard	cd (both lamps)	Nearest controlled point	Δlog cd from the best	lx @ 25 m	Δlog lx from 33
1	SAE	10,000	0.5D, 4R	0	16.0	-.31
2	Europe	3,750	1.25D, 3.75R	-.43	6.0	-.74
3.5	Japan	?	no nearby test point	N.A.	N.A.	N.A.
3.5	U.S.	?	no nearby test point	N.A.	N.A.	N.A.

Homogeneity of the beam (a comparison of 1.25D, 3.75R and 1.25D, V)

We planned to evaluate the homogeneity of the beam by comparing the illumination directed towards two foreground test points, one on the right side and one straight ahead. The selected test points were 1.25D, 3.75R and 1.25D, V. To perform the evaluation, we needed both the minima and maxima in this region of the beam, so that we could estimate the likely illumination. In the standards under review, that was not the case. Thus, homogeneity is simply not addressed by the standards and, therefore, cannot be evaluated.

Illumination prone to reflected glare from wet pavement (2D, 3.5L)

Geometry. One aspect of visual performance in adverse weather was evaluated by considering the amount of illumination reflected from the wet pavement in the direction of an oncoming driver in the adjacent lane at the distance of 50 m (i.e., the illumination *reflected* from the pavement towards the same point as the *direct* glare illumination considered above). The direct glare was evaluated for 0.5U, 3.5L. The calculated direction for the light to be reflected towards 0.5U, 3.5L at 50 m is 2D, 3.5L. This calculation assumes longitudinal separation between the oncoming driver and the headlamps of 50 m, lateral separation between the driver and the headlamps of 3 m, mounting height of headlamps of 0.6 m, and driver eye height of 1.1 m.

Criterion illuminance. The proportion of light reflected in the direction of interest depends on the type of the road surface and the extent to which the standing water fills the depressions in the road surface. Because the proportion of reflected light varies quite substantially with these two factors, no criterion illuminance was set.

Findings. None of the standards have a maximum near 2D, 3.5L. Consequently, we were not able to evaluate the standards on this function.

Lateral spread

The extent of the lateral spread of illumination (important for visual performance on sharp horizontal curves and at intersections) was evaluated by examining the widest controlled test points. Table 14 ranks all standards in terms of the decreasing lateral angle of the most extreme controlled test points. Within proposals with equivalent width of the coverage, the proposals are ranked in the decreasing order of the specified minimum luminous intensity.

Relation between seeing illumination and direct glare illumination (a comparison of 0.5D, 1.25R and 0.5U, 3.5L)

The relation between seeing illumination and direct glare illumination was evaluated by computing the ratio between illumination at 0.5D, 1.25R and 0.5U, 3.5L. Table 15 ranks all standards in terms of the decreasing ratio of luminous intensity directed towards 0.5D, 1.25R and 0.5U, 3.5L (column 3). It also lists the differences in logarithms between these two luminous intensities (column 5) and the difference between the logarithm of the ratio in question and the highest ratio from all standards (column 6). From the standpoint of visibility, the highest ratios are most desirable because they indicate high visibility with low glare to oncoming drivers.

Table 14. The extent of lateral spread.

1	2	3	4
Rank	Standard	Widest controlled points	cd (both lamps)
1	SAE	20L/20R (4D)	600
2	U.S.	15L/15R (2D)	1,700
3	Japan	15L/15R (2D)	1,400
4	Europe	9L/9R (1.7D)	2,500

Table 15. Ratio of seeing illumination (0.5D, 1.25R) and glare illumination (0.5U, 3.5L).

1	2	3	4	5	6
Rank	Standard	Ratio	Nearest controlled points	Δlog cd	Δlog ratio from the best
1	Europe	30.0:1	see Tables 10 and 7	1.48	0
2	SAE	15.4:1	see Tables 10 and 7	1.19	-.29
3	U.S.	10.0:1	see Tables 10 and 7	1.00	-.48
4	Japan	6.0:1	see Tables 10 and 7	0.78	-.70

Reliability of visual aiming

Vertical aiming is of primary concern here. The evidence indicates that reliability of vertical visual aiming is affected by the luminous-intensity contrast between vertically adjacent parts of the beam (Poynter, Plummer, and Donohue, 1989; Sivak, Flannagan, Chandra, and Gellatly, 1992). Contrast in these two studies was computed in steps of 0.1° from available candela matrices. However, such a computation of contrast is not possible for the standards under consideration because of the limited number of test point/regions.

In the absence of actual contrast measures, we estimated the gradient by using the method proposed by Kosmatka (1992b), which involves the following steps (see Table 16): (1) Select a point to the right of vertical and below horizontal that involves a minimum and is within 1° of horizontal. (2) Select a point to the right of vertical and above horizontal that involves a maximum, is within 1° of horizontal, and is at the same lateral position as the previously considered minimum. (3) Compute the ratio of these two values. Raise this ratio to the power that is the inverse of the number of 0.1° steps that separate the two points under consideration. Subtract 1.00 and multiply by 100. This yields the percent by which candela values change over each 0.1° step, assuming that the gradient is constant in terms of percent change over the entire interval (i.e., assuming that log candela values change linearly with angle). The results of these calculations are shown in Table 16. The computed gradients ranged from 37% to 8%.

Adverse effects of misaim

Of primary concern here are the potential adverse effects of vertical misaim: misdirecting seeing illumination to glare zones, and restricted glare illumination to seeing zones. As a first approximation, the effect of vertical misaim is likely to be *inversely* proportional to the steepness of the vertical gradient. Consequently, the *inverse* of the ranking in Table 16 represents our best prediction concerning the effects of misaim.

Table 16. Gradient to the right of vertical.

1	2	3	4	5	6	7
Rank	Standard	Minimum cd below horizontal for one lamp (controlled point)	Maximum cd above horizontal for one lamp (controlled point)	Vertical distance (°) between points in 3 and 4	Ratio of columns 3 and 4	Gradient (%)
1	Europe*	7,500 (0.6D, 1.1R)	438 (0.3U, 1.1R)	.9	17.1	37
2.5	SAE	10,000 (0.6D, 1.3R)	2,400 (0.5U, 1.3R)	1.1	4.2	14
2.5	U.S.	10,000 (0.5D, 1.5R)	2,700 (0.5U, 1.5R)	1.0	3.7	14
4	Japan	6,000 (0.5D, 2R)	2,700 (0.5U, 2R)	1.0	2.2	8

*The European standard allows two different types of cutoff (see Table 2). The calculations in this table are for the continuously inclining cutoff; for the horizontal cutoff, the resulting gradient is 27%.

DISCUSSION

There are three main problems in coming up with a single overall figure of merit. First, the visual performance functions are not equally important from the safety point of view. For example, strong arguments could be made that visibility and direct glare should be weighted more heavily than the other functions such as indirect glare via rearview mirrors (because of the existence of dual-position and variable-reflectance mirrors) and homogeneity of the beam (because it deals mostly with considerations of aesthetics and comfort). There is no general consensus, however, about the appropriate weights for all the different functions addressed here. Second, performance on certain visual performance functions could not be evaluated for some lamp standards under review. This happened because the selected critical points did not always coincide with, or fall near to, test points in the standards. Furthermore, in some instances where there was a coincidence of test points, the required minimum luminous intensity for seeing considerations was not included in the standards. Third, because of the lack of relevant empirical data, for several functions we were unable to determine criterion illuminance values against which to evaluate the standards.

In spite of the above limitations, we tried to integrate and summarize in Table 17 the rankings for the individual visual performance functions from Tables 5 through 16. (The inverse of the ranking of the reliability of visual aiming [Table 16] was used to estimate effects of misaim.) Table 17 also presents the mean rankings for all visual performance functions.

CONCLUSIONS

This study evaluated the SAE J1735 Draft Proposal for a low-beam headlighting pattern in relation to the current standards in the U.S., Europe, and Japan. The approach consisted of the following steps: (1) identifying a set of 15 important visual performance functions (including seeing and glare) for low-beam headlamps; (2) defining the relevant geometry relative to the visual performance functions; (3) setting criterion values of illumination for each of the visual performance functions based on the available empirical data; and (4) evaluating the standards relative to the criterion values by using the worst-allowed-case approach (evaluating the *minima* specified by the standards for seeing functions, and the *maxima* for glare functions).

The results indicate that the SAE J1735 Draft Proposal tended to require better performance than the current U.S., European, and Japanese standards.

ACKNOWLEDGMENT

This study was supported by the American Automobile Manufacturers Association (AAMA).

Table 17. Rankings of the standards on the individual visual performance functions in Tables 5 through 16.

Visual performance function	Standard			
	SAE	U.S.	Europe	Japan
Right traffic signs (Table 5)	1.5	1.5	3.5	3.5
Overhead traffic signs (Table 6)	1.5	1.5	3	4
Direct glare (Table 7)	2	3.5	1	3.5
Rearview-mirror glare (Table 8)	2	4	1	3
Fog, rain, and snow scatter (Table 9)	2	2	4	2
Intermediate-distance targets (Table 10)	1.5	1.5	3	4
Illumination on hills (Table 11)	2	2	4	2
Illumination on sags (Table 12)	1.5	1.5	3	4
Near-distance targets (Table 13)	1	3.5	2	3.5
Lateral spread (Table 14)	1	2	4	3
Seeing-to-glare ratio (Table 15)	2	3	1	4
Reliability of visual aiming (Table 16)	2.5	2.5	1	4
Effects of misaim (Table 16)	2.5	2.5	4	1
Mean ranking	**1.77**	**2.38**	**2.65**	**3.19**

REFERENCES

Alferdinck, J.W.A.M. (1984). *Deterioration of retroreflective properties of signs and pilons in use for road construction sites* (Report IZF 1984 C-2 [in Dutch]). Soesterberg, The Netherlands: TNO Institute for Perception.

Anderson, J.W. and Carlson, G.C. (1966). *Vehicle spray pattern study* (Investigation 338). St. Paul, MN: Minnesota Highway Department.

Bhise, V.D., Farber, E.I., and McMahan, P.B. (1976). Predicting target-detection distance with headlights. *Transportation Research Record, 611*, 1-16.

Bhise, V.D., Farber, E.I., Saunby, C.S., Troell, G.M., Walunas, J.B., and Bernstein, A. (1977). *Modeling vision with headlights in a system context* (SAE Technical Paper Series No. 770238). Warrendale, PA: Society of Automotive Engineers.

Burgett, A., Matteson, L., Ulman, M., and Van Iderstine, R. (1989). Relationship between visibility needs and vehicle-based roadway illumination. *Proceedings of the Twelfth International Technical Conference on Experimental Safety Vehicles.* Washington, D.C.: National Highway Traffic Safety Administration.

Cobb, J. (1990). *Roadside survey of vehicle lighting 1989* (Research Report 290). Crowthorne, England: Transport and Road Research Laboratory.

Cox, N.T. (1968). *The effect of dirt on vehicle headlamp performance* (Report LR 240). Crowthorne, England: Road Research Laboratory.

de Boer, J.B. (1967). Visual perception in road traffic and the field of vision of the motorist. In J.B. de Boer (Ed.), *Public lighting*. Eindhoven, The Netherlands: Philips Technical Library.

ECE. (1992). *Agreement concerning the adoption of uniform conditions of approval and reciprocal recognition of approval for motor vehicle equipment and parts* (ECE Regulation 20.02). Geneva: United Nations.

Fisher, A.J. (1970). A basis for a vehicle headlighting specification. *IES Lighting Review* (Australia), 32, No. 1 and 2.

FMVSS (Federal Motor Vehicle Safety Standard). (1993). Standard No. 108. Lamps, reflective devices, and associated equipment. In, *Code of federal regulations [Title] 49*. Washington, D.C.: Office of the Federal Register.

Helmers, G. and Lundquist, S.-O. (1991). *Visibility distance to road markings in headlight illumination* (Publication 657) (in Swedish). Linköping: Swedish Road and Traffic Research Institute.

Helmers, G. and Rumar, K. (1975). High beam intensity and obstacle visibility. *Lighting Research and Technology*, *7*, 35-42.

JASIC (Japan Automobile Standards Internationalization Center). (1993). Automobile Type Approval Handbook for Japanese Certification. Tokyo: JASIC.

Jenkins, S.E. and Gennaoui, F.R. (1992). *Terminal values of road traffic signs* (Special Report No. 49). Vermont South, Australia: Australian Road Research Board.

JIS (Japanese Industrial Standard) D 5500B1-1984. (1984). Tokyo: Japanese Standards Association.

Kosmatka, W.J. (1992a). *Obstacle detection rationale for upper beam intensity* (unpublished manuscript). Nela Park, Ohio: GE Lighting.

Kosmatka, W.J. (1992b). Personal communication, May 3.

Leibowitz, H.W. (1987). Ambient illuminance during twilight and from the full moon. In *Night vision: Current research and future directions* (pp. 19-22). Washington, D.C. National Academy Press.

Leibowitz, H.W. and Owens, D.A. (1991). Can normal outdoor activities be carried out during civil twilight? *Applied Optics*, *30*, 3501-3503.

Olson, P.L. (1985). *A survey of lighting equipment on vehicles in the United States* (SAE Technical Series No. 850229). Warrendale, PA: Society of Automotive Engineers.

Olson, P.L. and Sivak, M. (1984a). Discomfort glare from automotive headlights. *Journal of the Illuminating Engineering Society*, *13*, 296-303.

Olson, P.L. and Sivak, M. (1984b). Glare from automobile rear-vision mirrors. *Human Factors*, *26*, 269-282.

Owens, D.A., Francis, E.L., and Leibowitz, H.W. (1989). *Visibility distance with headlights: a functional approach* (SAE Technical Series No. 890684). Warrendale, PA: Society of Automotive Engineers.

Owens, D.A., Sivak, M., Helmers, G., Sato, T., Battle, D., and Traube, E. (1992). *Effects of light transmittance and scatter on nighttime visual performance* (Report No. UMTRI-92-37). Ann Arbor: The University of Michigan Transportation Research Institute.

Padmos, P. and Alferdinck, J.W.A.M. (1988). *Optimal light intensity distribution of the low beam of car headlamps* (Report No. IZF 1988 C-9/E). Soesterberg, The Netherlands: TNO Institute for Perception.

Poynter, W.D., Plummer, R.D., and Donohue, R.J. (1989). *Vertical alignment of headlamps by visual aim* (Report No. GMR-6693). Warren, MI: General Motors Research Laboratories.

SAE. (1991). *Low beam design guide and conformance requirements (Proposal 7A)*. Warrendale, PA: Society of Automotive Engineers, Headlamp Beam Pattern Task Force.

SAE. (1994). The SAE J1735 draft proposal for a harmonized low-beam headlighting pattern. Warrendale, PA: Society of Automotive Engineers, Headlamp Beam Pattern Task Force.

Schmidt-Clausen, H.J. and Bindels, J.T.H. (1974). Assessment of discomfort glare in motor vehicle lighting. *Lighting Research and Technology*, *6*, 79-88.

Sivak, M. and Ensing, M. (1989). *Human factors considerations in the design of truck lighting, signaling, and rearview mirrors* (Report No. UMTRI-89-9). Ann Arbor: The University of Michigan Transportation Research Institute.

Sivak, M. and Flannagan, M.J. (1993). Partial harmonization of international standards for low-beam headlighting patterns (Report No. UMTRI-93-11). Ann Arbor: The University of Michigan Transportation Research Institute.

Sivak, M., Flannagan, M., Chandra, D., and Gellatly, A.W. (1992). *Visual aiming of European and U.S. low beam headlamps* (SAE Technical Paper Series #920814). Warrendale, PA: Society of Automotive Engineers.

Sivak, M., Flannagan, M., and Gellatly, A.W. (1991). *The influence of truck driver eye position on the effectiveness of retroreflective traffic signs* (Report No. UMTRI-91-35). Ann Arbor: The University of Michigan Transportation Research Institute.

Sivak, M., Helmers, G., Owens, D.A., and Flannagan, M.J. (1992). Evaluation of proposed low-beam headlighting patterns (Report No. UMTRI-92-14). Ann Arbor: The University of Michigan Transportation Research Institute.

Sivak, M. and Olson, P.L. (1985). Optimal and minimal luminance characteristics for retroreflective highway signs. *Transportation Research Record*, *1027*, 53-57.

Sivak, M., Olson, P.L., and Pastalan, L.A. (1981). Effect of driver's age on nighttime legibility of highway signs. *Human Factors*, *23*, 59-64.

Woltman, H.L. and Szczech, T.J. (1989). Sign luminance as a methodology for matching driver needs, roadway variables, and signing materials. *Transportation Research Record*, *1213*, 21-26.

950599

Dual-Filament Replaceable Bulb Headlamp Using a Multi-Reflector Optimized with a Neural Network

Yutaka Nakata
Ichikoh Industries, Ltd.

1. ABSTRACT

A system has been developed that makes it possible to design headlamps having an ideal light distribution pattern. Using multi B-spline surfaces (MBSS) for the reflector and dual filament (c-8/c-8) for the light source, the shape and the light distribution pattern it generates are computer simulated. The characteristics of the light distribution are analyzed and, using a neural network, the reflector is partitioned. Each surface is continuously modified so as to obtain the optimum light distribution pattern.

2. INTRODUCTION

Recently, with the increase of older drivers, a headlamp having a light distribution to fit even for older drivers is needed. Older drivers are more sensitive to glare than younger drivers are, and older drivers need brighter light than younger drivers do. Considering the natural environment, what is needed is energy saving. From this point the merits and the demerits of a dual-filament, replaceable bulb headlamp are as follows;

In case of using H4 bulbs, the glare can be decreased because the cut line can be generated clearly. On the other hand, the produced luminous flux is decreased because the light emitted from sub filament is reduced by a shade in a bulb. In case of using #9007 or #9004 bulb, compared with H4 bulb, the volume of utilizable luminous flux is greater but the volume of glare is also greater because the cut line becomes obscure due to no shade. Therefore presently it is considered necessary to have a dual-filament replaceable bulb headlamp having low glare as with the H4 bulb and more effective luminous flux as with the #9007 bulb.

Light distribution can be made without lens prisms by giving the reflector's reflective surface a special curvature. For the lens one can use a transparent lens without prisms, eliminate the loss at the boundary part of the prism, and avoid bending of the light distribution pattern caused by the slant of the lens. But if only the low beam is considered when designing the reflector, the upper beam light distribution pattern will become scattered, so what is needed is a method for designing reflectors so as to optimize both upper beam and low beam light distribution patterns. This research has solved these problems by the following method.

1) A system was developed that generates a reflector surface automatically. This surface reflects sub-filament image having cut off line. The system analyzes the reflection characteristics of the reflector by comparing the reflected image of the main and sub-filament of each part of the reflector.

2) A system was developed that appropriately partitions the reflector based on the results of the above analysis and determines the light distribution role of each part.

3) A system was developed that optimally apportions the light intensity distribution of the low and upper beam light distribution patterns using a Neural Network algorithm and calculates the shape of the reflector.

4) The sub-filament and main filament are located close in the bulb tube. This system involves the effect of loss of light output caused from interference between filaments.

5) The reflector-generation program of this system enables control of the beam pattern for each part of reflector surface. (Horizontal scattering, vertical scattering, Lower or upper corner intensity gradient, Cut off line)

This system, given only the size of the reflector and the maximum light intensity required, can partition the reflector and automatically generate the shape of each part. The resulting light distribution pattern is properly balanced between the low beam and upper beam patterns. Also, good results were achieved in the evaluation of trial-manufactured reflectors made from designs using this system.

3. EXPLANATION OF THE MODEL

Figure 1 shows the flowchart of the MBSS dual-filament replaceable bulb headlamp automatic design program. When the program is started, a basic reflector is automatically generated using the default values (③) and displayed on the screen. The program is then ready for the interactive input of parameters (②). The following shows the content and output screen for each step using type A reflector of Table 1.

Start		Reflector	Screen Pattern
① ③	Generation of basic reflector using default values. Deformed parabolic reflector	One smooth surface 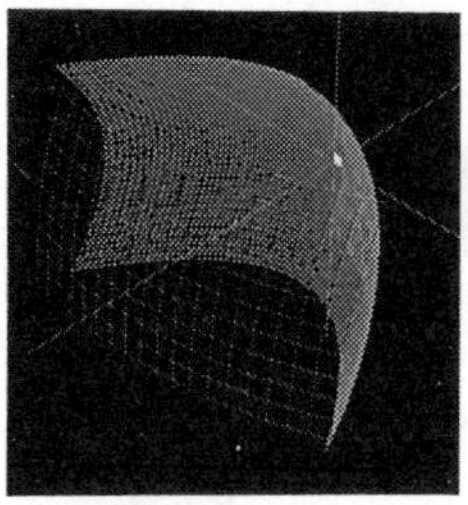	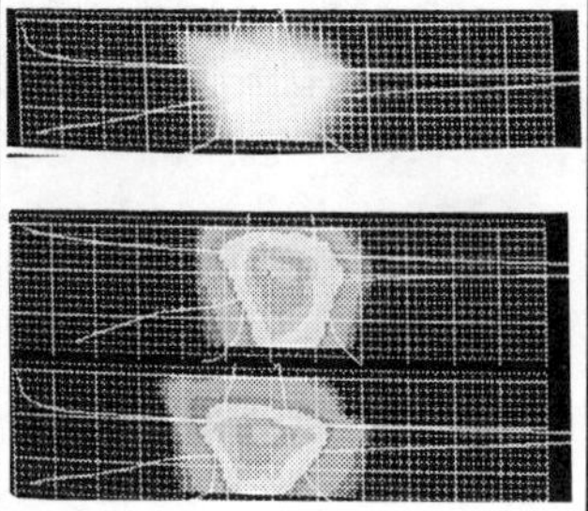
④	Generation of reflector having cut line The reflector is modified in such a way that all the reflected light (filament image) of the light incident on each point on the above reflector touch the lower side of the cut line.		
⑤	Generation of reflector having appropriate partial pattern The reflector is modified in such a way that the above individual patterns roughly intersect at the H-V point		
⑥	Light distribution characteristics analysis Classification (15 types) by differences in the shape and position of each of the above reflection patterns: See Figures 4-1 and 4-2		
⑦	Automatic partitioning Partitioned appropriately by the neural network 	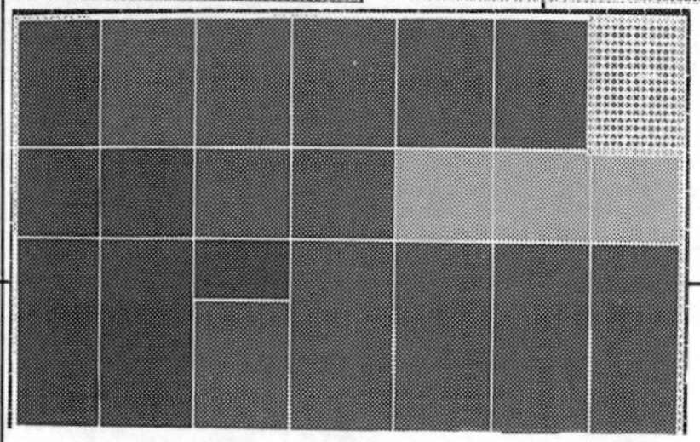	
⑧	Automatic surface calculation A surface (B-spline surface) is automatically generated so that the individual parts of the partition form the target pattern. See Figure 10		
⑨	Generation of the individual and overall light distribution pattern Display of surface diagram	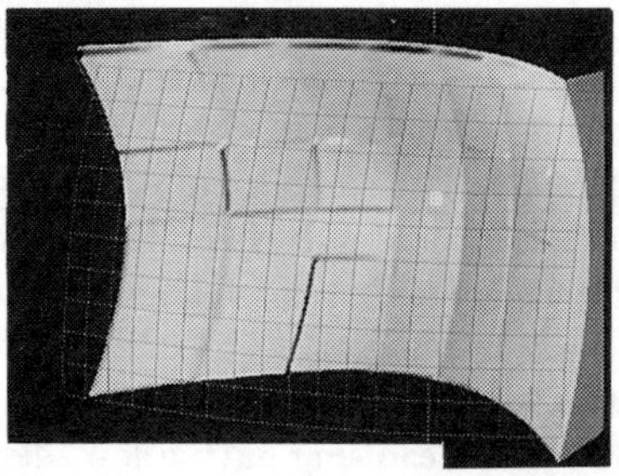	
⑩	Result image processing 1) Light distribution pattern 2) Solid model		
⑪	Saving of light distribution and reflector shape data in database		
⑫	Standards evaluation: SAE, ECE, JIS etc. Actual car simulation evaluation: SAE paper No.920816 Computerized Graphic Light Distribution Fuzzy Evaluation System for Automobile Headlighting Using Vehicle Simulation		
⑬ ⑭ End	Fine adjustment needed ? - Not needed → End - Needed → Input amount of shift and amount of diffusion while viewing the pattern on the screen, repeating steps ⑧、⑨、⑩	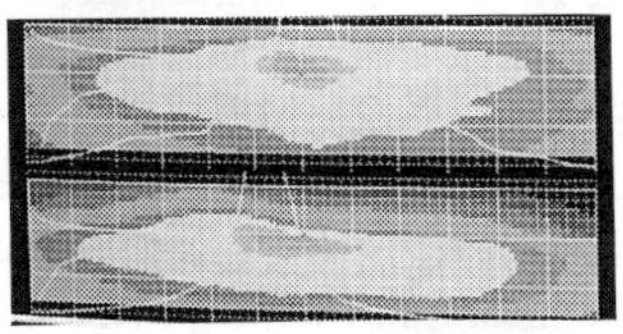	

Table 1 Process of MBSS Headlamp Automatic Design Program

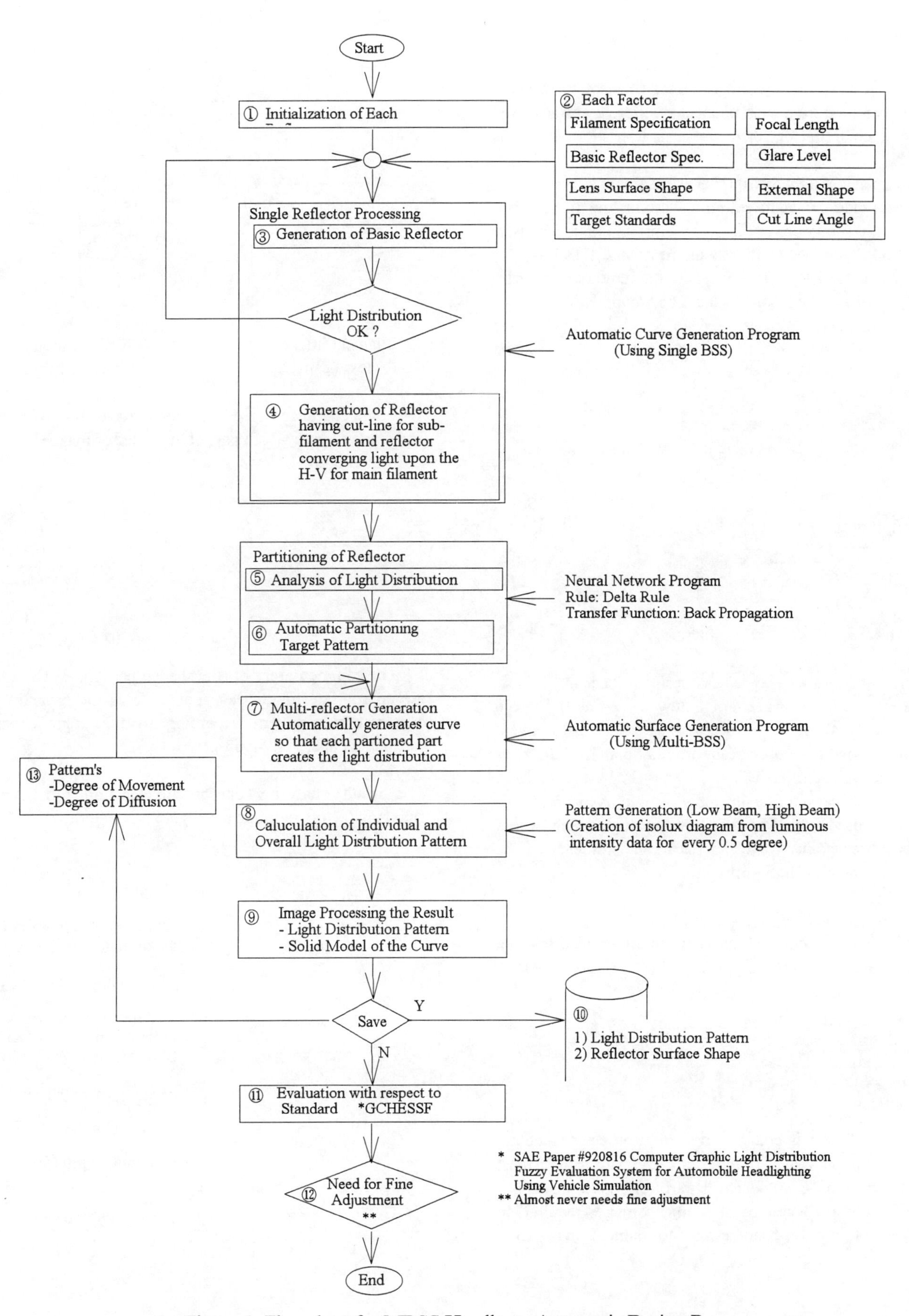

Figure 1 Flowchart for MBSS Headlamp Automatic Design Program

4. BASIC REFLECTOR AND FILAMENT LOCATION

Two lamp headlamp system use dual filament replaceable bulb. Light emitted from each filament is reflected and scattered by a common reflector then the characteristic pattern is generated.

Requirements of light distribution are as follows:

1) low beam should have its cut-off line around the line 0.5°D.

2) Low beam should have its maximum luminous intensity about 30,000cd and the location should be around 1°D-1°L.

3) High beam should have its maximum luminous intensity about 50,000cd and the location should be around H-V.

Single MBSS reflector is used to solve 1) above. This reflector is divided into three parts as shown below by the shape of low beam pattern.

Reflector

Filament image for low beam

Filament image for upper beam

The result analyzed by a computer using algorithm A, B and C to research which part of the reflector makes hot spot H is shown in Figures below.

Lower Beam Pattern

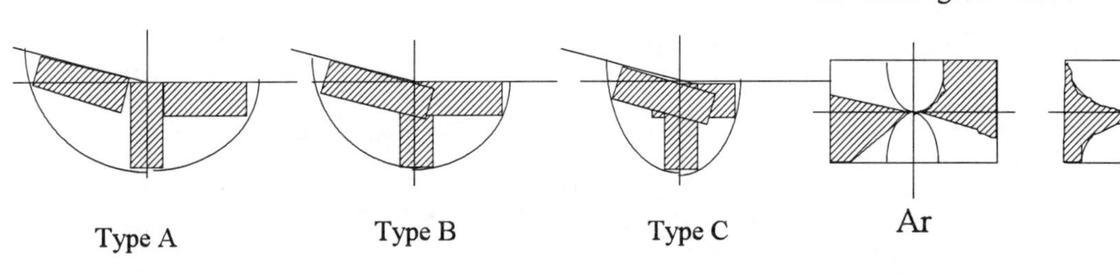

Type A Type B Type C

Ar Br Cr

To comply with 2) above, type A is desirable because the low beam pattern isn't converged just below the point H-V. To comply with 3) above, it is necessary to find the filament location that converges upper beam around point H-V as far as possible using type A reflector.

The pattern because of having longer length is inappropriate for the pattern used for hot zone of low and high beams at that time. From these points we will research the influence of the filament location on the light distribution.

4.1. Appropriate location of the main filament

Each low beam and each upper beam are divided into hot spot H and scattered pattern S. (Figure below)

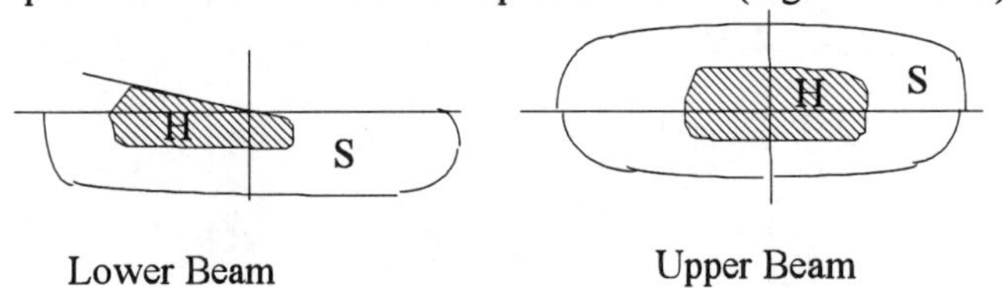

Lower Beam Upper Beam

Type A reflector is changed to a multi reflector to obtain the light distribution above. H and S of each beam pattern must be reflected patterns from identical parts of reflector surface. Described below are locations of filament images required to generate the hot spot. (a, b and c refer to filament image from reflector)

The common areas of the shaded portions in Br, and in Cr, are appropriate for the hot spot. Ar is the program used to construct sub pattern according to the approach in SAE Technical Paper No.940638.

The result analyzing appropriate location for filament by changing factors in following Figure 3 is as follows.

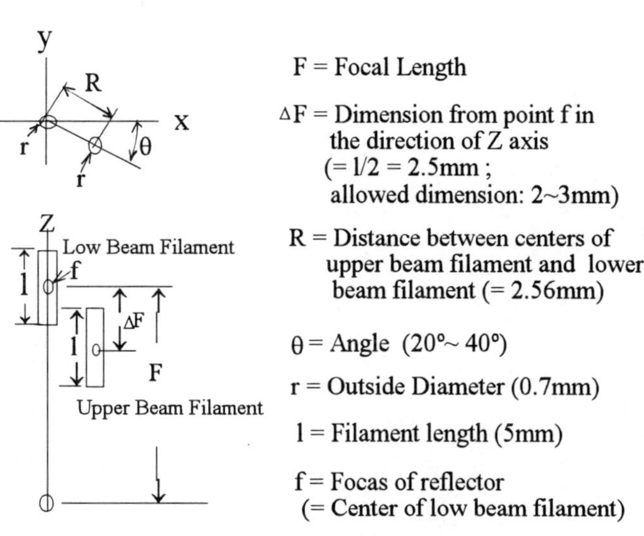

Figure 3

Type A:
Output screen of Algorithm (A) analyzing the distribution for low beam filament

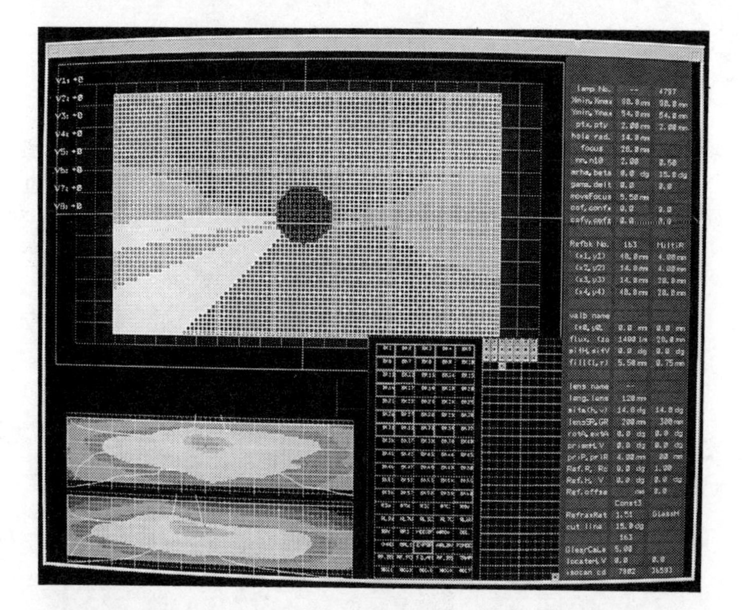

Type B:
Output screen of Algorithm (B) analyzing the distribution for upper beam filament

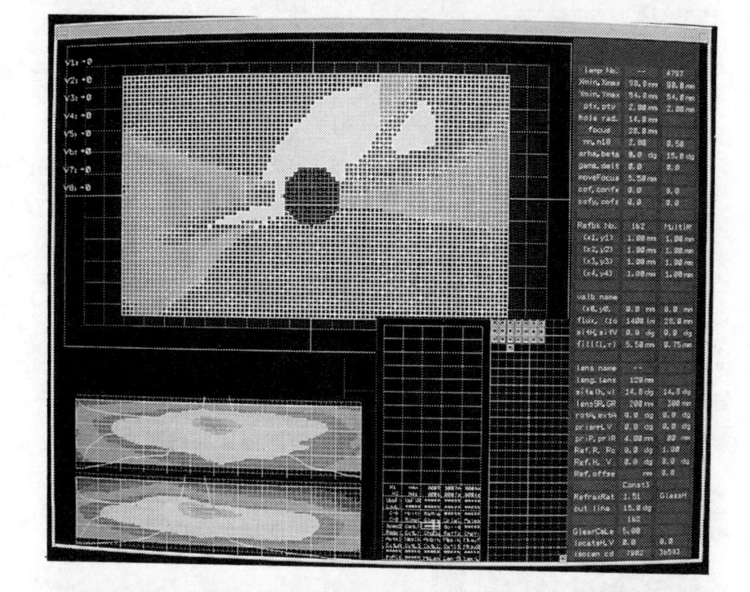

Type C:
Output screen of Algorithm (C) analyzing the distribution for low beam filament

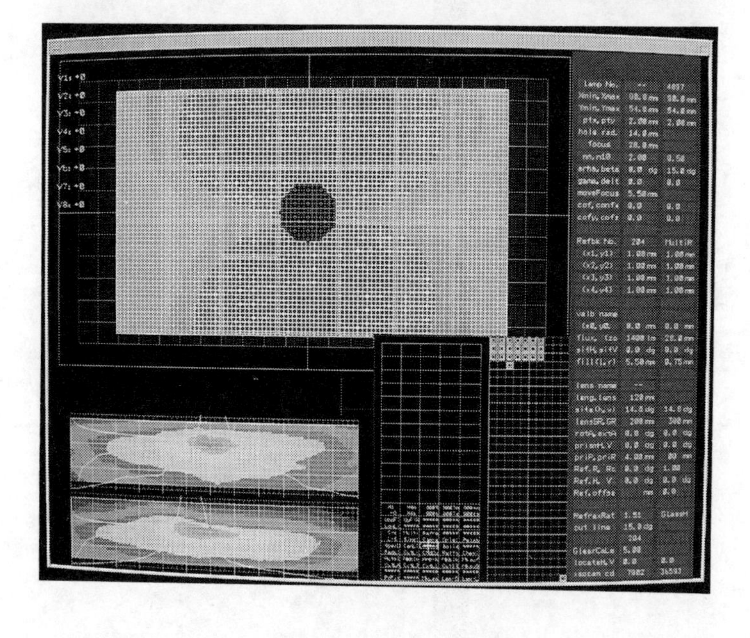

Figure 2 Photometric Characteristics of Reflector Before Partitioning

5. ALGORITHM

5.1 Analysis of reflector light distribution characteristic

A surface is created by replacing the reflector with a B-spline surface. Then the filament of the light source is modeled with cylindrical shape. When the light source is lit, the image of the cylindrical filament, as reflected by the reflector, appears on the screen. The algorithm for this is shown in Figures 4, 5 and 6.

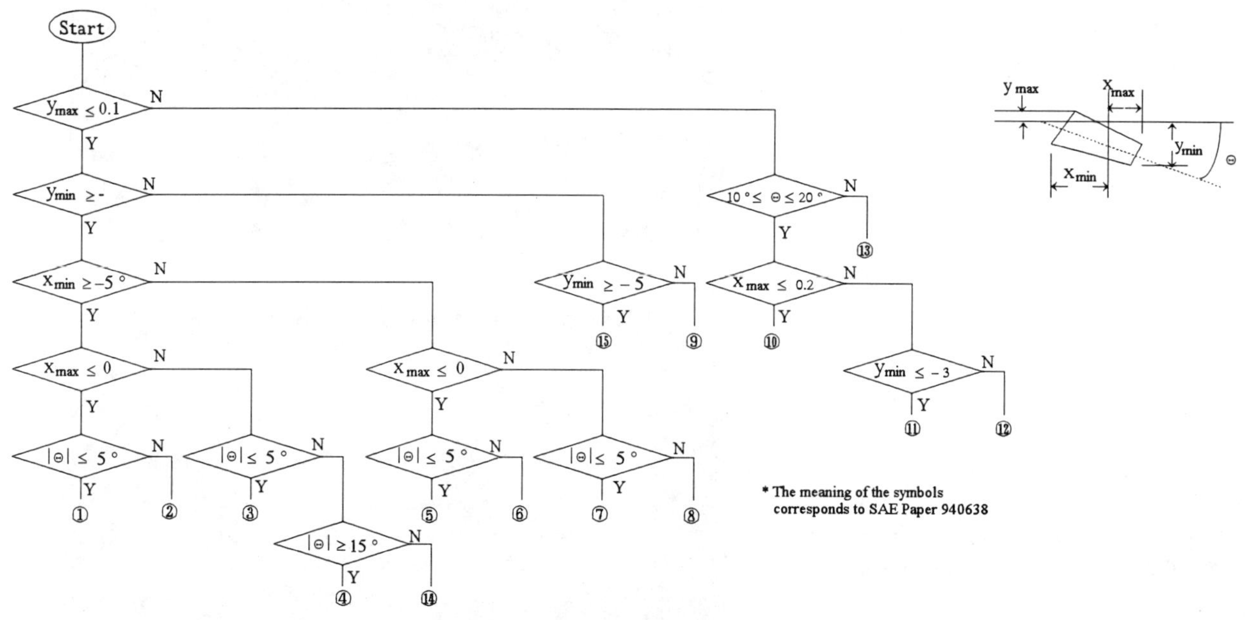

Figure 4 Classification algorithm(A) for pattern used in light distribution characteristics analysis for low beam filament

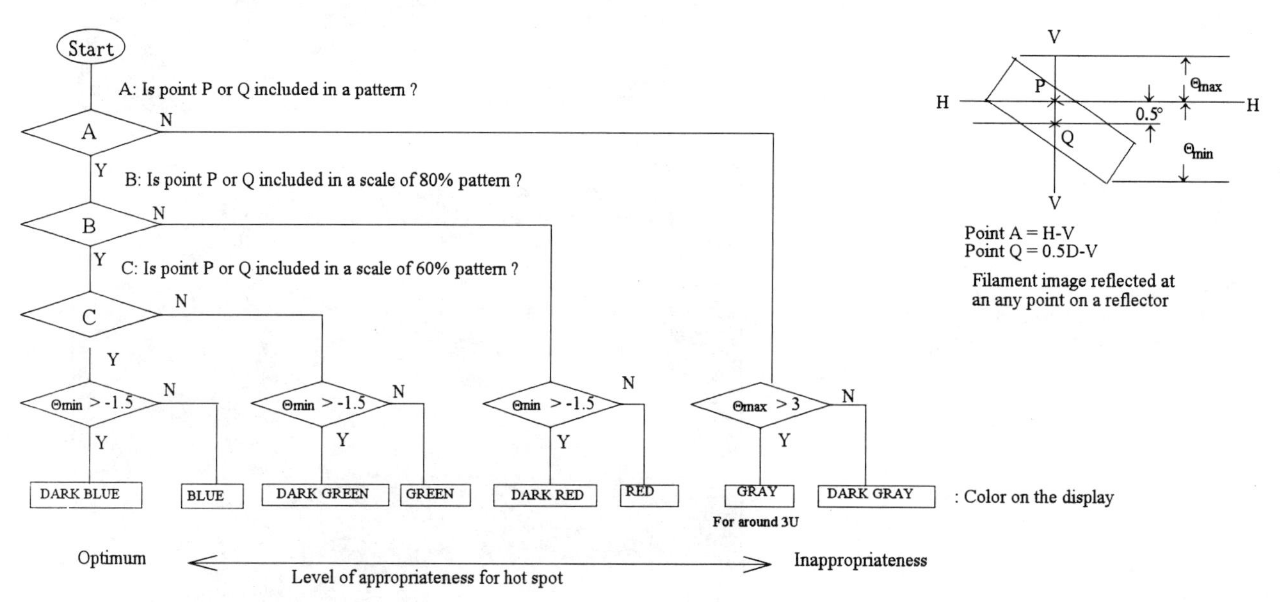

Figure 5 Algorithm (B) analyzing the distribution of hot spot area (For upper beam filament)

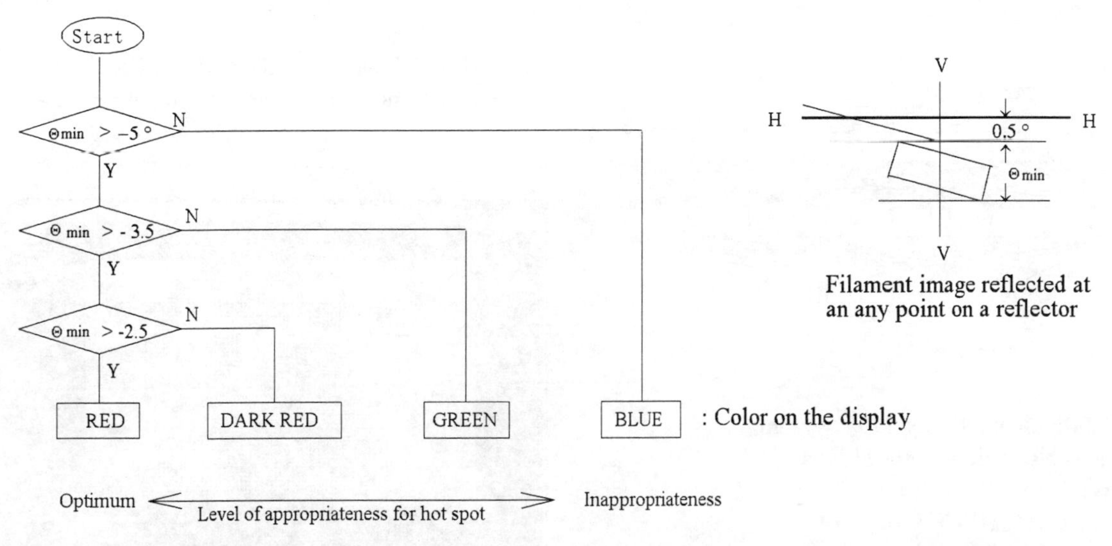

Figure 6 Algorithm a analyzing the distribution of hot spot area (For lower beam filament)

.2 Automatic partitioning of reflector

.2.1 Conditions for automatic analysis

1) Made into a lattice pattern in order to make NC processing easier when the mold is made.
2) The number of segments is kept to a minimum in order to keep down the cost of the mold as much as possible.
3) The size of an individual segment is made at least 10 mm each vertically and horizontally in front view.
4) The reflector is a rectangle having a height of 80 to 120 mm and width of 180 to 220 mm.
5) The light source is of C-8/C-8 type similar to #907 bulb.

.2.2 The minimum number of partitions required is 3 ertically and 7 horizontally, as in Figure 6. The required atterns are broadly classified into:

- Hot zone of upper and lower beam (maximum light intensity, diagonal cut line) ( )
- Horizontal cut line ( )
- Delta zone (below the diagonal cut line) ( )
- Diffusion (large, medium, small) ( )

When analyzed with the results of light distribution analysis, can be graphically represented as shown in Figure 7.

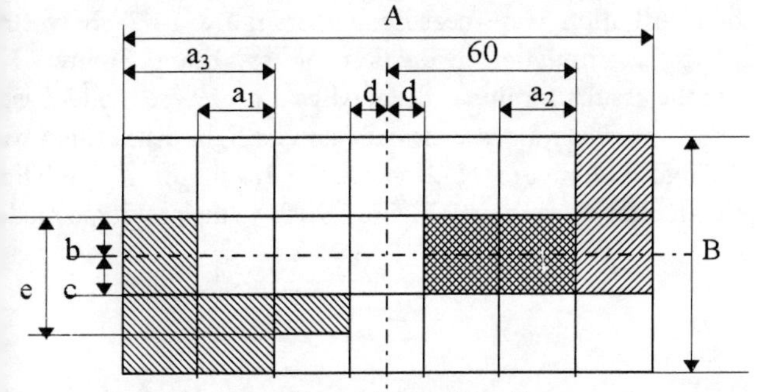

Figure 7 Partition of Reflector

A : Horizontal length of reflector 180 - 220 mm
B : Vertical length of reflector 80 - 120 mm
a_1, a_3 : Length determined by the intensity of hot zone
a_2 : Length determined by the intensity of small diffusion
b : Length determined by the intensity of horizontal cut off line
c : Upper limit that can be used as diagonal cut off line
d : Left and right limit that can be used only for large diffusion
e : Lower limit that can be used as delta zone

A, B, a_2, c, d, e determine the results of the reflector light distribution characteristics.

5.2.3 Concerning the values of a_1, a_3 and b;

The values of a_1, a_3, b are obtained from the neural network. The neural network changes a_1, a_3 and b and repeatedly sets and calculates Θ_H, Θ_V, Θ_W, Θexp, $\Theta_{W\Theta}$ so as to obtain the optimum light distribution. The neural network is created as in Figure 8 which takes as instructional input the resulting light intensity (IL, IU) and A, B. When the required light intensity is given by using this neural network, the values of a_1, a_3 and b are automatically calculated.

5.3 Concerning symbol attached to Θ ;

H : amount of shift of pattern in horizontal direction
V : amount of shift of pattern in vertical direction
exp : amount of diffusion of pattern
w : target cut line angle when optimized after shifting and diffusion of pattern
wΘ : amount of diffusion vertical shift of the horizontal cut off line when pattern is shifted

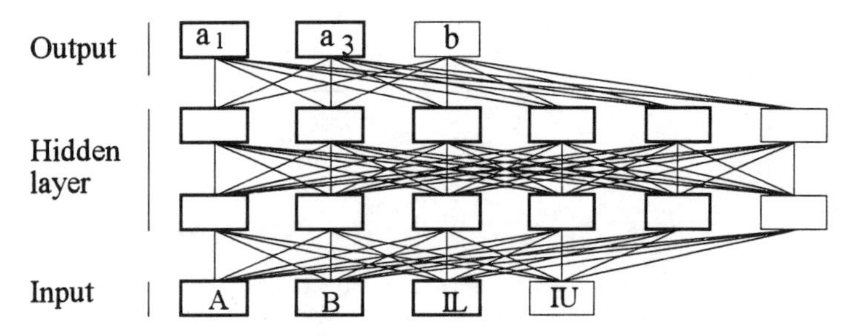

Figure 8 Neural Network

A : horizontal length of reflector 120 -180 mm
B : vertical length of reflector 60 - 110 mm
IL : luminous intensity (Low Beam)
IU : luminous intensity (High Beam)
a_1 : hot zone
b : horizontal cut off line

5.4 Generation from basic reflector surface to BSS surface

A B-spline surface P(u, w) in matrix form corresponding to the position vectors given for the lattice pattern consisting of Q_{00}, Q_{01}, Q_{0n} --- Q_{n0} --- Q_{nm} is given by equation (1).

$$P_{ij}(u,w) = \left[N_{0,4}(u)\; N_{1,4}(u)\; N_{2,4}(u)\; N_{3,4}(u)\right] x \begin{bmatrix} Q_{i-1,j-1} & Q_{i-1,j} & Q_{i-1,j+1} & Q_{i-1,j+2} \\ Q_{i,j-1} & Q_{i,j} & Q_{i,j+1} & Q_{i,j+2} \\ Q_{i+1,j-1} & Q_{i+1,j} & Q_{i+1,j+1} & Q_{i+1,j+2} \\ Q_{i+2,j-1} & Q_{i+2,j} & Q_{i+2,j+1} & Q_{i+1,j+2} \end{bmatrix} \begin{bmatrix} N_{0,4}(w) \\ N_{1,4}(w) \\ N_{2,4}(w) \\ N_{3,4}(w) \end{bmatrix} \quad \text{--- (1)}$$

Here $N_{i,4}(u)$ and $N_{i,4}(w)$ are, respectively, fourth-order and third-degree normalized B-spline functions for defining the B-spline curves.

In case of the inverse transform from the surface vector $P_{i,j}$ to the surface definition vector $Q_{i,j}$. $V_{i,j}$ is replaced with $P_{i,j}$ by using equations (2) and (3) and the loop is repeated so as to reduce the error with the true value of $P_{i,j}$, thereby determining $V_{i,p}$, and $Q_{i,p}$ is determined from $V_{i,p}$ in the same way.

$$P_{i,j} = \frac{1}{6}V_{i-1,j} + \frac{2}{3}V_{i,j} + \frac{1}{6}V_{i+1,j} (1 \le i \le m-1; 1 \le j \le m-1) \quad \text{--- (2)}$$

$$V_{i,j} = \frac{1}{6}Q_{i-1,j} + \frac{2}{3}Q_{i,j} + \frac{1}{6}Q_{i+1,j} (1 \le i \le m-1; 1 \le j \le m-1) \quad \text{--- (3)}$$

The ideal reflector pattern is generated by modifying the entire reflector surface over numerous interactions using the above equations (2) and (3), considering the reflector pattern and the shift volume obtained from comparing the initially created default pattern and the target pattern.

6. DESCRIPTION OF OUTPUT SCREEN

Figure 9 shows a sample of an output screen.

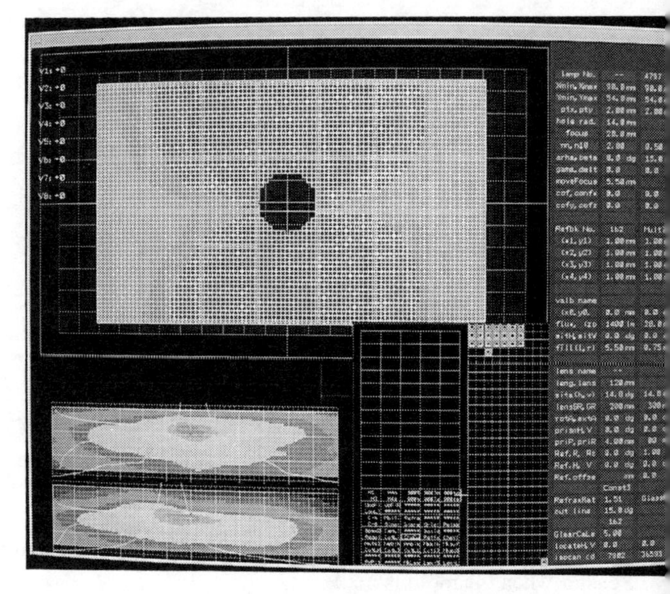

Figure 9 Output Screen

7. DESIGN

We used this system to design an MBSS reflector du filaments replaceable light source headlamp. Extern dimensions are 100 mm height x 180 mm width, and the targ maximum luminous intensities of low and upper beams a 35,000 cd and 50,000cd, respectively. The light distributic pattern of each reflector is shown in Figure 10 for low beam ar for upper beam.. Figure 11 show the light distribution patte combined according to the required pattern shown in Figure The predicated light distribution for low beam is shown Figure 12-1, and the predicated evaluation results for low bea are shown in Figure 12-2. The predicated light distribution f upper beam is shown in Figure 13-1, and the predicate evaluation results for upper beam are shown in Figure 13-2.

Next, Figure 14-1 shows the results for low beam when prototype model based on the numerical values was made and i light distribution was measured, and Figure 14-2 shows th results of a simulation evaluation for low beam. Figure 15- shows the results for upper beam when a prototype model base on the numerical values was made and its light distribution wa measured, and Figure 15-2 shows the results of a simulatic evaluation for upper beam. Figure 16 shows the prototype mode

1	2	3	4	5	6	7
11	12	13	14	15	16	17
21	22	23	24	25	26	27
	31					

Number means the segment No. of partitioned reflector

Upper : Pattern by basic reflector
Lower : Pattern by final reflector

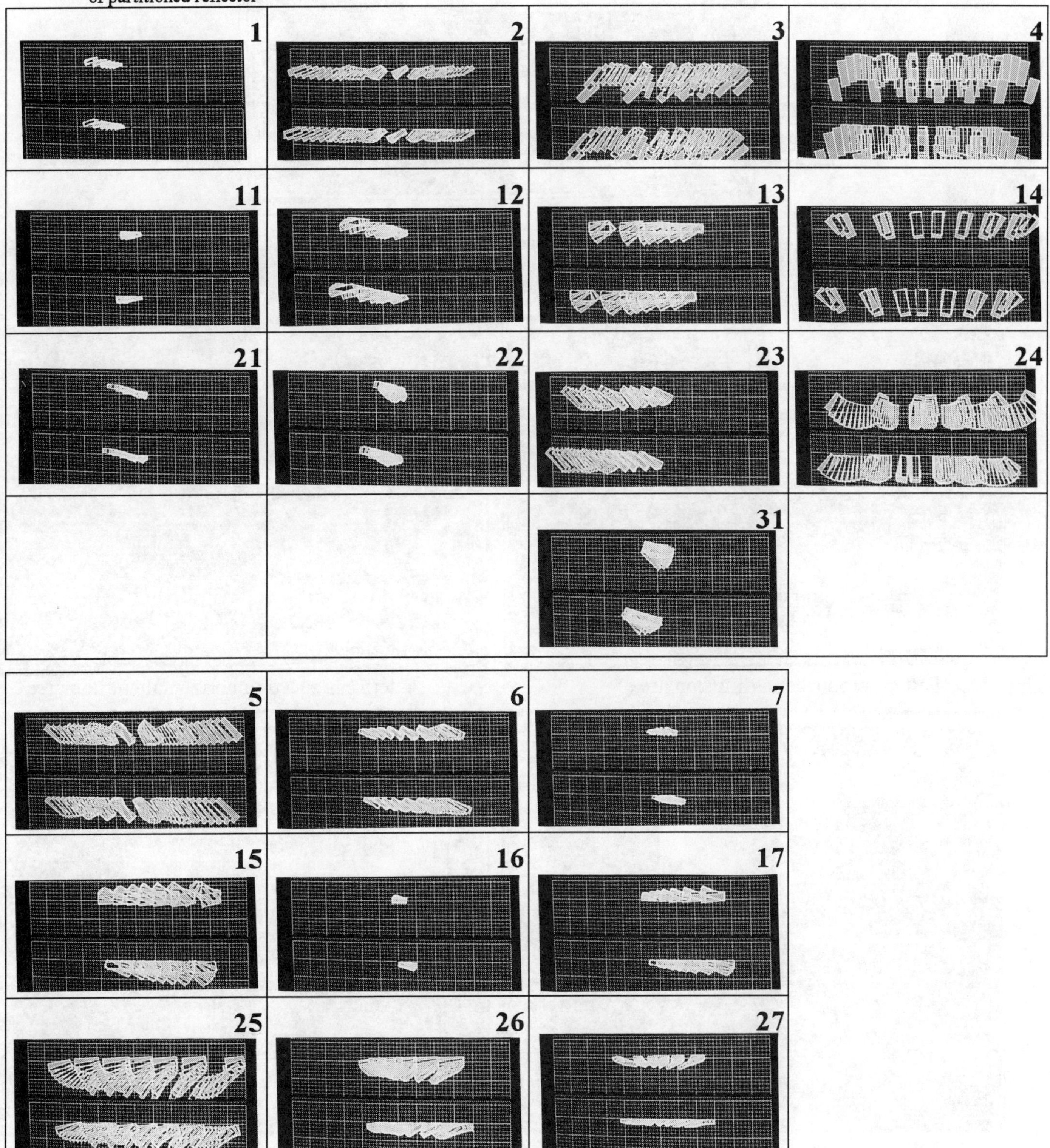

Figure 10 Individual Segment Patterns

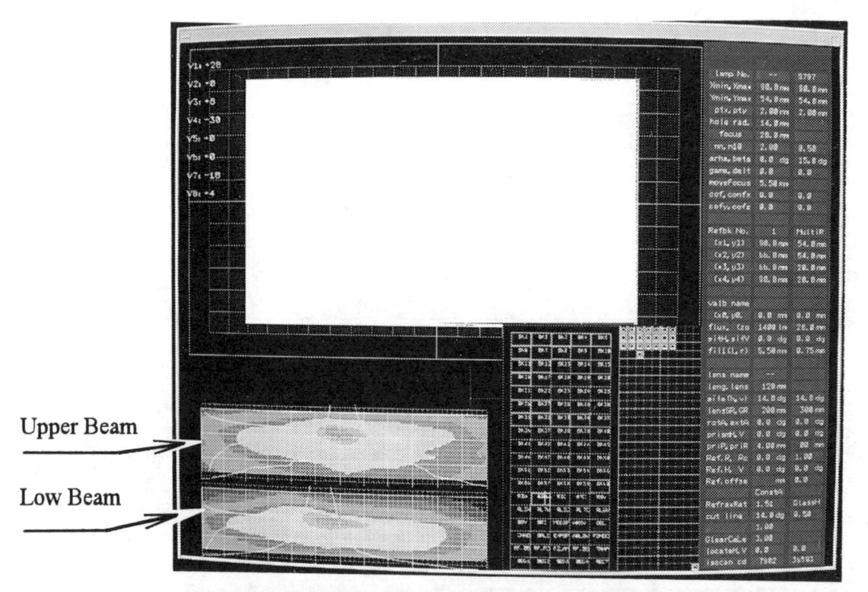

Pattern produced in whole reflector

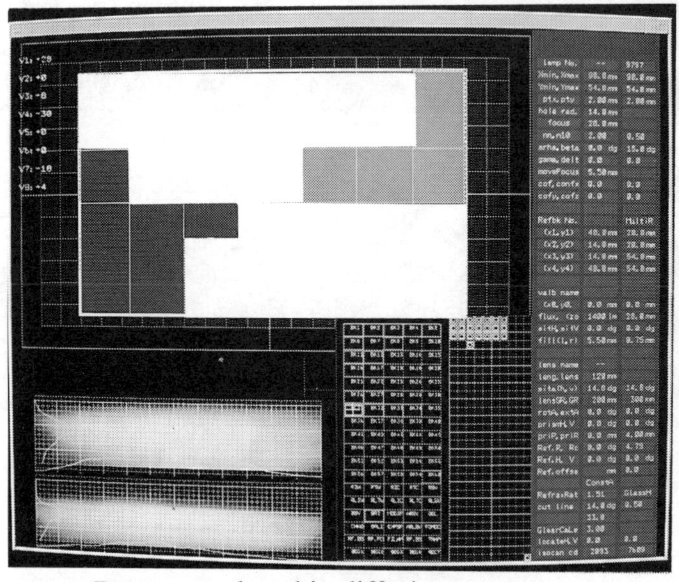

Pattern produced in diffusion area

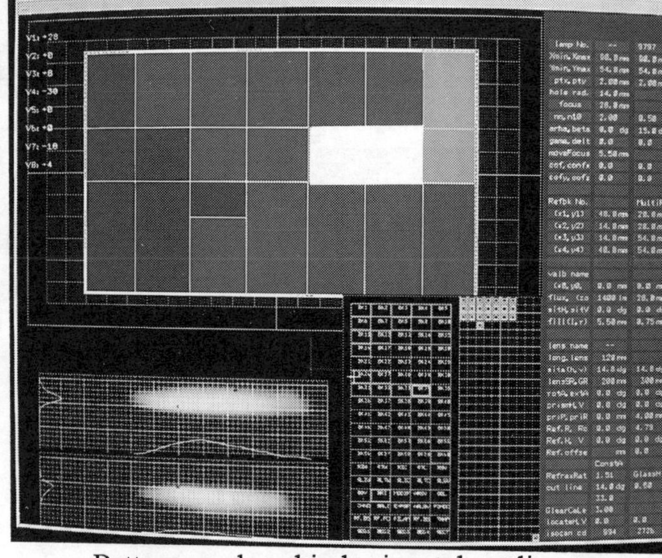

Pattern produced in horizontal cut line area

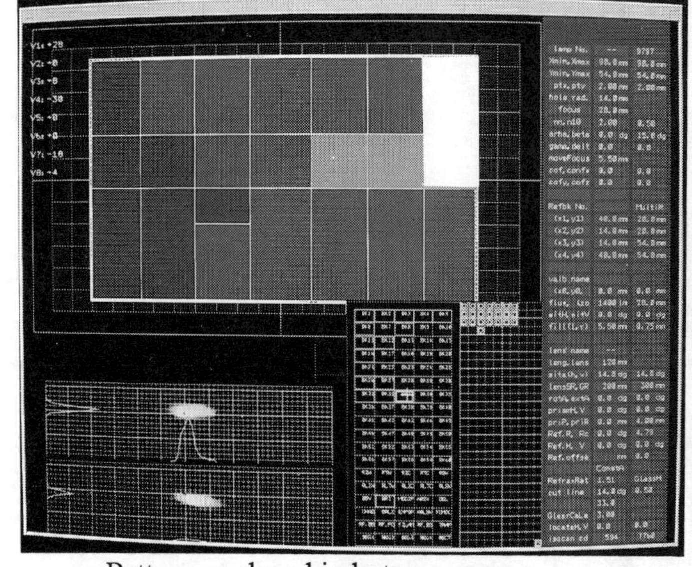

Pattern produced in hot zone area

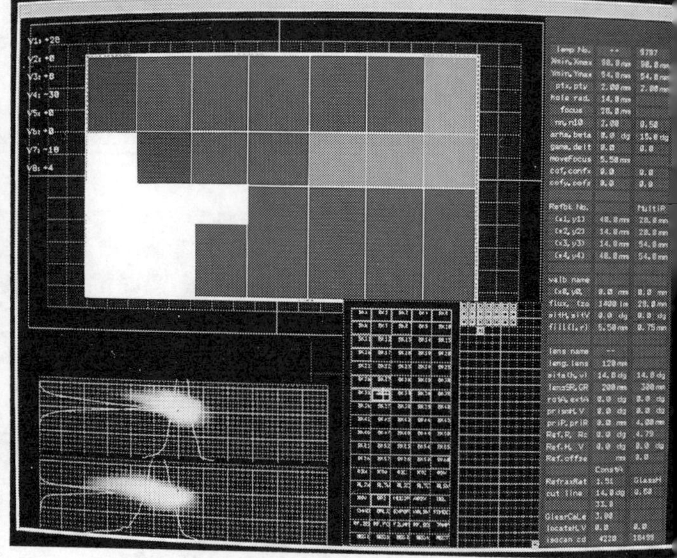

Pattern produced in delta zone area

Figure 11 Light Distribution Pattern Combined according to Required Pattern

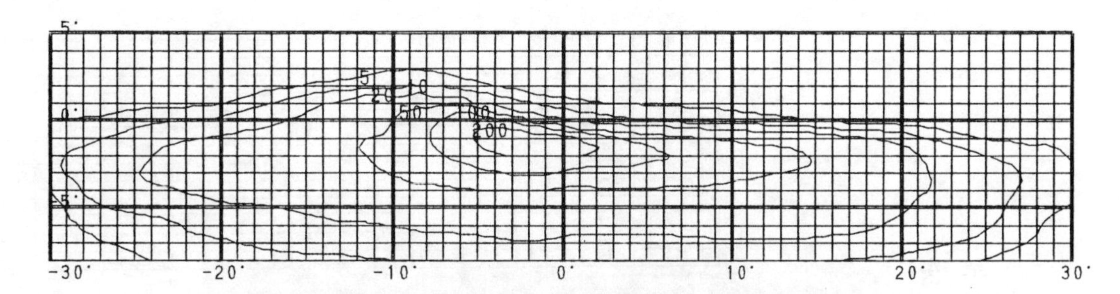

Figure 12-1
Theorical light distribution
for low beam

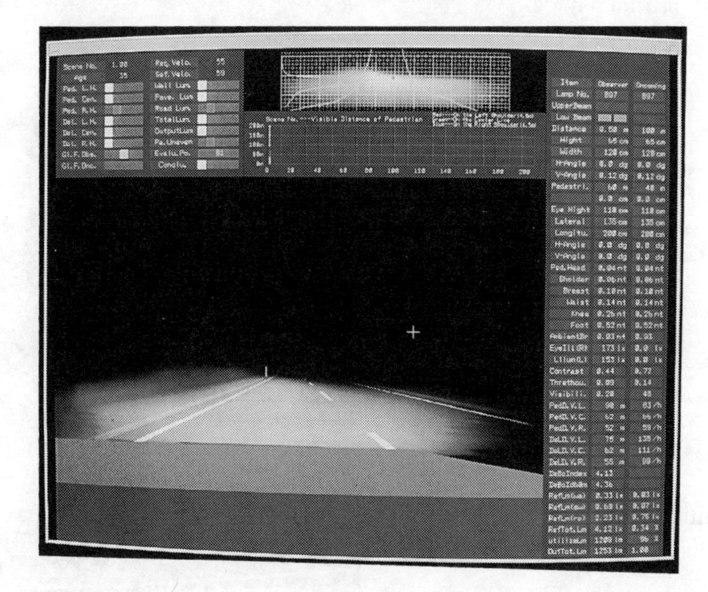

Figure 12-2
Theorical evaluation results
for low beam

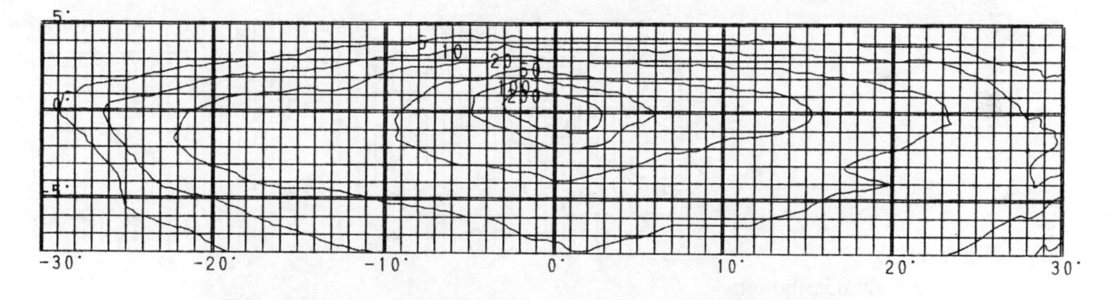

Figure 13-1
Theorical light distribution
for upper beam

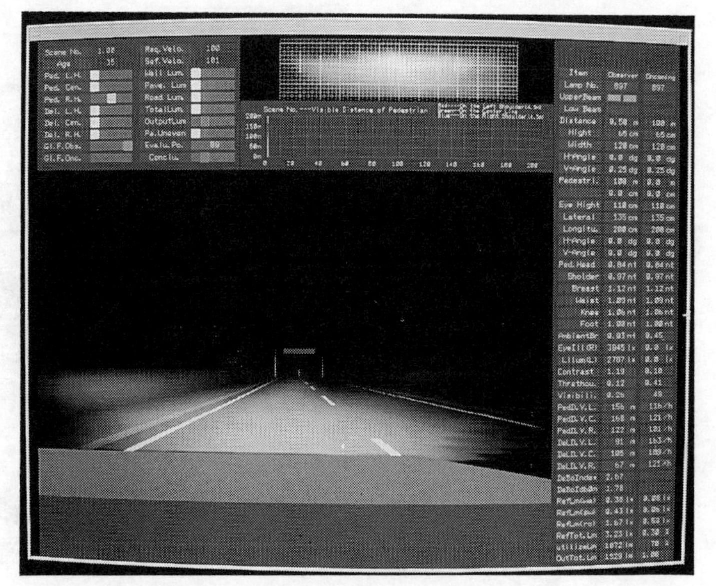

Figure 13-2
Theorical evaluation results
for upper beam

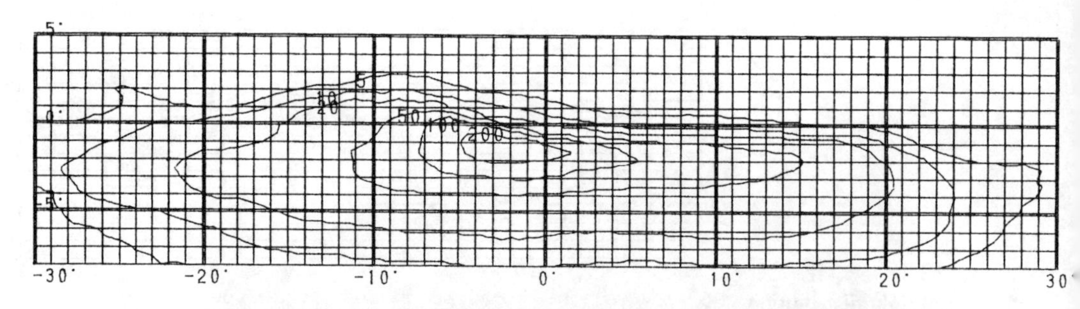

Figure 14-1
Prototype light distribution
for low beam

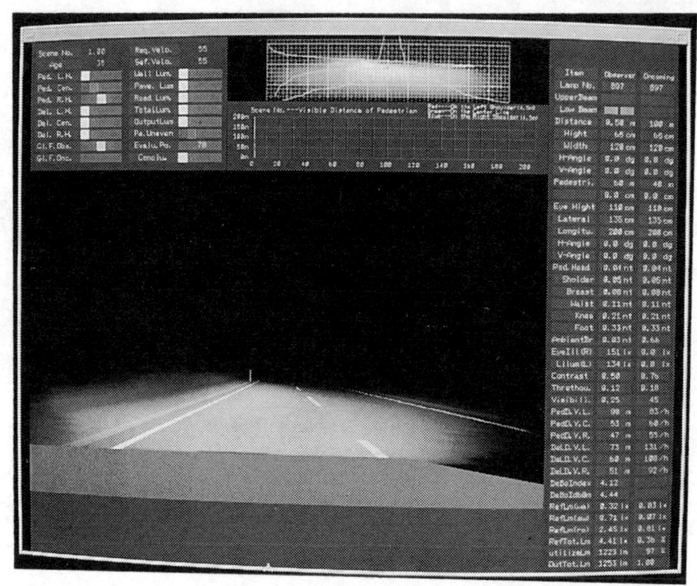

Figure 14-2
Prototype evaluation results
for low beam

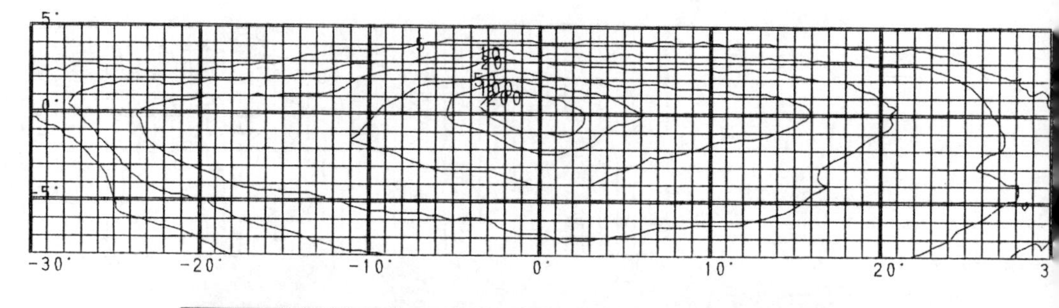

Figure 15-1
Prototype light distribution
for upper beam

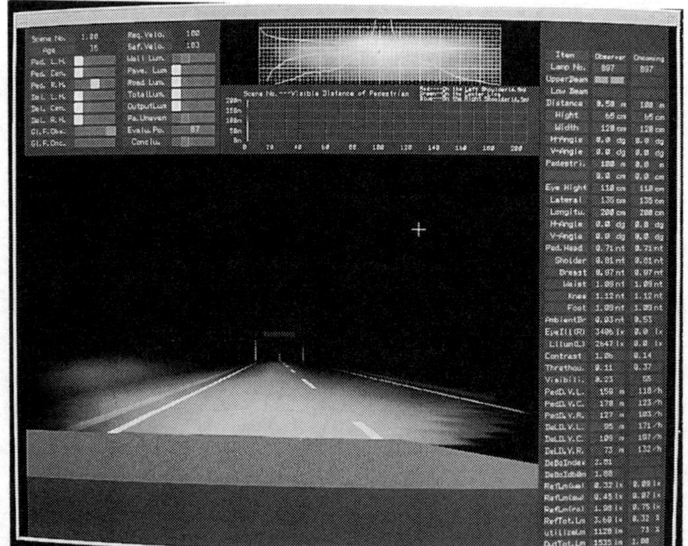

Figure 15-2
Prototype evaluation results
for upper beam

Assembly

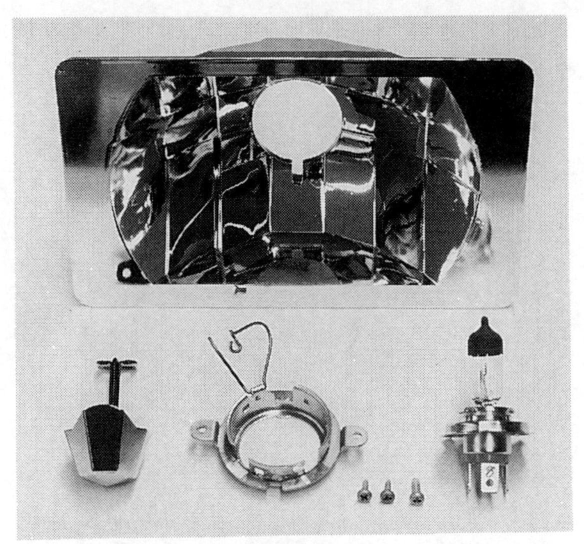

Parts

Figure 16 Prototype Model

8. RESULTS AND CONCLUSIONS

1. The appropriate reflector shape used for a dual-filament replaceable bulb headlamp is 100mm or more height x 180mm or more width. The shape and the location of filament at that time is shown in Figure 3.

2. It was confirmed, as described above, that the design predicted light distribution and the light distribution of the prototype model is very similar. This is evidence that the algorithm for generating the surface of each reflector segment is also correct.

3. This design system is capable of designing in a short period of time, because it uses a neural network rather than a feedback algorithm for its optimization design technique. {about 100 minutes, using IBM RS6000(590)}

By combining this automatic MBSS reflector design program for a dual-filament replaceable bulb headlamp using a neural network with the fuzzy light distribution evaluation simulation system announced at the 1992 SAE, it has become possible to build a comprehensive system that combines a nearly ideal design system with a nearly human evaluation system.

Future planned improvements include increasing the precision in order to make it a better system. We would like also like to apply this approach to reflectors having different reflectance characteristics..

REFERENCE

1. Nakata Y., et al, (1992) Computerized graphic light distribution fuzzy evaluation system for automobile headlighting using vehicle simulation, SAE Paper No. 920816

2. Nakata Y. (1994) Multi B-Spline surface reflector optimized with neural network, SAE Paper No. 940638

950600

On-the-Road Visual Performance with Electrochromic Rearview Mirrors

Michael J. Flannagan, Michael Sivak, Masami Aoki, and Eric C. Traube
University of Michigan Transportation Research Institute

Abstract

This study was part of a series of studies on variable-reflectance rearview mirrors. Previous work included laboratory studies of human visual performance, field collection of photometric data, and mathematical modeling of the visual benefits of variable-reflectance mirrors. We extended that work in this study by collecting photometric and human-performance data while subjects drove in actual traffic.

Three mirror conditions were investigated: (1) fixed-reflectance mirrors in the center and driver-side positions, (2) a variable-reflectance mirror in the center with a fixed-reflectance mirror on the driver side, and (3) variable-reflectance mirrors in both positions. The fixed and variable reflectivities were produced by the same mirrors by overriding the circuitry that normally controlled reflectance in the variable mode.

Results indicated that variable-reflectance mirrors provided a substantial reduction in discomfort glare without a measurable reduction in subjective ratings of rearward seeing ability. They did not cause major improvements in forward seeing (in agreement with previous laboratory and modeling results).

The present study is inconclusive with respect to the benefits of a variable-reflectance driver-side mirror relative to a fixed-reflectance mirror with 50% reflectance. The reason for this is that the particular driver-side mirror used in this study became noticeably green in the low-reflectivity state, and thus low reflectivity was confounded with a color change. The effect of the driver-side mirror should be clarified by further research.

Introduction

Over the past several years the growing availability of variable-reflectance rearview mirrors has offered drivers a new option for limiting glare from rearview mirrors that is more flexible than that provided by two-state, mechanically switched prism mirrors. Much of the past work on the effects of rearview mirror reflectivity on driver vision suggests that variable-reflectance mirrors offer drivers improved visual performance, primarily because the low-reflectance level of prism mirrors (4%) is too low for most conditions (Mansour, 1971; Olson, Jorgeson, & Mortimer, 1974; Ueno & Otsuka, 1988).

The potential refinement in glare control offered by variable-reflectance mirrors relative to prism mirrors has increased the need to understand the visual factors that might determine the optimal level of mirror reflectivity for a given set of lighting conditions. We have undertaken a program of research to identify, quantify, and comprehensively model those factors (Flannagan & Sivak, 1994). The present study is a field test of some of the conclusions that have emerged from that work.

Our previous work has been based on a working hypothesis that optimal rearview mirror reflectivity is determined by a tradeoff among three factors: (1) subjective discomfort caused by glare from mirrors, (2) forward seeing ability, and (3) rearward seeing ability. The first two factors are improved by reducing mirror reflectivity; the necessity to make a tradeoff arises because the third is harmed by reduced reflectivity. In previous studies, we have argued for a number of tentative conclusions about the effects of mirror reflectivity, including that 4% reflectivity is in fact needlessly low for glare protection in most circumstances (Flannagan & Sivak, 1990), that the limits on glare required to preserve forward seeing ability are normally dominated by the limits needed to reduce subjective discomfort (Flannagan, Sivak, & Gellatly, 1992), and that flexible control of reflectivity is needed for both the center and driver-side mirrors to achieve optimal control of discomfort (Flannagan & Sivak, 1994).

Those conclusions have been based on measurements made in the laboratory and under controlled field conditions, combined with mathematical models of visual performance. In the present study we had drivers make assessments of the three factors involved in the central tradeoff (discomfort from rearview-mirror glare, forward seeing ability, and rearward seeing ability) while they drove an instrumented vehicle in actual nighttime traffic. The vehicle was equipped with variable-reflectance mirrors in the center and driver-side positions. Control circuitry allowed the mirrors to be set to fixed reflectivity levels that are typical of the high-reflectivity levels of conventional mirrors (80% for the center mirror and 50% for the driver-side mirror), or to vary in response to prevailing lighting conditions. In the variable-reflectivity mode the reflectivity levels of the mirrors were determined by control circuitry, supplied by the mirror manufacturer, that was responsive to both ambient light level, as detected by a forward-oriented sensor, and glare from the rear, as detected by a rearward-oriented sensor.

The vehicle was instrumented to collect photometric information about the luminance of the forward pavement and about illuminance from rearward sources of glare (primarily the headlamps of following vehicles).

All measurements were made in actual traffic at night. Subjects drove a planned course that included a mix of conditions: city streets, expressways, and rural roads. The levels of traffic and fixed lighting varied considerably.

Method

Subjects

There were two age groups, each with an equal number of males and females. There were 6 people in the younger group, ranging from 18 to 25 with a mean age of 22.0, and 6 in the older group, ranging from 59 to 66 with a mean age of 63.5. All were licensed drivers. None had previous experience with variable-reflectance mirrors.

Instrumented vehicle

The vehicle was a 1993 Nissan Altima. It was equipped with photometers to measure simultaneously the luminance of the pavement in front of the vehicle and illumination from the rear that could cause rearview-mirror glare (primarily from the headlamps of following vehicles). We had used it previously, in a slightly different configuration, to collect photometric data in a rearview mirror study that did not involve human subjects (Flannagan & Sivak, 1994). The configuration of photometers on the vehicle is shown in Figure 1.

The luminance of the pavement in front of the vehicle, as viewed from approximately the driver's eye position, was measured with a Minolta LS-100 luminance meter with a 1-degree field of view and a nominal sensitivity limit of 0.001 cd/m^2. The meter was positioned so that the far edge of the field of view was aligned with a point on level pavement 50 m in front of the driver's eye position, and directly ahead of the driver. The patch of pavement measured by the luminance meter extended from that point toward the vehicle for about 21.7 m, with a maximum width of 0.7 m, as is shown in Figure 2. This area contains the mean visual fixation point for drivers on straight roads using U.S. low beam headlamps as measured by Graf and Krebs (1976).

The luminance meter was mounted near the front passenger-seat eye position, as show in Figure 1. It was turned slightly to the left (about 1 degree) so that it was aimed at the pavement directly in front of the driver. It was aimed through the windshield, with approximately the same line of sight as the driver, except for being displaced laterally to the passenger seat.

Illuminance from the rear that could potentially cause rearview mirror glare was measured by a Minolta T-1 illuminance meter, fitted with a standard cosine receptor. The meter has a nominal sensitivity limit of 0.01 lx. It was installed inside a set of baffles on the roof of the vehicle, just above the rear window, as shown in plan view in Figure 1. In the rooftop position the photometer assembly did not interfere with the driver's field of view. The baffles excluded light that originated from locations that would not be visible to a driver through the interior rearview mirror. In order to check the angles of acceptance of this configuration, we made a series of static tests in which we compared lux readings from the photometer in the roof position to lux readings taken at the position of the center rearview mirror. We positioned a glare vehicle behind the instrumented vehicle at distances of 1, 2, 3, 4, 5, 7.5, 10, and 50 m. The glare vehicle was located either directly behind the instrumented vehicle or laterally displaced to the left or right by the width of a typical lane (3.7 m). We took measurements for both the low- and high-beam headlamps of the glare vehicle. Although there was not perfect absolute agreement, largely because of rear-window transmittance, the overall correlation between readings taken at the two positions was very high, even for the shorter separations at which the parallax difference between the two photometer positions was greater. The readings are shown in Figure 3.

Mirrors

The rearview mirrors in the center and driver-side (left) positions were electrochromic mirrors. They were supplied specifically for the test vehicle by a mirror manufacturer. They were configured with standard control circuitry including both a forward-oriented light sensor for measuring the level of ambient illumination and a rearward-oriented sensor for measuring potential glare light from the rear. The sensors were both mounted on the center mirror and in normal operation the reflectance of the driver-side mirror was controlled by the same signal as the center mirror. The passenger-side mirror was a conventional convex mirror. For half the subjects it was effectively eliminated by covering it with an opaque shield. For the other subjects it was used normally.

Additional control circuitry could override the normal operation of the electrochromic mirrors. Overall there were three control modes: (1) the driver-side mirror was at a fixed reflectance of 50% and the center mirror was at 80% (hereafter designated the 50/80 condition), (2) the driver-side mirror was fixed at 50% and the reflectance of the center mirror varied according to the normal control signal (the 50/Var condition), or (3) the reflectance of both the driver-side and center mirrors varied with the normal control signal (the Var/Var condition). All reflectance levels were measured with a standard tungsten-halogen headlamp bulb as the light source.

The center and driver-side mirrors differed somewhat in color in their low-reflectivity states. Spectral reflectance functions for the two mirrors are shown in Figure 4.

Test route

The main test route consisted of a mix of expressways, city streets, and rural roads in and around the city of Ann Arbor. It was 96 km long. A separate, shorter route (13 km) was used for practice trials.

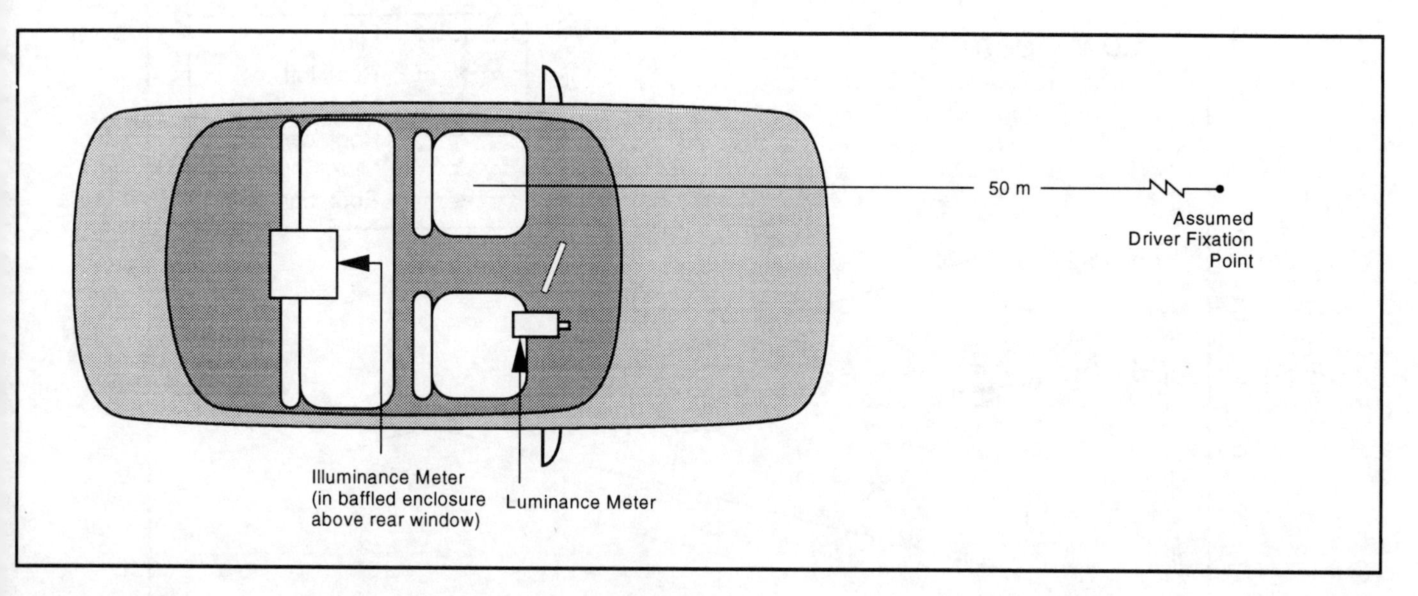

Figure 1. Plan view of the on-board photometric instrumentation.

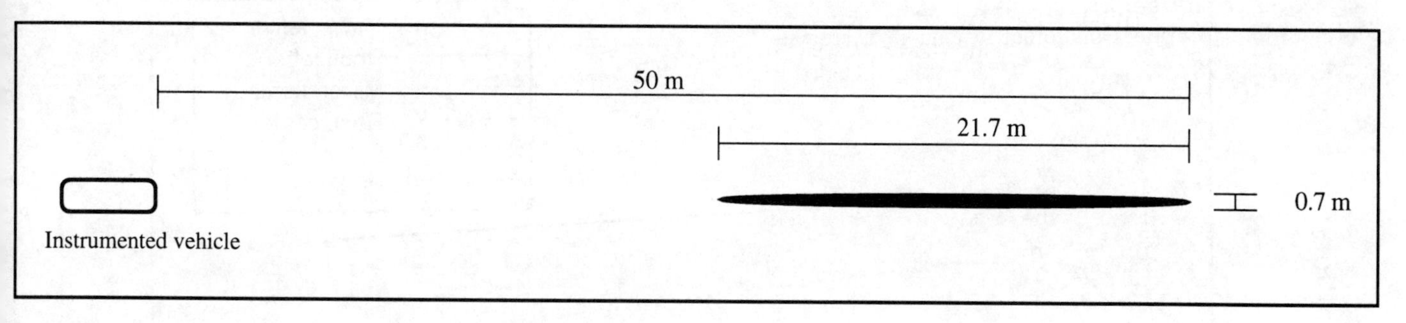

Figure 2. Plan view of the area of pavement from which luminance was measured by the spot photometer. The thin black oval is the projection of a round, 1-degree field of view directed approximately along the driver s line of sight.

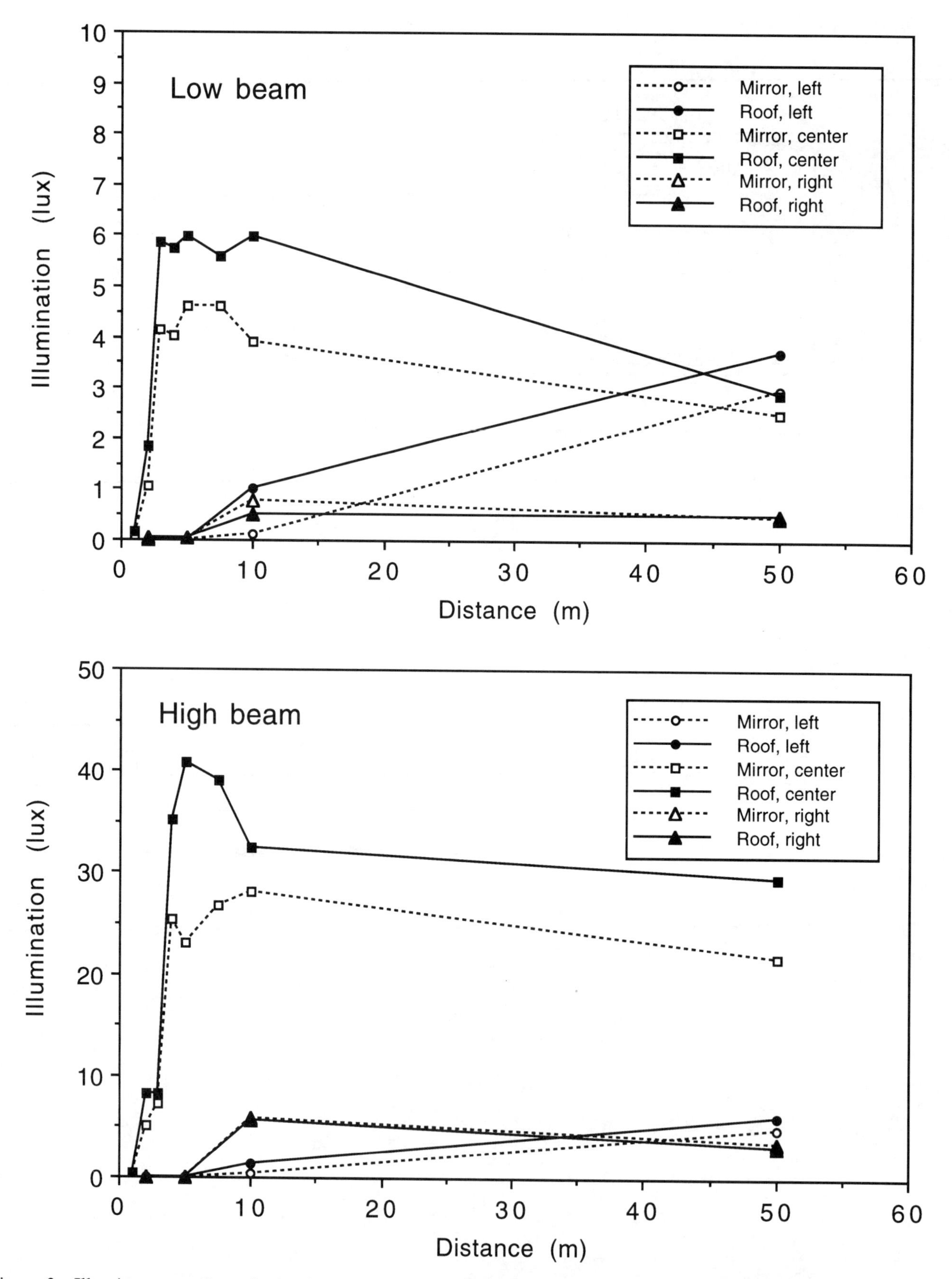

Figure 3. Illuminance readings (lux) taken by the baffled photometer at the position of the center rearview mirror and at the rooftop position that was used so as not to block the driver s view of the center mirror. Illumination was provided by a single vehicle using either low or high beams, at various distances behind the instrumented vehicle, and in the same lane or displaced one lane to the left or right. The labels in the legend refer to the location of the photometer on the instrumented vehicle and the lateral location of the glare vehicle relative to the instrumented vehicle. The upper panel shows data for the low beams of the glare vehicle and the lower panel shows data for the high beams.

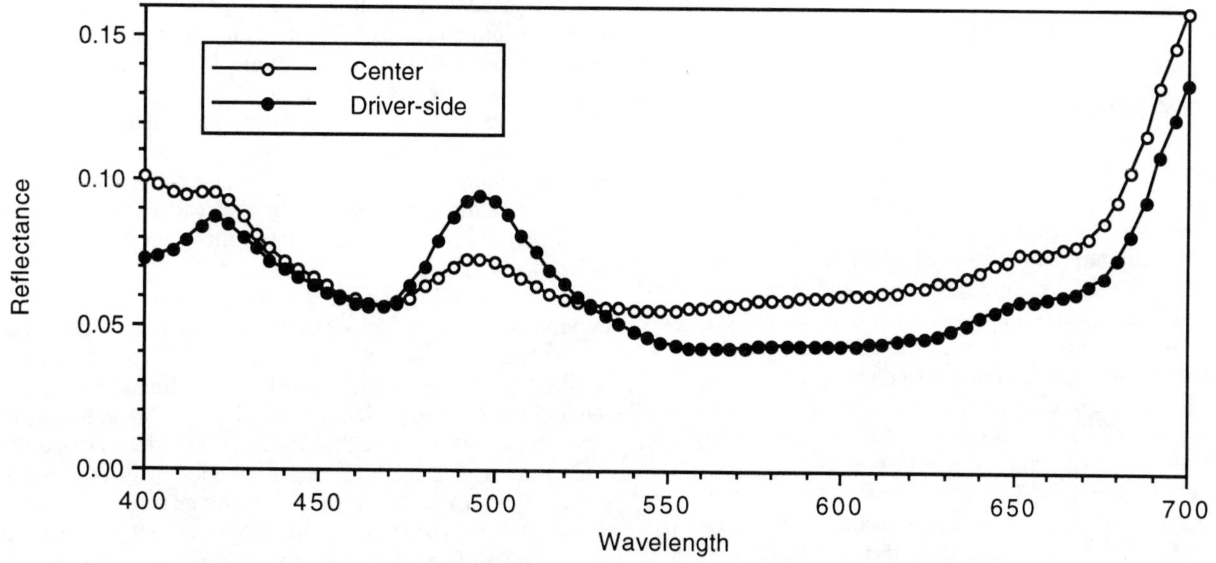

Figure 4. Spectral reflectance functions in the visible region for the center and driver-side rearview mirrors in low-reflectivity states.

Procedure

Subjects were run individually. Sessions began about a half hour after sunset and lasted about two and a half hours. The subjects drove the instrumented vehicle on a predetermined route. Two experimenters rode with them to collect data and to instruct them about the route. Subjects first drove a short practice route for about 20 minutes and then the main test route for about 90 minutes.

Subjects were informed that the study concerned an innovative rearview mirror system, but otherwise were not told about the purposes of the study or the nature of the mirror system.

Every two minutes during the practice and main driving periods, the photometers collected the current forward pavement luminance and the current level of illumination from the rear. At the same times, the subjects made ratings of three visual quantities: (1) the level of discomfort that they were currently experiencing from rearview-mirror glare, (2) their ability to see the road ahead of them, and (3) their ability to see to the rear. They made their three ratings in that order, using numerical scales from 1 to 10 that had been described to them in the initial instructions. Each of the scales had two verbal anchors, for the values 1 and 10. The anchors for the discomfort scale were ϶Very little (1) and ϶Very much (10). For both of the seeing-ability scales the anchors were ϶Very poorly (1) and ϶Very well (10). The instructions emphasized that for discomfort glare subjects should, as much as possible, rate the combined effect of glare from all the mirrors and disregard any discomfort that they believed was attributable to glare from oncoming traffic.

The main driving period was divided into nine 10-minute blocks. Because trials occurred every two minutes there were 5 trials per block. The mode of mirror operation (50/80, 50/Var, or Var/Var; as described above) was changed between blocks. The order of mirror modes was balanced across subjects, and the ordering was always such that each mirror condition occurred once in each third of the main driving period (once among the first three 10-minute blocks, once among the second three, and once among the third three).

The transitions between mirror modes were not identified for the subjects. The subjects were not even informed that the mirrors would be operated in different modes. Our intent was that on each trial the subjects should simply rate their current experience, with no explicit knowledge of the experimental manipulation of the mirrors.

For half of the subjects the convex passenger-side (right) mirror was covered with an opaque shield. For the other half it was uncovered and (like the other mirrors) was adjusted for normal aim by the subjects before they began driving.

After they finished driving, the subjects were systematically debriefed with a written questionnaire that asked progressively more specific questions about mirror performance. The questionnaire consisted of two pages, with a single general question on the first page and three more specific questions on the second. Subjects were not allowed to see the second page until they had answered the first, more general question. The questions are shown in Table 1.

Table 1.
Questions on the written questionnaire given to subjects after they finished the driving task.

First page:
During the study you were using new advanced mirror systems on the car you were driving. Did you notice anything interesting about the mirrors? If so, what did you notice?
Second page:
The mirrors that you were using used new technology to change reflectivity during the study, so that at times they were more reflective than at others. Did you notice anything interesting about this reflectivity change? If so, what did you notice?
Did you have any preferences for different reflectivities? If so, what did you like? Dislike?
What further comments do you have on the mirror systems?

Results

Photometric Data

Figure 5 shows histograms of the photometric data, combined over all subjects. Individual panels of the figure show illuminance values (in log lux) and luminance values (in log cd/m^2), summarized in terms of which mirror mode was activated at the times the measurements were taken. As would be expected because of the counterbalancing of order of the mirror modes across subjects, the distributions are approximately the same for the three different mirror modes. The relationship between illuminance and luminance values, combined over all subjects and mirror modes, is shown in Figure 6. The pattern is very similar to previous data we have collected (Flannagan & Sivak, 1994). The overall correlation coefficient is .37. The nature of the relationship is that low values of glare illuminance and high values of pavement luminance do not occur together.

Subject ratings of visual performance

Figures 6, 7, and 8 show the main effects of mirror mode on the three visual-performance ratings. There is a substantial and statistically significant effect on discomfort glare ratings, $F(2,16) = 6.75$, $p < .01$. In contrast to the mode in which both mirrors are of fixed reflectance and the center mirror reflectance is relatively high (50/80), both of the variable modes result in substantially reduced discomfort glare. However there is no evidence for a further reduction in discomfort glare when the driver-side mirror is variable rather than fixed at 50%. In fact, there is a slight trend (not statistically significant) for more discomfort when that mirror is variable.

Neither the effect of mirror mode on forward seeing, $F(2,16) = 2.43$, nor the effect of mirror mode on rearward seeing, $F(2,16) = 0.80$, was statistically significant.

No interactions involving age or sex were significant, indicating that the pattern of results was similar for all combinations of age and sex.

There were no significant differences depending on whether the passenger-side mirror was covered or not.

Discussion

These field results support a number of the conclusions about mirror performance that our previous laboratory and analytical work has suggested. Most importantly, they indicate that variable-reflectance mirrors are capable of substantially reducing discomfort glare without a significant reduction in the principal function of rearview mirrors allowing drivers to see to the rear. There is a slight tendency, evident in Figure 9, for drivers to rate rearward vision lower when rearview-mirror reflectance varies rather than remaining fixed at a high level. However, the trend is not significant, indicating that any such difference is substantially less salient to drivers than the reduction in discomfort glare, which was significant when measured under the same conditions with the same statistical power.

Also in agreement with our previous work, there was no evidence for a perceived increase in forward seeing ability when glare was reduced by variable-reflectance mirrors. Modeling of the disability effects of glare based on a veiling luminance from light scattered in the eye predicts that to be the case (Flannagan et al., 1992). Although there should be some effect of rearview mirror glare on forward seeing ability, it should be much smaller than the effects of oncoming glare because of the much larger angles between the glare sources (i.e., the mirrors) and the stimuli to be seen (e.g., a distant pedestrian on the road ahead). For both the driver-side and center mirrors those angles are on the order of 45 degrees.

A discrepancy between these results and our previous work is the lack of evidence for a difference in rated discomfort glare depending on whether the driver-side mirror is fixed at 50% or variable. Our modeling, based on the work of Schmidt-Clausen and Bindels (1974) indicates that there should be a substantial further reduction in discomfort glare if a variable-reflectance driver-side mirror is added to a system with a variable-reflectance center mirror and a 50% driver-side mirror (Flannagan & Sivak, 1994). There are several possible explanations for the failure to find an effect. It is possible that subjects simply weight glare from the center mirror more highly. The Schmidt-Clausen and Bindels model weights glare sources equally if they are at the same visual eccentricity (as is approximately true for the driver-side and center mirrors). The instructions emphasized that subjects should rate glare from all mirrors together, but that may be a difficult instruction to follow.

Alternatively, the color of the driver-side mirror used in this study may have had an effect on drivers discomfort glare ratings. On the written debriefing forms, which did not explicitly ask about color, eight of the twelve subjects spontaneously mentioned the green color of the driver-side mirror (which can be inferred from the data in Figure 4). Most of those comments were neutral, but two subjects explicitly connected the green color to an increased perception of glare. The green tint becomes more noticeable as the mirror reflectance goes down, and thus it may have partially counteracted what otherwise would have been a perceived reduction in discomfort resulting from the lower reflectance. Therefore we do not consider the present results conclusive on the issue of fixed versus variable driver-side reflectance. The issue of how much additional benefit a variable-reflectance driver-side mirror adds to a system with a variable-reflectance center mirror should be studied further.

Acknowledgment

We wish to thank Ichikoh Industries, Ltd. for their generous support of this research.

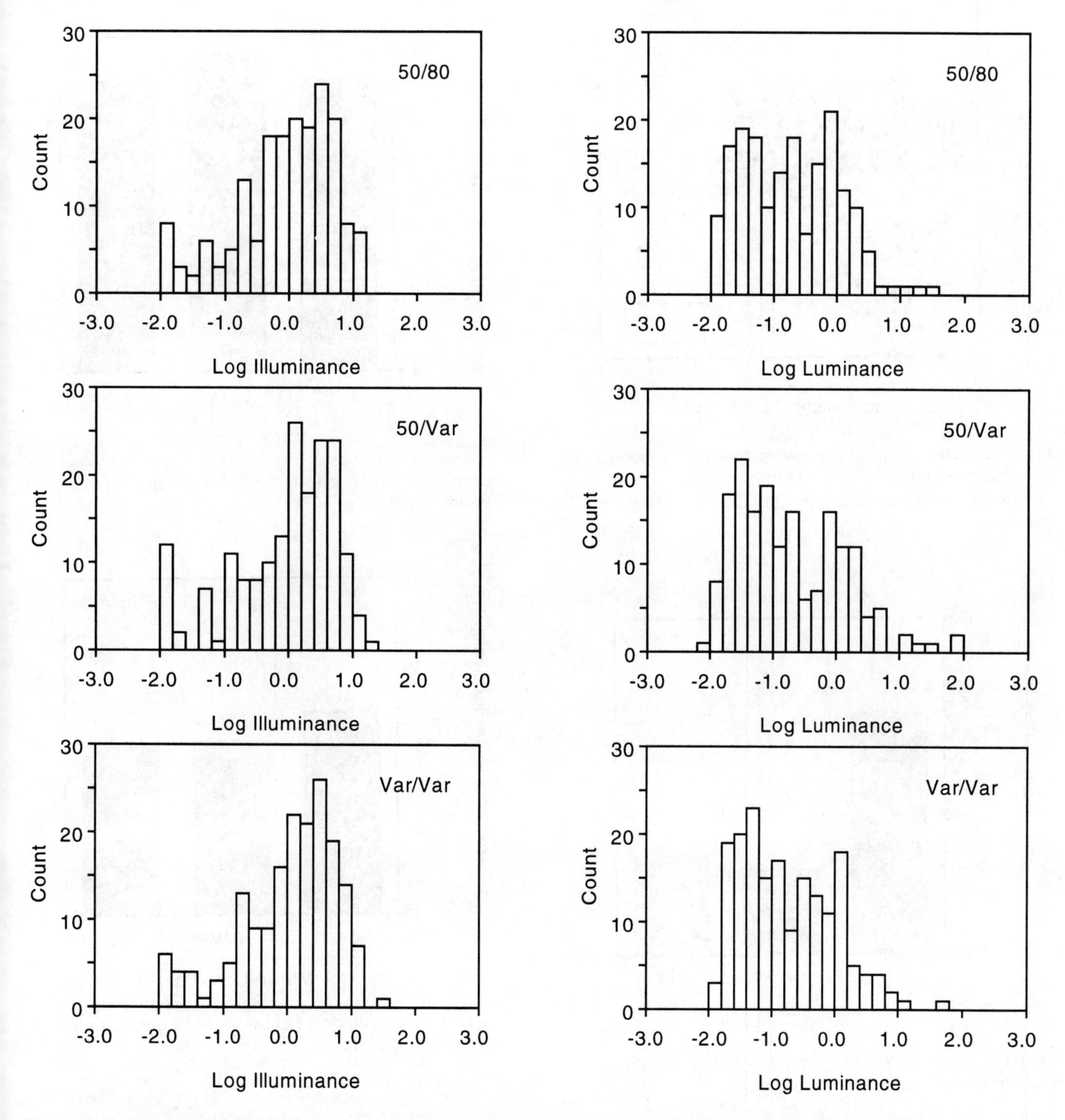

Figure 5. Histograms of photometry values (log illuminance [lux] from rearward glare sources and log luminance [cd/m^2] of forward pavement) for the three mirror conditions.

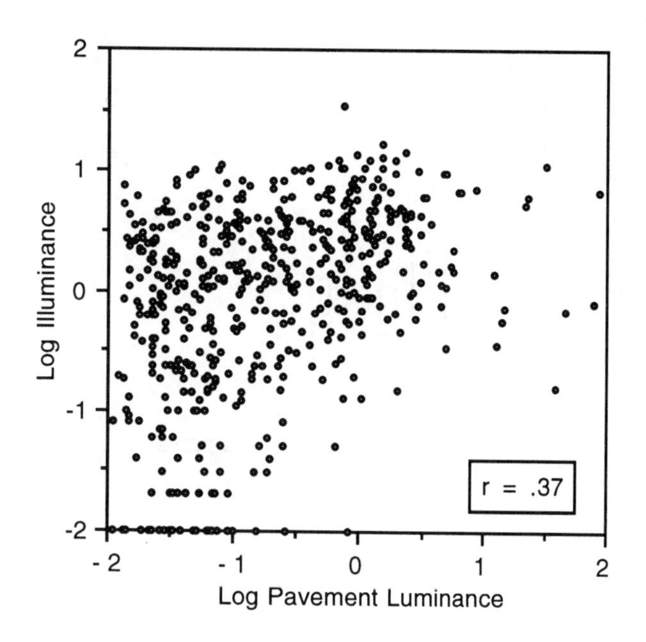

Figure 6. The relationship between log illuminance (lux) from the rear and log pavement luminance (cd/m2) over all photometric measurements made in the study (540 pairs of values).

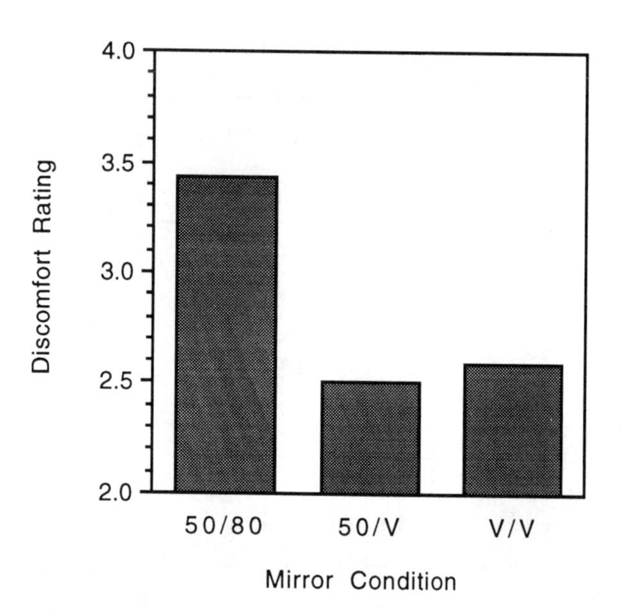

Figure 7. The main effect of mirror condition on ratings of discomfort from rearview mirror glare.

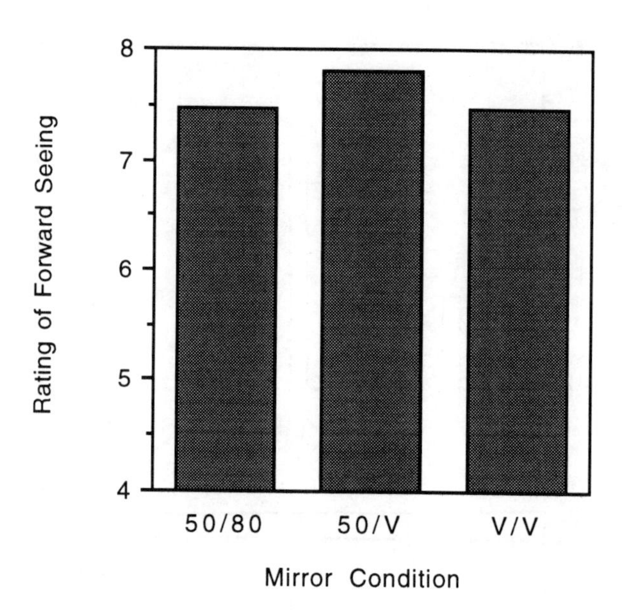

Figure 8. The main effect of mirror condition on ratings of forward seeing ability.

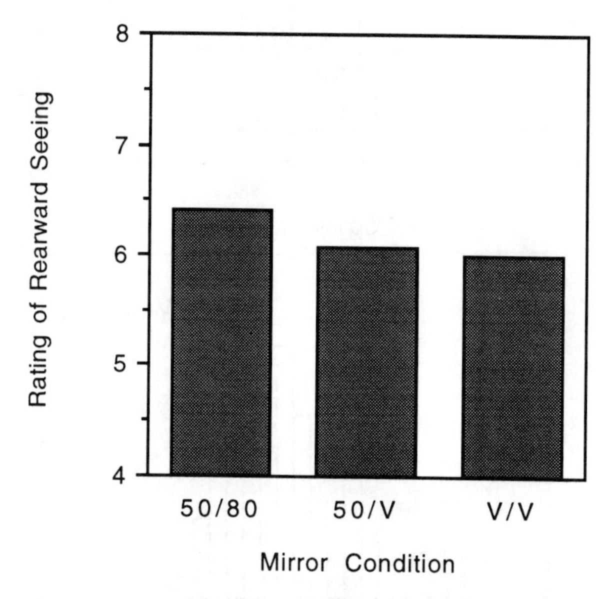

Figure 9. The main effect of mirror condition on ratings of rearward seeing ability.

References

Flannagan, M. J., & Sivak, M. (1990). Nighttime effectiveness of rearview mirrors: Driver attitudes and behaviors. In *Vehicle lighting and driver visibility for the 1990s, SP-813* (pp. 69-79). Warrendale, Pennsylvania: Society of Automotive Engineers.

Flannagan, M. J., & Sivak, M. (1994). *Quantifying the benefits of variable reflectance rearview mirrors* (SAE Technical Paper Series No. 940641). Warrendale, Pennsylvania: Society of Automotive Engineers.

Flannagan, M. J., Sivak, M., & Gellatly, A. W. (1992). *Rearview mirror reflectivity and the tradeoff between forward and rearward vision* (SAE Technical Paper Series No. 920404). Warrendale, Pennsylvania: Society of Automotive Engineers.

Graf, C. P., & Krebs, M. J. (1976). *Headlight factors and nighttime vision* (76SRC13). Minneapolis: Honeywell.

Mansour, T. M. (1971). *Driver evaluation study of rear view mirror reflectance levels* (SAE Technical Paper Series No. 710542). New York: Society of Automotive Engineers.

Olson, P. L., Jorgeson, C. M., & Mortimer, R. G. (1974). *Effects of rearview mirror reflectivity on drivers' comfort and performance* (Report No. UM-HSRI-HF-74-22). Ann Arbor: The University of Michigan Highway Safety Research Institute.

Schmidt-Clausen, H.-J., & Bindels, J. T. H. (1974). Assessment of discomfort glare in motor vehicle lighting. *Lighting Research and Technology*, *6*, 79-88.

Ueno, H., & Otsuka, Y. (1988). *Development of liquid crystal day and night mirror for automobiles* (SAE Technical Paper Series No. 880053). Warrendale, Pennsylvania: Society of Automotive Engineers.

950601

The Geometry of Automotive Rearview Mirrors - Why Blind Zones Exist and Strategies to Overcome Them

George Platzer
Consultant

ABSTRACT

Equations are derived which describe and relate the magnification, viewing angle and reflected illuminance of convex mirrors as used in automotive applications. The derived equations are compared to those for plane mirrors. Using these equations, the viewing angles of automotive rearview mirrors are calculated and depicted. The blind zones are defined in terms of the viewing angles, obstructions to vision, perceptibility limitations, and the lateral separation of vehicles. Various strategies for overcoming the blind zones are discussed.

INTRODUCTION

The blind zones produced by automotive rearview mirrors have long been of concern. A variety of ways of coping with them are in use, and new ways are being proposed. Blind zones are an important factor in accidents caused by lane changing maneuvers. The National Highway Traffic Safety Administration (NHTSA) Crash Avoidance Research Program has targeted Lane Change/Merge (LCM) crashes as one of five categories of crashes potentially suitable for high technology Intelligent Vehicle Highway System (IVHS) crash avoidance countermeasures (Knipling,1993). Wang and Knipling (1994) estimate that LCM crashes account for 4.0% of passenger car crashes, 225 fatalities and 630,000 crashes annually. About 50% of these occur on urban divided highways, and about 75% occur during daylight hours. Involvement by direction of lane change is about equally split between right to left and left to right changes. An analysis of LCM crashes by Najm et. al.,1994 shows that in 61.2% of crashes, the driver did not see the other vehicle. In 29.9% of crashes, the driver misjudged the position and /or speed of the other vehicle.

Mirror blind zones are not responsible for all of the LCM type crashes. However, they are extremely important in that they are not well understood by the average driver, and yet they are an integral part of the data acquisition system used by drivers in LCM maneuvers. Understanding the blind zones is important, and key elements in understanding them are the viewing angles of the mirrors and where the views are directed.

Equations will be derived which quantify the viewing angles in terms of the relevant parameters. Much of the derivations are by way of review, but some new relationships are shown. Graphical methods can be used to show viewing angles, but analytical methods bring out insights not obtainable graphically. Hence the analytical approach.

The derived equations are next used to calculate the viewing angles of the mirrors on a vehicle using their dimensions and their positions relative to the driver. Then the blind zones are defined and depicted. Several factors in addition to the viewing angles determine the extent of the blind zones, including obstructions to vision due to the vehicle, the driver's ability to perceive objects in both the mirror and his or her peripheral vision, and the lateral separation of the vehicles on the roadway.

Once the blind zones have been established, various strategies which have been developed to overcome the blind zones will be reviewed.

GEOMETRY OF REARVIEW MIRRORS

On US passenger cars, the inside mirror and left outside mirror are plane, and the right outside mirror is convex. Europe and Japan allow the use of convex mirrors on the driver's side. Since a plane mirror is a special case of a convex mirror with an infinite radius, the convex mirror equations will be derived first.

The spherical convex mirror presents an image at a magnification less than unity. Most introductory physics books give the equations relating the mirror radius, the object distance, the image distance and the magnification. For the spherical convex mirror of Figure 1, it is easily shown that:

$$\frac{1}{p}+\frac{1}{q}=\frac{2}{r}=\frac{1}{f}. \qquad \text{EQ(1)}$$

and,

$$m = \frac{q}{p}, \qquad \text{EQ(2)}$$

where, p = distance of the object from the mirror
q = distance of the image from the mirror
r = radius of the mirror
f = focal length of the mirror
m = magnification.

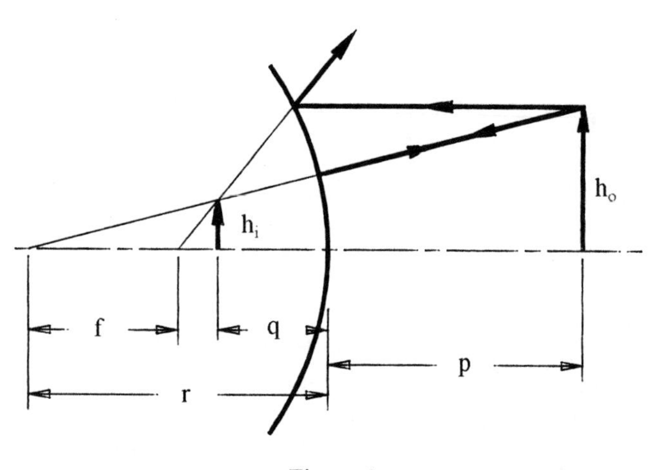

Figure 1

In Figure 1, distances to the right of the mirror are positive and distances to the left are negative. The image distance and the radius will be negative for the convex mirror. The image of the convex mirror is erect and virtual, i.e., an object in the mirror appears standing right side up, and the image cannot be focused on a screen. Concave mirrors have inverted real images which may be focused on a screen.

Since we want to know what an image in the mirror looks like to the driver, lets begin by calculating the image distance and height of an object from EQ(1). The image distance is;

$$q = \frac{rp}{2p - r}. \qquad \text{EQ(3)}$$

The image height is;

$$h_i = mh_o = \frac{q}{p} h_o, \qquad \text{EQ(4)}$$

where h_o is the height of the object. Substituting EQ(3) into EQ(4), shows that;

$$h_i = \frac{r}{2p - r} h_o. \qquad \text{EQ(5)}$$

Note that r has negative values for the convex mirror and that h_i does not go to infinity when $p = r/2$. The magnification is;

$$m = \frac{h_i}{h_o} = \frac{r}{2p - r}. \qquad \text{EQ(6)}$$

The magnification is seen to be a function of the object distance. As a numerical example, choose $r = -0.5$ ft. and $p = 5.0$ ft. Then $m = -0.5$. How does this compare to a plane mirror? For a plane mirror $r = \infty$, and m is always -1.

Now we know that the convex mirror produces a virtual image smaller than that of a plane mirror and that the image size is a function of the object distance from the mirror. The eye forms an image on it's retina of the mirror's virtual image. We could calculate the overall magnification from the object height to the retinal image height if we assumed a "standard" eye with an associated "standard" focal length lens and lens to retina distance. However, as drivers or mirror engineers, we really only want to know how the eye sees the convex mirror image compared to a plane mirror image, i.e., how much smaller is the image from a convex mirror than the image from a plane mirror. To do this, we only need to compare the angles subtended from the eye to the mirror virtual images. This is because the retinal image height is proportional to the subtended angle.

Figure 2 shows the subtended angle, φ , for a convex mirror. For simplicity, the eye and the object being viewed are on the axis of the mirror. The eye is shown at a distance s from the mirror. The construction lines show the image of the object and the angle subtended at the eye by the image.

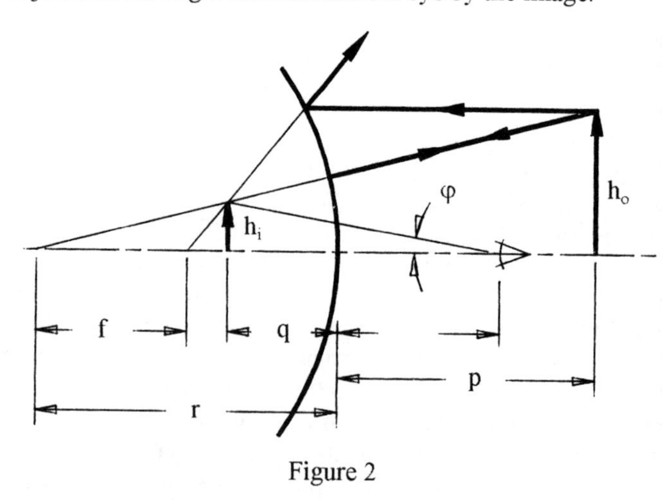

Figure 2

From Figure 2 it is seen that the subtended angle φ is;

$$\varphi = \tan^{-1} \frac{h_i}{s - q} = \tan^{-1} \frac{rh_0}{(s - q)(2p - r)}$$

$$\varphi = \tan^{-1} \frac{h_0}{\frac{2sp}{r} - (s + p)}. \qquad \text{EQ(7)}$$

EQ(7) is the subtended angle for a convex mirror, and letting r go to infinity in EQ(7) gives the subtended angle for a corresponding plane mirror. The relative magnification of the convex mirror compared to the plane mirror will be denoted by m_R, and it is:

$$m_R = \frac{\tan^{-1} \dfrac{h_0}{\dfrac{2sp}{r} - (s+p)}}{\tan^{-1} \dfrac{h_0}{-(s+p)}}. \qquad \text{EQ(8)}$$

While precise, EQ(8) is difficult to interpret at a glance. This can be helped by recalling that ;

$$\tan^{-1} x = x - \frac{x^3}{3} + \frac{x^5}{5} - \cdots \qquad \text{EQ(9)}$$

For an automotive mirror, r $\cong$ - 5 ft, s $\cong$ 4 ft, and h_0 is at most 1/3 of p. In this case, the higher order terms are extremely small and they may be ignored. Then;

$$m_R \cong \frac{1}{1 - \dfrac{2sp}{r(s+p)}}. \qquad \text{EQ(10)}$$

Finally, we have a simple understandable equation that compares what we would see in a convex mirror with what we would see in a like plane mirror.

Figure 3 is a graph of m_R vs p for r = - 5 ft and s = 4 ft. It is seen that m_R goes to unity at $p = 0$ and to .38 at $p = \infty$. As a car approaches, it appears to increase in size at a faster rate than would a car in a plane mirror.

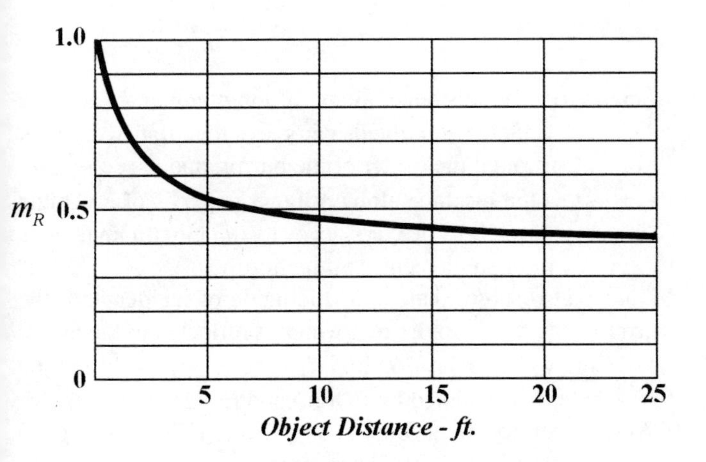

Figure 3

Next, an expression for the viewing angle of a convex mirror will be derived. Figure 4 shows a convex mirror of radius, r, and width, w, in a horizontal plane to depict the horizontal viewing angle. An observer is shown at a distance s from the mirror. The incoming ray shown defines the widest angle that can be seen for the mirror width and eye position depicted. The total viewing angle of the mirror will

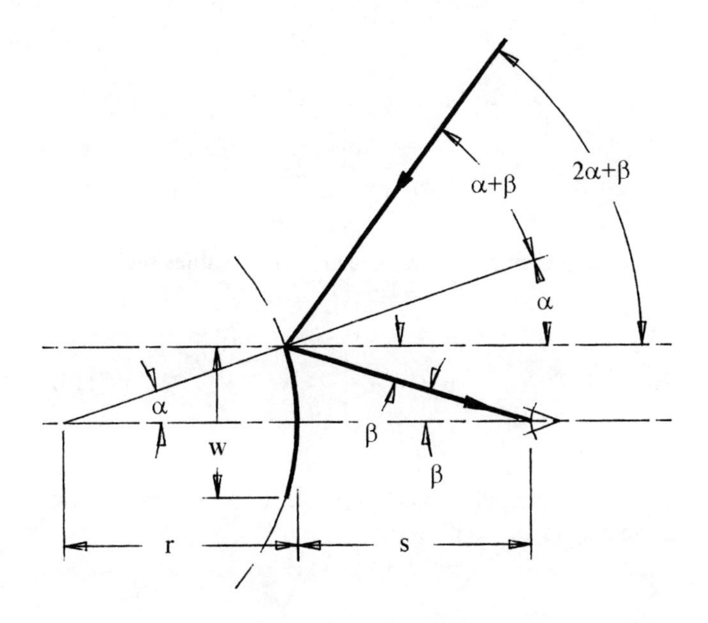

Figure 4

be twice the angle that the incoming ray makes with the axis line. This angle is obviously;

$$\theta = 2(2\alpha+\beta). \qquad \text{EQ(11)}$$

Noting that,

$$\alpha = \tan^{-1} \frac{w}{2r}, \qquad \text{EQ(12)}$$

and,

$$\beta = \tan^{-1} \frac{w}{2s}, \qquad \text{EQ(13)}$$

then,

$$\theta = 2\left[2 \tan^{-1} \frac{w}{2r} + \tan^{-1} \frac{w}{2s}\right]. \qquad \text{EQ(14)}$$

It is of interest to compare this angle with the viewing angle of a plane mirror of the same width at the same position. The ratio of the convex mirror viewing angle to the plane mirror viewing angle will be denoted by θ_R, and it is;

$$\theta_R = \frac{2\left[2\tan^{-1}\frac{w}{2r} + \tan^{-1}\frac{w}{2s}\right]}{2\tan^{-1}\frac{w}{2s}},$$

$$\theta_R = 1 + 2\frac{\tan^{-1}\frac{w}{2r}}{\tan^{-1}\frac{w}{2s}}. \qquad \text{EQ(15)}$$

For automotive mirrors, w, s and r have values such that;

$$\theta_R \cong 1 + \frac{2s}{r}. \qquad \text{EQ(16)}$$

In EQ(10), when the object distance goes to infinity the relative magnification becomes;

$$|m_R|_{p=\infty} \cong \frac{1}{1 - \frac{2s}{r}}. \qquad \text{EQ(17)}$$

EQ(17) is evaluated with negative values for r, since r was defined as being negative in Figure 1. Hence, we can write ;

$$|m_R|_{p=\infty} \cong \frac{1}{1 + \frac{2s}{|r|}}. \qquad \text{EQ(18)}$$

Then;

$$\theta_R \cong \frac{1}{|m_R|_{p=\infty}}. \qquad \text{EQ(19)}$$

That is, the viewing angle of a convex mirror is greater than the plane mirror viewing angle by a factor equal to the reciprocal of the relative magnification evaluated at $p = \infty$.

Figure 4 shows the eye on the axis of the mirror, and of course, this is not the way the mirror is used. Figure 5 shows the eye off the mirror axis as it would be for a right side convex mirror. The mirror is at an angle λ to the line from the eye to the center of the mirror. To the eye, the mirror appears to be reduced in width by a factor of $\cos\lambda$. Then

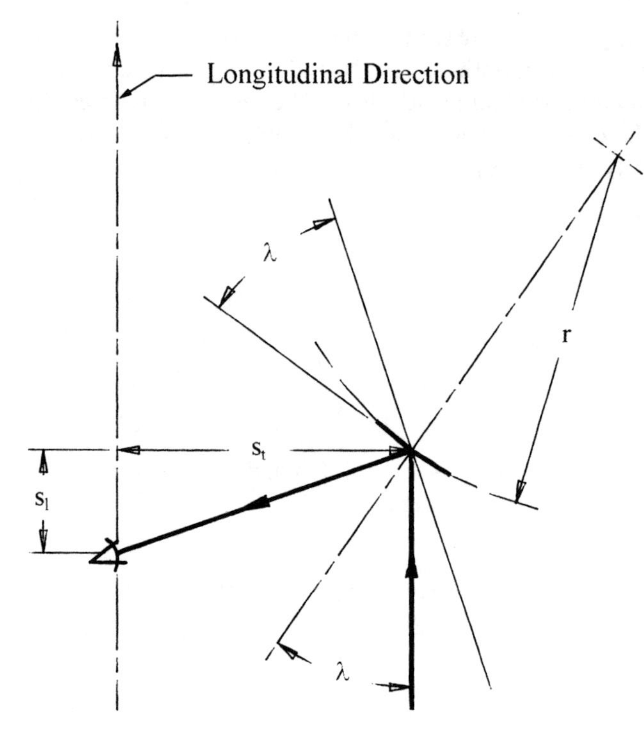

Figure 5

EQ(14) becomes;

$$\theta \cong 2\left[2\tan^{-1}\frac{w}{2r} + \tan^{-1}\frac{w\cos\lambda}{2s}\right]. \qquad \text{EQ(20)}$$

It is easily shown that,

$$\lambda = \frac{1}{2}\tan^{-1}\frac{s_t}{s_l}, \qquad \text{EQ(21)}$$

where s_l is the distance along a longitudinal axis of a vehicle from the driver's eye to a transverse axis through the center of the mirror, and

s_t is the distance along a transverse axis of a vehicle from the driver's eye to a longitudinal axis through the center of the mirror.

Figure 5 also shows that λ is the angle of incidence to the mirror of the ray from the rear which is reflected to the eye.

Using values of $r = 60$ in., $s_t = 50$ in., $s_l = 20$ in. and $w = 6$ in., θ is calculated by EQ(20) and EQ(21) to be 16.73°. CADAM shows θ to be 16.75°. Hence, EQ(20) is a good approximation for automotive mirrors.

One other characteristic of the convex mirror is of interest here, and that is the intensity of light reflected from it. Figure 6 shows a point source of light at a distance p from a convex mirror, forming a virtual image of the source at distance q behind the mirror. An eye looks at the mirror from a distance s . Distances r, p and q are defined as in EQ(1).

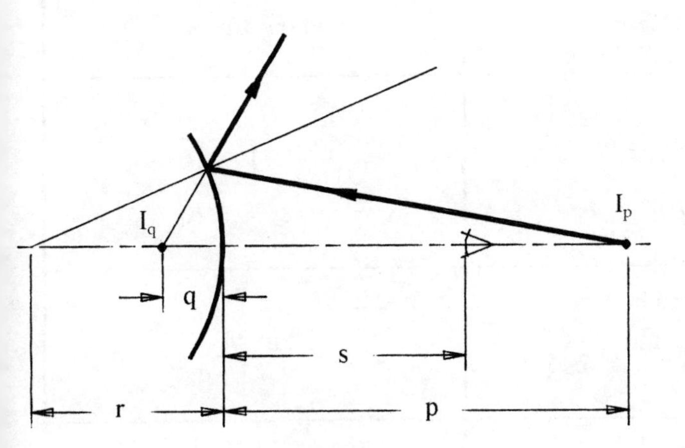

Figure 6

The point source of light has a luminous intensity of I_p. Then the illuminance at the mirror surface due to I_p will be;

$$E_1 = \frac{I_p}{p^2}. \qquad \text{EQ(22)}$$

The eye sees a point source of light at a distance q behind the mirror having a luminous intensity of I_q. If I_q were a real illuminant, the illuminance at the mirror due to I_q would be;

$$E_2 = \frac{I_q}{q^2}. \qquad \text{EQ(23)}$$

If the reflectance of the mirror is ρ ,

$$E_2 = \rho E_1, \qquad \text{EQ(24)}$$

and,

$$\frac{I_q}{q^2} = \rho \frac{I_p}{p^2}. \qquad \text{EQ(25)}$$

Then,

$$I_q = \rho I_p \frac{q^2}{p^2}, \qquad \text{EQ(26)}$$

and substituting EQ(3) into EQ(26),

$$I_q = \rho I_p \left(\frac{r}{2p - r} \right)^2. \qquad \text{EQ(27)}$$

The illuminance seen by the eye is,

$$E_3 = \frac{I_q}{(s-q)^2}. \qquad \text{EQ(28)}$$

If the convex mirror were a plane mirror, the illuminance at the eye due to I_p would be,

$$E_4 = \frac{\rho I_p}{(s+p)^2}. \qquad \text{EQ(29)}$$

The ratio of E_3 to E_4 is the relative illuminance from a convex mirror compared to a plane mirror, and it will be denoted by E_R.

$$E_R = \frac{I_q}{\rho I_p} \left(\frac{s+p}{s-q} \right)^2, \qquad \text{EQ(30)}$$

Substituting EQ(27) into EQ(30),

$$E_R = \left[\frac{(s+p)r}{(s-q)(2p-r)} \right]^2, \qquad \text{EQ(31)}$$

and substituting EQ(3) into EQ(31),

$$E_R = \left[\frac{(s+p)r}{(s - \frac{pr}{2p-r})(2p-r)} \right]^2,$$

$$E_R = \left[\frac{1}{1 - \frac{2sp}{r(s+p)}} \right]^2 = m_R^2. \qquad \text{EQ(32)}$$

E_R is a measure of the glare potential of a convex mirror. If the mirror has a 55% chromium coating and r = - 5 ft, s = 4 ft. and p = 100 ft, the mirror has as effective reflectivity of 8.5%.

Table 1 summarizes the results of EQ(1) through EQ(32), giving a concise statement of the characteristics of both the convex and plane mirror.

Summary of Mirror Characteristics		
	Convex Mirror	***Plane Mirror***
Relative Magnification	$\dfrac{1}{1-\dfrac{2sp}{r(s+p)}} = m_R$	1
Viewing Angle *(monocular, on axis)*	$2\left[2\tan^{-1}\dfrac{w}{2r} + \tan^{-1}\dfrac{w}{2s}\right]$	$2\tan^{-1}\dfrac{w}{2s}$
Relative Viewing Angle *(monocular, on axis)*	$\left.\dfrac{1}{m_R}\right\vert_{p=\infty}$	1
Reflected Illuminance	$\dfrac{\rho I_p}{\left[\dfrac{2sp}{r}-(s+p)\right]^2}$	$\dfrac{\rho I_p}{(s+p)^2}$
Relative Reflected Illuminance	m_R^2	1

Table 1

The equations of Table 1 alone are not enough to define the viewing angles on a car. Most of us see with two eyes, and our eyes are off the axis of the mirror.

Figure 7 shows the viewing angles in a horizontal plane generated by two eyes spaced equidistantly about the axis of

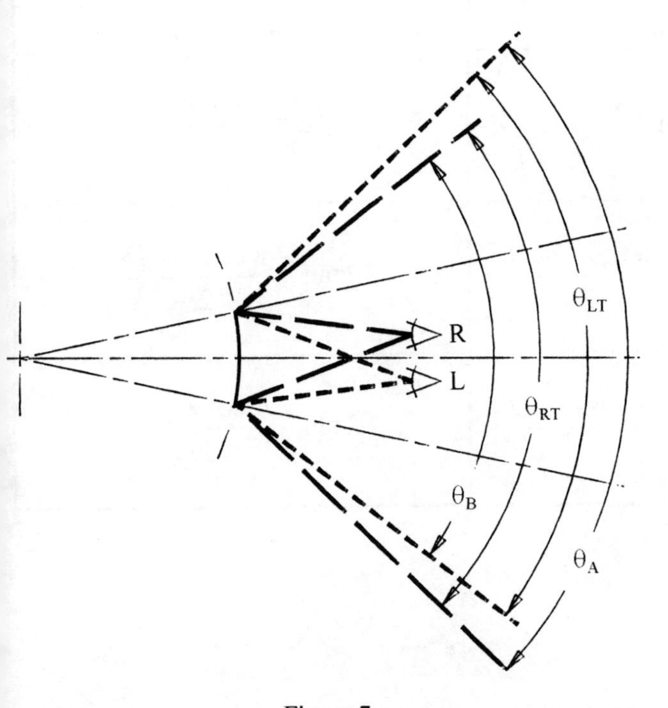

Figure 7

the mirror. The left and right eye viewing angles are designated by θ_{LT} and θ_{RT} respectively. The total viewing angle generated by both eyes separately is designated θ_A. SAE Recommended Practice J1050a calls this total view the ambinocular view. The view where θ_{LT} and θ_{RT} overlap is marked θ_B, and it is designated by J1050a as the binocular view. It will be stereoscopic.

The total viewing angle is easily derived from Figure 8 (half of Figure 7), noting that the widest angle is obtained

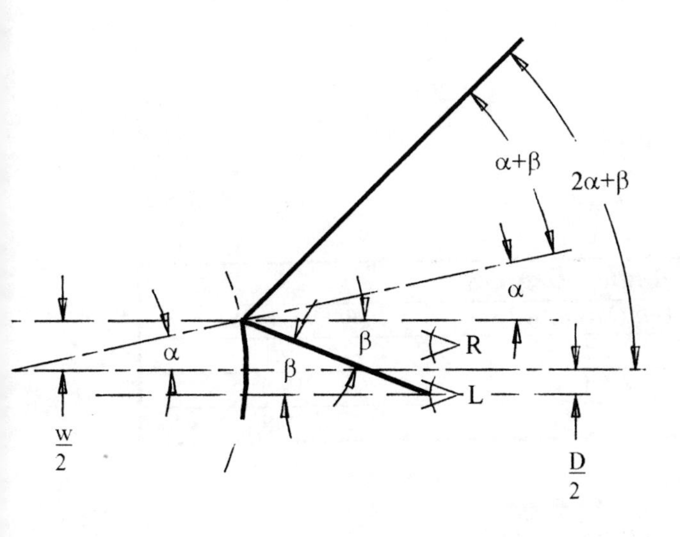

Figure 8

when the left eye looks at the right edge of the mirror and the right eye looks at the left edge of the mirror. We would like to know what angle the incoming ray makes with the mirror axis. If we displace a line parallel to the axis and going through the point where the incoming ray hits the mirror, we can easily calculate the angle which the incoming ray makes with this displaced axis. This angle is equal to the angle the incoming ray makes with the original axis. Twice this angle is obviously the viewing angle of the mirror. In Figure 8,

$$\beta = \tan^{-1}\frac{\frac{w}{2}+\frac{D}{2}}{s} = \tan^{-1}\frac{w+D}{2s}, \qquad \text{EQ(33)}$$

and the total viewing angle will be,

$$\theta_A = 2\left[2\tan^{-1}\frac{w}{2r} + \tan^{-1}\frac{w+D}{2s}\right]. \qquad \text{EQ(34)}$$

Again, for off axis viewing,

$$\theta_A \cong 2\left[2\tan^{-1}\frac{w}{2r} + \tan^{-1}\frac{w\cos\lambda + D}{2s}\right], \qquad \text{EQ(35)}$$

where λ is the angle between a line perpendicular to the center of the mirror and a line from the center of the mirror to a point midway between the eyes.

A similar analysis for the binocular viewing angle shows that,

$$\theta_B \cong 2\left[2\tan^{-1}\frac{w}{2r} + \tan^{-1}\frac{w\cos\lambda - D}{2s}\right]. \qquad \text{EQ(36)}$$

This is the same as EQ(35) except for the sign of D. Table 2 summarizes the viewing angle equations.

Mirror Viewing Angles		
	Convex Mirror	***Plane Mirror***
Monocular	$2\left[2\tan^{-1}\frac{w}{2r}+\tan^{-1}\frac{w\cos\lambda}{2s}\right]$	$2\tan^{-1}\frac{w\cos\lambda}{2s}$
Ambinocular	$2\left[2\tan^{-1}\frac{w}{2r}+\tan^{-1}\frac{w\cos\lambda+D}{2s}\right]$	$2\tan^{-1}\frac{w\cos\lambda+D}{2s}$
Binocular	$2\left[2\tan^{-1}\frac{w}{2r}+\tan^{-1}\frac{w\cos\lambda-D}{2s}\right]$	$2\tan^{-1}\frac{w\cos\lambda-D}{2s}$

Table 2

Representative Eye to Mirror Distance - Inches						
	Mirror					
	Inside w = 8.0 in., r = ∞		*Left Outside* w = 5.75 in., r = ∞		*Right Outside* w = 5.75 in., r = 60 in.	
Seat Position	s_l	s_t	s_l	s_t	s_l	s_t
Forward	9	14	16	18	16	48
Center	13	14	20	18	20	48
Back	17	14	24	18	24	48

Table 3

Ambinocular Viewing Angle - Degrees			
Seat Position	*Inside Mirror*	*Left Outside Mirror*	*Right Outside Mirror*
Forward	31.9	18.3	19.1
Center	28.9	16.6	19.0
Back	25.7	15.1	18.9

Table 4

WHY BLIND ZONES EXIST

Using the equations of Table 2, the viewing angles of automotive rearview mirrors are easily calculated, and with the angles so determined we can make scale drawings to accurately show the blind zones. First, we must select values of the independent variables in the Table 2 equations which will provide the viewing angles of most interest. This was done by taking a mid-size passenger car and measuring the mirror widths, the radius of curvature of the convex mirror, and the eye to mirror longitudinal and transverse distances. The measurements were made for three conditions; driver's seat full forward, centered, and full back. This particular vehicle had a manual seat and the measurements were made with the recliner full forward. Table 3 shows the measurements taken, with distances rounded to the nearest inch.

Table 4 shows the viewing angles obtained by using the measurements of Table 3. Note that short drivers have a considerably better rearview than tall drivers. The viewing angles of the Table 4 center position will be used to depict the viewing angles attained by an average height driver. These viewing angles are shown in Figure 9, which is a scale drawing of a mid-size passenger car in a standard 12 ft. wide lane. The outside mirrors are adjusted to just see the side of the car. The shaded regions are blind zones which are bounded by the outer vision limits of the outside mirrors and the lines where the driver's peripheral vision begins.

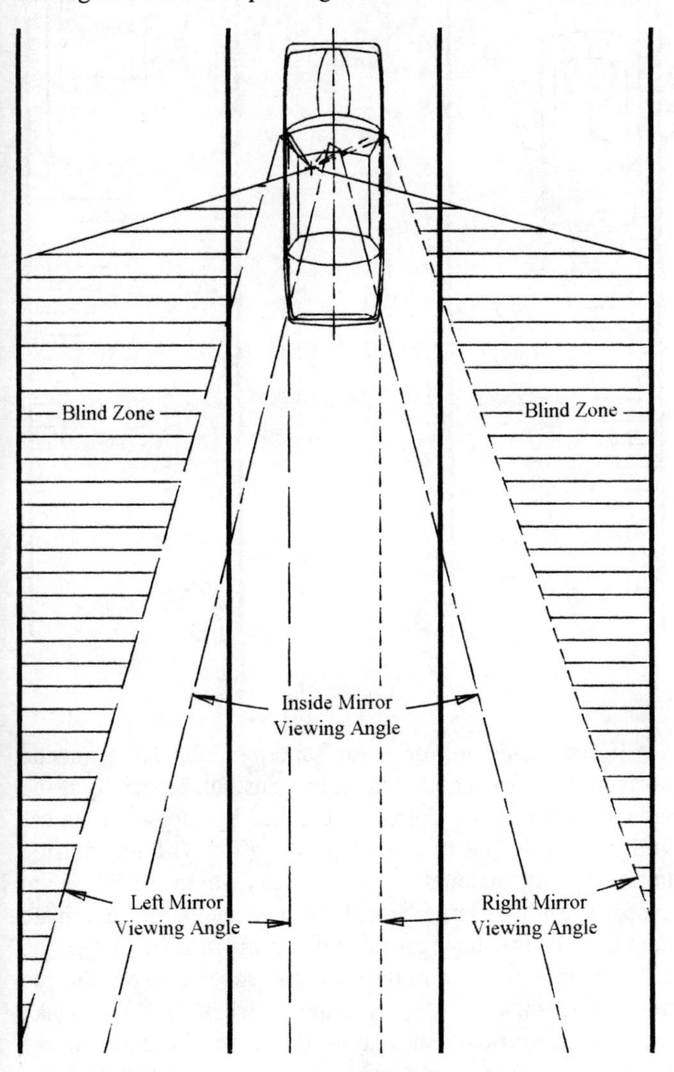

Figure 9

Let's examine the driver's side blind zone more closely, beginning with the peripheral vision line. Figure 10 shows a driver with head turned slightly and eyes directed at the outside rearview mirror. If driver's had 180° of forward vision, the peripheral vision line would be perpendicular to a line from the driver's left eye to the mirror. However, as shown by Burg (1968), driver's do not have 180° of vision. The horizontal peripheral vision angle measured from the axis of the eye to the widest angle at which a target is perceived was found by Burg to be a function of age. Burg measured this angle with an apparatus which momentarily introduced a white illuminated object into the subject's field of view. The object was contrasted against a black background, and it would suddenly appear at discrete angular intervals. Thus conditions for detecting the object where good. Burg found that the peripheral vision angle peaked at about 88° at age 20 and diminished to about 75° at age 70.

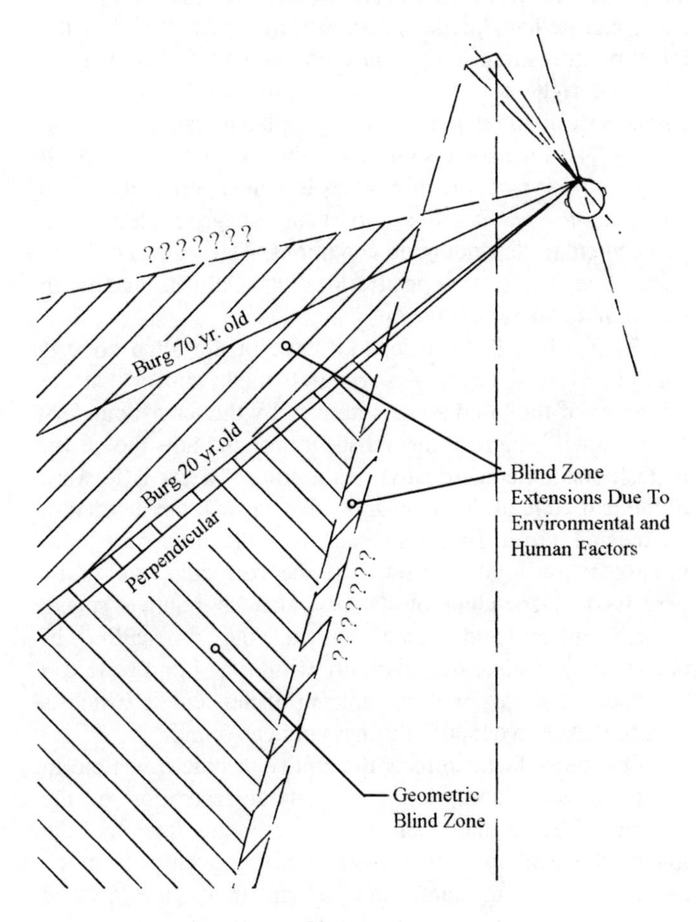

Figure 10

Ball et al (1988) show that the useful field of view, which involves detection and identification of targets, varies with factors such as the conspicuity of the target, the central task being performed, any distracting influences and age. Significant error rates in target identification were observed at relatively low eccentricities from the fixation point of the eye, as low as 30°.

When the left rearview mirror is used to observe the distance and closing velocity of a following vehicle, the driver's foveal vision is fixed on that mirror. Objects outside the foveal region may or may not be detected for any of the reasons set out by Ball.

Another restriction on the driver's peripheral vision may be the vehicle itself. A tall driver with the seat all the way back, and perhaps partially reclined, may well find that the left B pillar restricts peripheral vision to the left. On the right side, the right seat back, right B pillar and a right side passenger can all restrict the right peripheral vision.

As far as the peripheral vision line is concerned then, it can vary from a couple of degrees less than perpendicular to the *eye to mirror line*, to many degrees less than perpendicular, depending on conditions. Thus, the peripheral vision line is not at a specific location. Only its maximum extent of about 88° is known.

So, we find that the forward edge of the blind zone is variable. How about the rear edge? It might appear that the rear edge of the blind zone is defined by the outer boundary of the viewing angle of the outside mirror and how the mirror is positioned. Geometrically, this is true. If a car is forward of the rear edge of the blind zone, that car will not be seen in the outside mirror. But, if a vehicle is in the blind zone, does it have to be totally forward of the rear edge not to be perceived? Depending on factors such as relative speed, color, contrast and scene content, the probability of perceiving a vehicle only partially within the blind zone can be expected to vary with the amount of the vehicle which is exposed. Data to quantify this has not been found.

One other factor effects the ability to hide a vehicle in the blind zone, and that is the lateral separation of the vehicles. The farther apart the vehicles are laterally, the longer the available space is in which a vehicle can be hidden. Observing cars traveling on multi-lane divided highways, it is observed that vehicles in close proximity tend to separate laterally. Typically, cars separate an additional 2 to 4 feet beyond what the separation would be if they were centered in their lanes.

Summarizing, we can define a geometric blind zone determined by the outer edge of the outside mirror's viewing angle and a line from the driver's eyes which is a couple of degrees less than perpendicular to a line from the eyes to the outside mirror. The actual blind zone is larger than the geometric blind zone because of environmental and human factors and cannot be precisely defined. This is true for both the left and right outside mirrors.

Now let's apply what we know about the blind zones to actual driving conditions. Figure 11 is a scale drawing showing three mid-sized cars on standard 12 ft. wide lanes. The two outer cars are hidden in the center car's blind zones. The hidden vehicles are displaced three feet laterally. The left mirror peripheral vision line is shown at 85° to the *eye to mirror line*, corresponding to a 50 year old driver in Burg's data. The mirror viewing angles are as shown in Figure 9. No other environmental or human factors have been included to reduce the peripheral vision line below 85°. The reality of course, is that it can be far less than the 85° shown, which would result in a larger blind zone.

From Figure 11 three facts are apparent.

1. The blind zones can be large enough to easily conceal a vehicle.

2. The outside mirrors add little additional visibility beyond that provided by the inside mirror. The addition is only a strip about four feet wide.

3. The inside mirror provides by far the best view and the most information. It allows the driver to accurately judge the distance and speed of vehicles approaching from the rear.

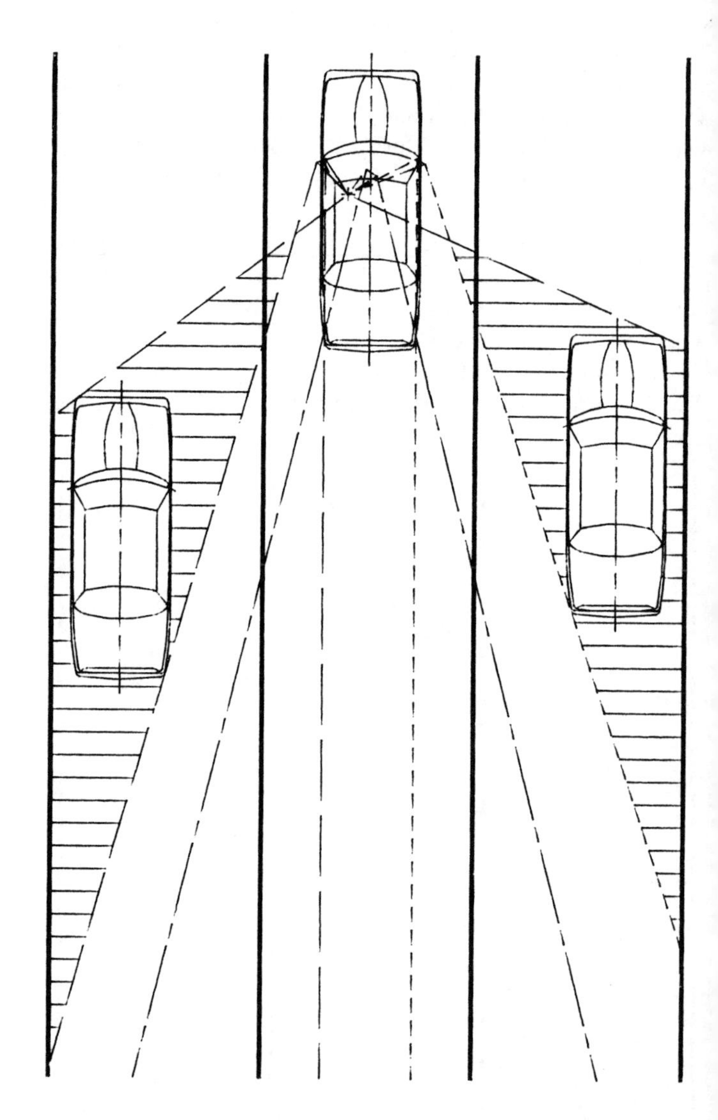

Figure 11

If the inside mirror is so superior, why have outside mirrors? Well, outside mirrors can be useful, especially if the view from the inside mirror is blocked by cargo or by a car immediately to the rear in stopped traffic. The left outside mirror is large enough and close enough to be of value, except it can't be large enough or close enough to eliminate the blind zone without causing other problems. Given that the left mirror is worth putting on a passenger car, despite the blind zone, then we seem to require a right mirror to make the car symmetrical. But putting the same size plane mirror on the right side would produce a nearly useless viewing angle. In fact, the viewing angle for a right side plane mirror inches wide would be only about 5.5°. So instead of a plane nirror, a convex mirror with about a 20° viewing angle is sed. The only problem is that the driver can't accurately ıdge either the distance or approach speed of vehicles seen ı it, and it still has a blind zone. However, the stylists are ontent because the car is symmetrical, and the mirror people ppear satisfied because they have more than a 5.5° viewing ngle.

TRATEGIES TO OVERCOME THE BLIND ZONES

Given then that we have inside and outside mirrors, how o we overcome the blind zones produced by them? The ıost common procedure on passenger cars is to set the utside mirrors to just see the side of the car. When changing ınes, the driver first looks in the outside mirror to determine ıe position and approach speed of vehicles to the rear and ıen turns to look for vehicles in the blind zone. There are roblems that exist with this procedure. First, the driver must emember to turn and look before changing lanes. Failing to o so can prove disastrous. Second, turning to look into the lind zone results in a loss of forward vision for times in the rder of a second. At highway speeds this translates into bout 100 feet of travel with the driver's eyes off the road. 'hird, turning one's head can get tiresome. Fourth, some rivers, especially older ones, have difficulty turning their eads.

One alternative to turning to look into the blind zones vould be to increase the horizontal viewing angles of the utside mirrors to about 35°. At such an angle, it is mpossible to hide a car in the remaining blind zone. On the ight side, a plane mirror with a 35° viewing angle would be 6 inches wide, so that's out. A right side convex mirror 6.5 nches wide would require a radius of curvature of 28 inches, nd that is less than the 35 inch minimum required by 'MVSS 111. So, we can't get 35° with a reasonably sized ight side mirror.

On the left side, a plane mirror with a 35° viewing angle vould be 16 inches wide, also unacceptable. A left side onvex mirror 6.5 inches wide would require a radius of urvature of 44 inches, but any convex mirror on the driver's ide is prohibited by FMVSS 111.

Volvo has developed an interesting solution to the blind one problem which was first sold in Sweden in 1979 Pilhall,1981). They use a mirror 6.7 inches wide with a adius of curvature of 82.7 inches on the inner 2/3's of the nirror and then a decreasing radius in the outer 1/3, going own to about a 10 inch radius. The magnification of the nner 2/3's is about 0.7 and the viewing angle is about 20°. 'he remaining 1/3 captures any vehicle in the blind zone.

Another approach car owners can use to address the lind zone problem is to add *stick on* high curvature mirrors ɔ their existing mirrors, since owners are not restricted by 'MVSS 111. *Stick ons* have the disadvantages of reducing he viewing angle of a plane mirror by effectively reducing ts width, and of marring the styling of the mirrors. However, large number of people are sufficiently dissatisfied with heir mirrors to resort to *stick ons*. Olson and Winkler (1985) found in a study of 620 vehicles that 6.6% had *stick on* mirrors on the driver's side mirror.

A simple and logical strategy to overcome the blind zones is suggested by further contemplation of Figure 11. As previously stated, the inside mirror provides by far the best view to the rear. Again, as seen in Figure 11, the outside mirrors actually provide very little additional viewing capability. This being the case, the blind zones can be effectively eliminated simply by rotating the outside mirrors to look into the Figure 11 blind zones as shown in Figure 12. The visible area now increases dramatically, and the remaining blind zones are incapable of hiding a car.

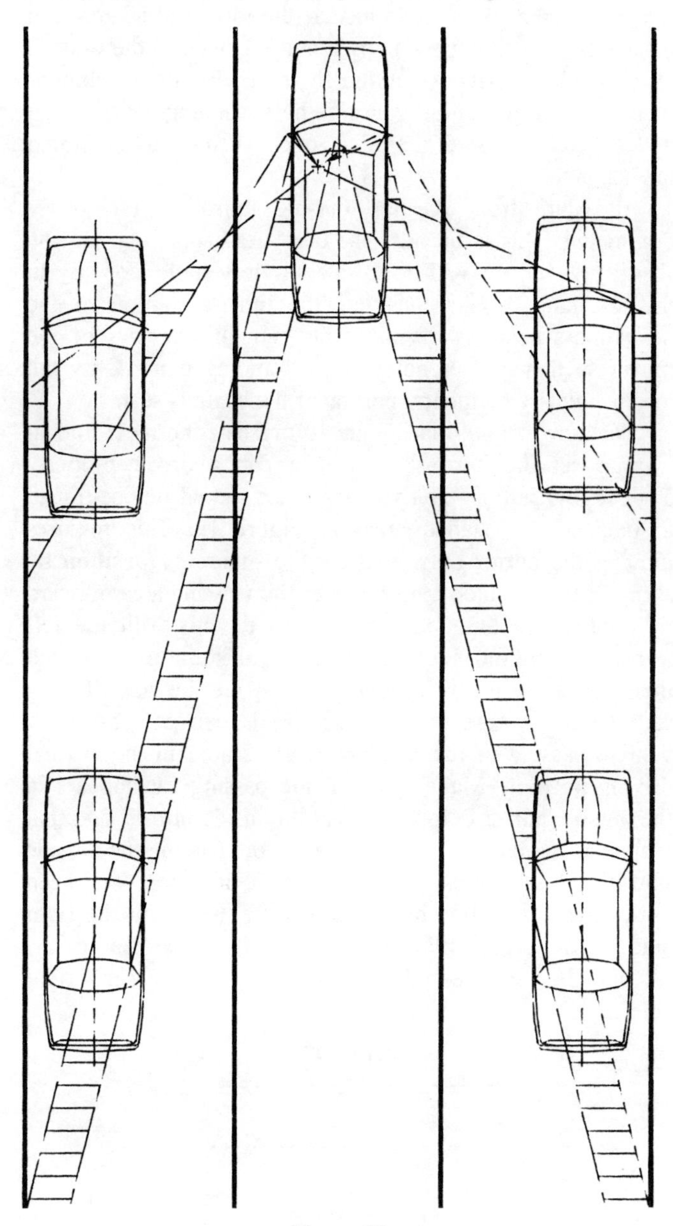

Figure 12

Driving with the mirrors positioned as in Figure 12 calls for a different lane change strategy. With this setting, the driver ***first*** looks in the ***inside*** mirror to check the position and approach speed of vehicles to the rear. ***Then,*** the driver glances at the outside mirror to see if a car is in what used to be the blind zone.

There are many advantages to the Figure 12 setting.

1. It is no longer necessary to turn to look into the former blind zones since they are now visible in the outside mirrors.

2. The forward driving scene is always within the driver's peripheral vision when glancing at the outside mirrors, as opposed to the loss of this forward vision when turning to look into the blind zones.

3. Less time is required for a glance to the outside mirror as compared to turning to look into the blind zone.

4. Older or physically restricted drivers no longer have the difficulty of turning to look into the blind zones.

5. It is now possible to include the former blind zones in a scan of the driving scene by a quick glance to the outside mirrors. Good drivers continually scan. By being able to include the former blind zones in their scanning, drivers will be less likely to forget the blind zones when contemplating a lane change.

6. Glare from the left outside mirror is effectively eliminated. This is because the only headlamps that can be seen it that mirror will be from a single vehicle close by in the left lane. In that position, the intense portion of the headlamp's beam is displaced far enough sideways of the mirror so that it does not produce blinding glare. Only the lower intensity peripheral portion of the beam is seen.

Acknowledging that Figure 12 presents the most logical way to set the mirrors, how then can a driver properly achieve this setting. The viewing areas should be positioned to balance the shaded areas in Figure 12. This requires moving the mirror outward from the Figure 11 position by about 7° to 8°, which in turn moves the viewing area outward by double that amount. One way to do this with the left mirror is to put the side of your head against the window, and then adjust the mirror so that you just see the side of your car. To make sure that this setting is correct, observe a vehicle passing in the left lane while seated in the normal driving position. Make sure that the passing car appears in the outside mirror before it leaves the inside mirror, and that it appears in your peripheral vision before leaving the outside mirror. This assures you that the blind zone has been eliminated. The same procedure can be used with the right side mirror, except, place your head at the center line of the car and then set the right mirror to just see the car's right side.

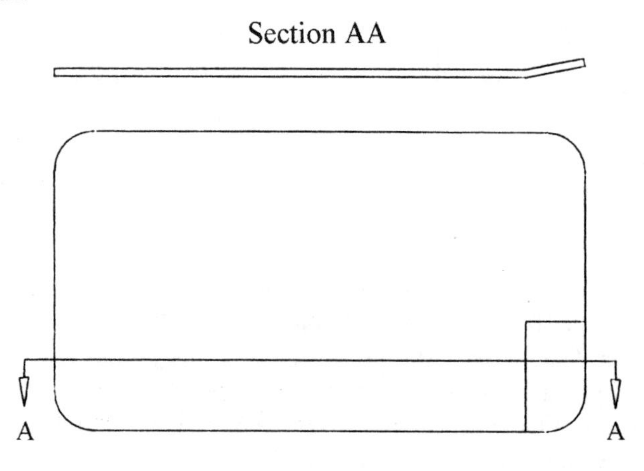

Figure 13

This procedure is cumbersome, and furthermore it is annoying to shift your head back and forth every time you want to see if the setting is correct. A solution to this problem is to have a mirror with a small portion of the lower inner corner canted as shown in Figure 13. The angle of the canted portion relative to the main mirror is 1/2 of the angle by which the viewing area is to be moved out away from the Figure 11 position and into the Figure 12 position. The Figure 13 mirror is set by adjusting it so that the side of the car is just seen in the canted portion.

Driving with the mirrors set in the Figure 12 position may require an adjusting period for some drivers. Acclimation to the Figure 12 setting will be easier once the driver fully realizes and accepts that the inside mirror provides all of the rearview information required except for revealing the presence of a vehicle in the former blind zones. The outside mirrors in the Figure 12 setting become simply "go, no-go" devices to tell a driver if there is a car in the former blind zone. Just a quick glance at the outside mirror reveals this.

The Figure 12 setting is already recognized by a small percentage of drivers as being a means of eliminating the blind zones. Olson and Winkler (1985) made measurements on their previously mentioned sample of 620 passenger cars to determine how drivers set their outside mirrors. Figure 14 shows their data for the driver's side outside mirror.

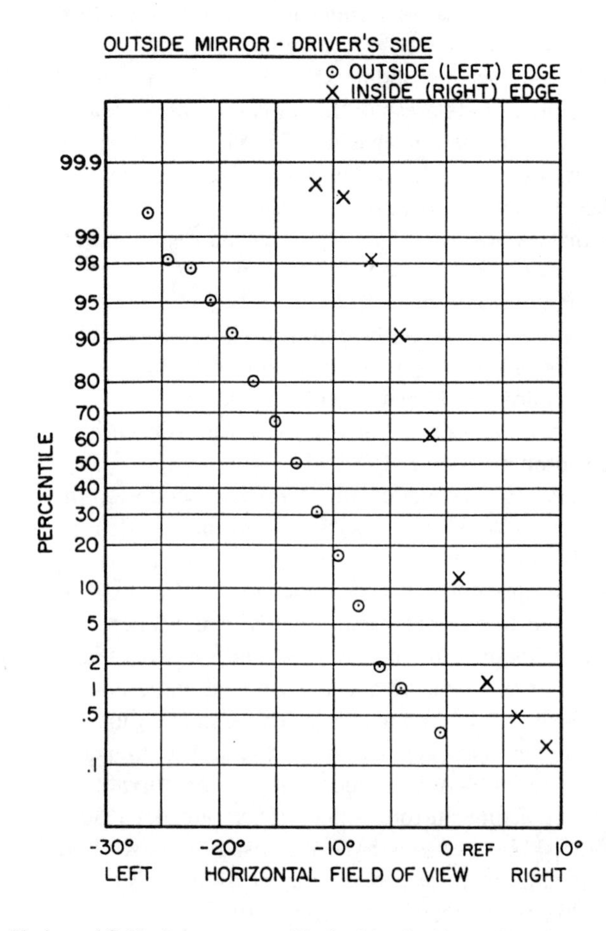

Horizontal field of view measured in the driver's side outside mirrors.

Figure 14

n Figure 14, O_{REF} is the left side of the car. This data shows hat about 50 % of drivers have the left outside mirror set so hat the left side of the car is not visible. About 5 % have the nirror turned out so that the inner edge of the field of view is ° or more away from the side of the car. A fraction of a percent appear to have their mirrors set to the full Figure 12 position. Thus, even in the absence of having the blind zone clearly defined and lacking a precise way to overcome the blind zone, many drivers are naturally leaning in the Figure 12 direction. Olson and Winkler in their conclusions state, *"As typically aimed, there is a great deal of overlap in the fields of view provided by the three mirrors with which most vehicles in the sample were equipped, and coverage to adjacent lanes close to the vehicle is limited. There would seem to be some merit in encouraging drivers to aim their outside mirrors so as to provide better coverage in these important areas."*

We have examined several low technology ways of overcoming the blind zones which could contribute to a reduction in LCM crashes. Chovan et al (1994) have developed high technology IVHS crash avoidance concepts for LCM crashes. These concepts include passive information displays to assist the driver, driver warning systems which provide warning in time for the driver to respond, control intervention systems which partially control the vehicle but do not eliminate driver control, and finally fully automatic control systems which react when the response times required are beyond the driver's capability. All of these systems require sensors which can determine position, velocity and acceleration of the vehicles in the zone of action. Most of the sensors required are available, but implementation revolves around issues of appropriate displays, intervention algorithms and cost. Some high technology counter measures such as detection of vehicles in the blind zones are feasible today. They will be evaluated in terms of cost and effectiveness compared to the lower technology alternatives.

There is one other aspect of overcoming the blind zones worth looking at, and that is the car manufacturer's perspective on this issue. One source of information on their perspective is the vehicle owner's manual. What do we find in these manuals regarding the blind zones? Absolutely nothing. It is as if the blind zones do not exist. Typically, an owner's manual will provide about 100 highlighted and specific warnings on a variety of subjects, but there are no warnings pertaining to the blind zones.

Most manuals instruct the driver to set the outside mirrors so that the side of the car is just visible. Having so instructed the driver, it would appear incumbent on the manufacturer to pictorially show the blind zone generated by this setting. This is not done.

One manufacturer instructs its owners to adjust the outside mirror to be able to see directly behind the car. This implies turning the mirror farther inboard than just seeing the side of the car, and that of course makes the blind zone larger.

Another manufacturer instructs its owners to adjust the *right* side mirror to just see the side of the car, and to adjust the *left* side mirror to center on the adjacent lane with a slight overlap of the view obtained on the inside mirror. This appears to be a step in the right direction, since it seems that the intent is to reduce the size of the left blind zone, i.e., don't set the left outside mirror to see the side of the car. However, the wording is unclear and more than likely leaves the owner confused.

In view of the information presented here it is apparent that the car manufacturers are not adequately addressing the blind zone problem. The one or two pages of the owner's manual usually devoted to mirrors needs to be expanded to include an explanation and graphical depiction of the blind zones on the car purchased, along with strategies to overcome them.

CONCLUSIONS AND RECOMMENDATIONS

1. The relative reflected illuminance of a convex mirror is equal to the square of the relative magnification, and the relative viewing angle is equal to the reciprocal of the relative magnification evaluated with the object at infinity.
2. The set of equations shown in Tables 1 and 2 can be used to accurately quantify the pertinent characteristics of automotive rearview mirrors.
3. A vehicle can be easily hidden in the blind zones created by setting the outside rearview mirrors to just see the side of the car.
4. The blind zones can be effectively eliminated if the outside rearview mirror's viewing angles are turned outward away from the side of the car.
5. Glare from the left outside mirror is effectively eliminated when it is turned outward to eliminate the blind zone.
6. An outside rearview mirror with a small canted portion at the lower inside edge can provide a means of properly setting the mirror to view the former blind zones.
7. It is recommended that the car manufacturers use the owner's manual, and any other means available, to explain to car owners why blind zones exist and to offer strategies for overcoming them.

REFERENCES

Ball, K.K., Beard, B.L., Roenker, D.L., Miller, R.L., and Griggs, D.S., "Age and Visual Search: Expanding the Useful Field of View," J. Opt. Soc. Am. / Vol.5 / No.12 / p.2210-2219 / Dec.1988

Burg, A., "Lateral Visual Field as Related to Age and Sex", Journal of Applied Psychology / Vol. 52 / No.1 / p.10-15 / 1968

Chovan, J.D., Tijerina, L., Alexander, G., and Hendricks, D.L., "Examination of Lane Change Crashes and Potential IVHS Countermeasures", DOT- VNTSC-NHTSA-93-2 / 1994

REFERENCES (CONTINUED)

Flannagan, M.J. and Sivak, M., "Indirect Vision Systems", Automotive Ergonomics, Taylor and Francis, London-Washington, D.C., p.205-217 / 1994

Knipling, R.R., "IVHS Technologies Applied to Collision Avoidance: Perspective on Six Target Crash Types", Technical Paper, IVHS America Annual Meeting / 1993

Morrow, I.R.V. and Salik, G., "Vision in Rearview Mirrors", The Optician, Part 2, 144(3732), p.340-345 / 1962

Najm, W.G., Koziol Jr., J.S., Tijerana, L., Pierowicz, J.A.,and Hendricks, D.L., "Comparative Assessment of Crash Causal Factors and IVHS Countermeasures", IVHS America Annual Meeting / 1994

Olson, P.L. and Winkler, C.B., "Measurement of Crash Avoidance Characteristics of Vehicles in Use", DOT HS 806 918 / 1985

Pilhall, S., "Improved Rearward View", SAE Technical Paper 810759 / 1981

Platzer Jr., G.E., US Patent No. 5,050,977

Rowland, G.E., Silver, C.A., Volinsky, S.C., Behrman, J.S., Nickols, N.F., and Clisham, W.F., "A Comparison of Plane and Convex Rearview Mirrors for Passenger Automobiles", DOT FH-11-7382 / 1970

Walraven, P.L., and Michon, J.A., "The Influence of Some Side Mirror Parameters on Decisions of Drivers", SAE Technical Paper 690207 / 1969

Wang, J., and Knipling, R.R., "Lane Change / Merge Crashes: Problem Size Assessment and Statistical Description", NHTSA Technical Report / DOT HS 808075 / 1994

Vehicle Owner's Manual Collection, Detroit Main Library, National Automotive History Collection

ABOUT THE AUTHOR

Automotive rearview mirrors have long been of interest to me, and I am a sole or co-inventor in 9 mirror patents.

I have a BSEE and an MS in Physics from Wayne State University and a Master of Automotive Engineering degree from the Chrysler Institute of Engineering.

I am an engineering consultant working in the areas of research and development in electronics and electro-mechanical devices.

Comments or questions on this paper may be addressed to me at 424 Cypress Road, Rochester Hills, MI 48309

950602

Automotive Field of View Analysis Using Polar Plots

E. J. McIsaac and V. D. Bhise
Ford Motor Co.

ABSTRACT

This paper describes how polar plots are constructed and used to evaluate fields of view from vehicles. A polar plot presents a driver's three dimensional view of the vehicle structure, such as the window openings or mirrors, and the objects outside of the vehicle, such as other vehicles in adjacent lanes, in a two dimensional angular (or polar) field.

These plots are simple and effective in understanding and visualizing complex visibility problems. Since the plot is made in angular space, a Human Factors Engineer can use the plots for direct assessment of drivers' visual problems, such as sizes of monocular and binocular obscurations. Location of visual targets in the driver's peripheral vision, and magnitude of eye and head turn angles, can be easily determined by measuring coordinates of details shown in a polar plot.

A PC-based computer program has been developed to generate polar plots under a wide variety of options, including choice of vehicles, eye points, views, targets, and mirrors. This paper includes examples of polar plots and illustrates how they can be used by an automotive designer to evaluate many specific problems. Example problems illustrated include: 1) angular locations of pillars, 2) monocular and binocular obscurations caused by pillars, 3) hood visibility, 4) visibility of adjacent vehicles, and 5) fields of view available from planar and convex mirrors.

INTRODUCTION

Most engineers involved in the automotive design process currently use computer graphics methods to determine what a driver can see from inside the vehicle. The outputs illustrating the driver's field of view generally involve projecting sightlines drawn from the driver's eye to items of interest and then projecting them onto horizontal and/or vertical planes. In such planar projections, the projected size of an object depends on the relative location between the eye, object, and the projection plane. This information is useful but not complete for a full vision analysis. Further defining the projected size of an object by its angular coordinates offers additional information that is useful in assessing visibility. This paper illustrates these advantages.

APPLICATIONS OF POLAR PLOTS TO AUTOMOTIVE DESIGN

Comparison of the visual field available to drivers of a proposed vehicle design can be made relative to drivers of benchmark vehicles by using polar plots. Both situations can be modeled and the views overlaid to emphasize certain characteristics of either vehicle.

Positioning items of interest, such as mirrors and other aids to indirect vision, gauges and switches on the instrument panel, and other controls including the steering wheel and gear shift lever, can be done using polar plots. Utilizing the polar plotting tools allows the automotive designer to take the customer's visibility needs and desires into consideration, making the vehicle "easier to use" or more ergonomic.

Driver specific situations can be modeled to determine if there will be problems with a particular design for a special group of drivers. For example, a tall driver may have less difficulty seeing over the hoodline of a truck than a shorter driver. However, the same tall driver may experience obstruction around the inside rearview mirror of a passenger car that the shorter driver does not. Polar plots can be used to evaluate these situations, resulting in a superior design.

Different variations exist in polar plotting to assist users to better interpret and display situations of interest. For instance, binocular plots can be created to emulate vision from two eye positions simultaneously. Polar plots can be used to show the objects that become easier or more difficult to see when the driver turns his or her head. For example, aiming the sideview mirrors in real life is typically done by the driver with the head turned directly toward the mirror in question. Emulating this scenario using polar plots and then turning the driver's sightline and eyes to view the forward position will model the same scene the driver would see in the mirror in the peripheral field when viewing forward traffic situations.

The polar plot is a very useful tool in communicating and understanding complex visibility issues and problems, as views seen in the polar plot have a high resemblance to the views seen by drivers. Polar plots allow simultaneous determination of what can be seen and what is obstructed to the driver of the design being analyzed. Either hardcopy or computer monitor output of the data gives support to the merits or flaws of a specified design. The full 360° field for the driver can be emulated in one take.

According to SAE J985, it is recommended that "the entire rear vision mirror is best located within the area 60 degrees to the left or 60 degrees to the right of the driver's forward field of view". Additional limitations are given above and below the forward line of sight. A polar plot representation of a proposed mirror configuration can be used to determine if a proposed system meets the criteria specified in the standards. Illustrations of these and other uses of polar plots are provided throughout the paper.

DEVELOPMENT OF POLAR PLOTS

Vehicle data in most CAD packages is stored in rectilinear (X,Y,Z) coordinates. In order to represent the field available to the human eye, angular (θ,ϕ) locations can be computed to define zones of vision. To create a polar plot, three dimensional (x,y,z) Cartesian Coordinate data for the locations of the driver's eye(s) and points defining objects in the driver's view relative to the vehicle origin and ground are needed.

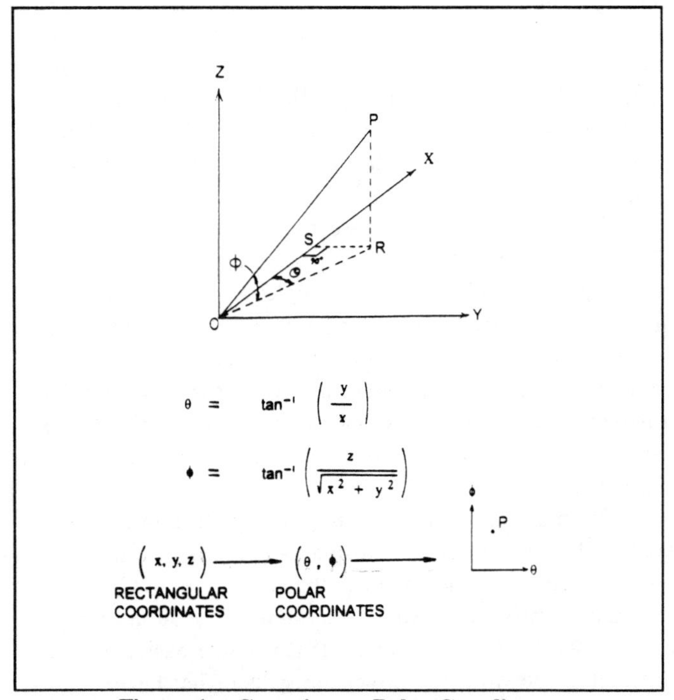

Figure 1 - Cartesian to Polar Coordinates

Figure 1 shows the (X,Y,Z) Cartesian Coordinate system with it's origin "O" at the driver's eye point. The point "P" with coordinate (x,y,z) can be described by the vector OP whose location can be defined by the angular coordinates (θ,ϕ) in a two dimensional space defined by a horizontal axis, giving the azimuth angle (θ) and a vertical axis defining the elevation angle (ϕ). The location of the point "P" in polar plot then shows its polar location.

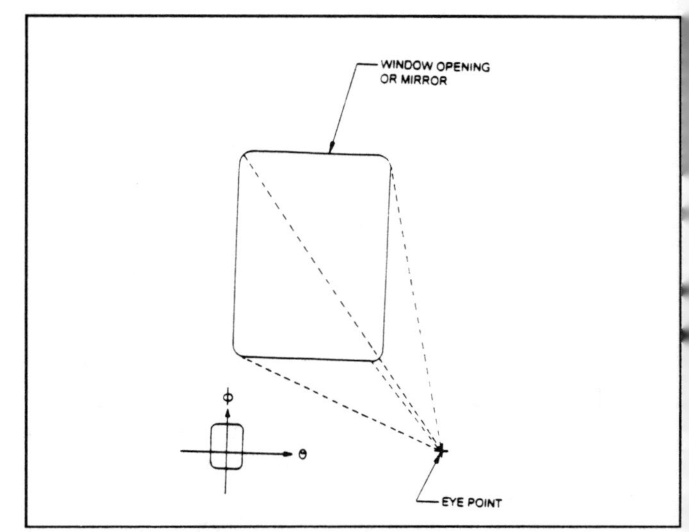

Figure 2 - Polar Application

Figure 2 illustrates how a polar plot can be generated by plotting angular locations of objects such as a window or mirror by transforming coordinates of all (x,y,z) points to angular coordinates.

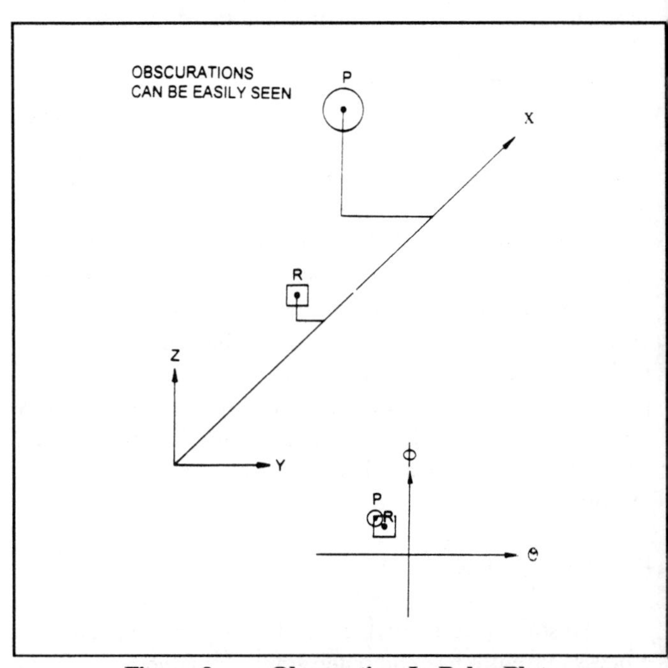

Figure 3 – Obscuration In Polar Plots

A polar plot becomes very useful in evaluating obscurations. Figure 3 illustrates the problem of determining whether the target "R" would obscure the target "P" when viewed from the eyepoint located at "O". A polar plot drawn with its origin "O" will show whether the target "R" is superimposed on the target "P".

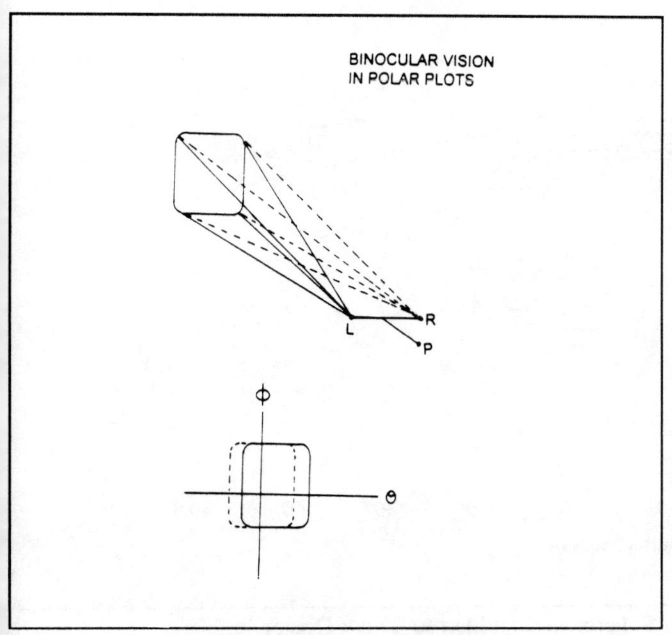

Figure 4 - Binocular Polar Plot

To determine a binocular view, polar plots generated from each eye are superimposed with a common origin in the polar space, as shown in Figure 4. If the object plotted is a window, then the observer's binocular view will include the view contained in both outlines of the window. If the object is opaque, the area common to the two outlines of the object in the polar plot will define the binocular obscuration caused by the object.

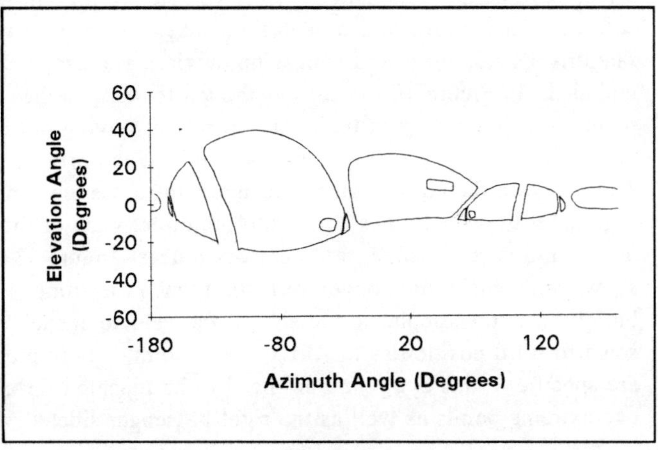

Figure 5
360° Polar Plot of Daylight Openings of Passenger Car With Cyclopian Eye

USING POLAR PLOTS FOR VISION ANALYSIS OF PROPOSED VEHICLE DESIGNS

When the daylight openings on a prototype vehicle have been defined, it is an opportune time to do a vision study to determine the difficulties that may or may not be present within the design being proposed. One starting point might be a polar plot of all of the daylight openings from the Middle Eye Centroid position or a 5th or 95th percentile driver eye position as defined in SAE J941 and J1050. Figure 5 is an illustration of this type of plot. It shows a full 360° view of the daylight openings including the rear window of a passenger car from a cyclopian eye location. Other targets can then be added to obtain further information about the design.

DEFINING TARGETS FOR FIELD OF VIEW ANALYSIS - In automotive design practices, target planes have been defined representing volumes that drivers should be able to see without obstruction. Some of these targets may represent forward, direct view regions where a sign, signal, or pedestrian might be located relative to drivers approaching and stopping at an intersection (e.g. Ford Forward Direct Field of View Targets)*. Another target might describe a region where the headlights or centerline of vehicles would be located when passing another vehicle. A driver might be expected to see every part of this target either directly, or indirectly with a visual aid, such as a mirror. Many targets of this type have been proposed by various governmental, industry, corporate and professional organizations for vision analysis.

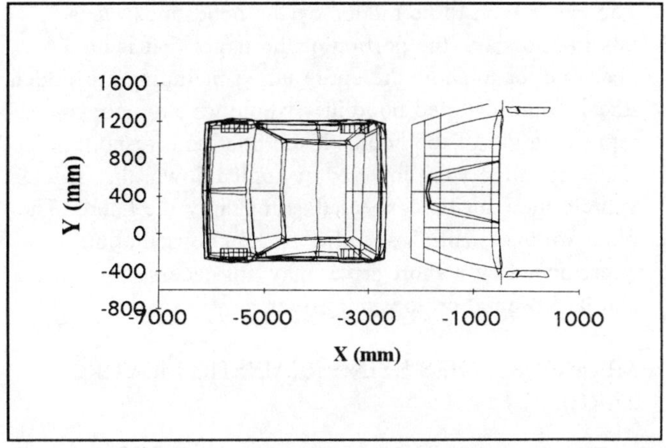

Figure 6 - Bird's-Eye View: Subcompact Forward of Truck

Representations of real three dimensional objects (e.g. a car) can also be used in polar plots. These targets can be used by themselves or in conjunction with target planes in vision analysis. When a design engineer sees the portion of a plane that is obstructed by some component of the vehicle, he or she is given measurable data that can be used to determine what targets are visible and what portions are not visible to the driver. To have more impact on feasibility or planning, a realistic target plotted in a polar plot can be used for demonstrations. Perhaps a small car can be hidden from the driver of a large truck by a raised hoodline (Figures 6 and 7). Figure 6 gives a bird's-eye view of the situation with the small car directly in front of the truck.

_ * Field of View From Automotive Vehicles, FORD Motor Company, SAE Report No. SP-381, pp. 17-48, May, 1973.

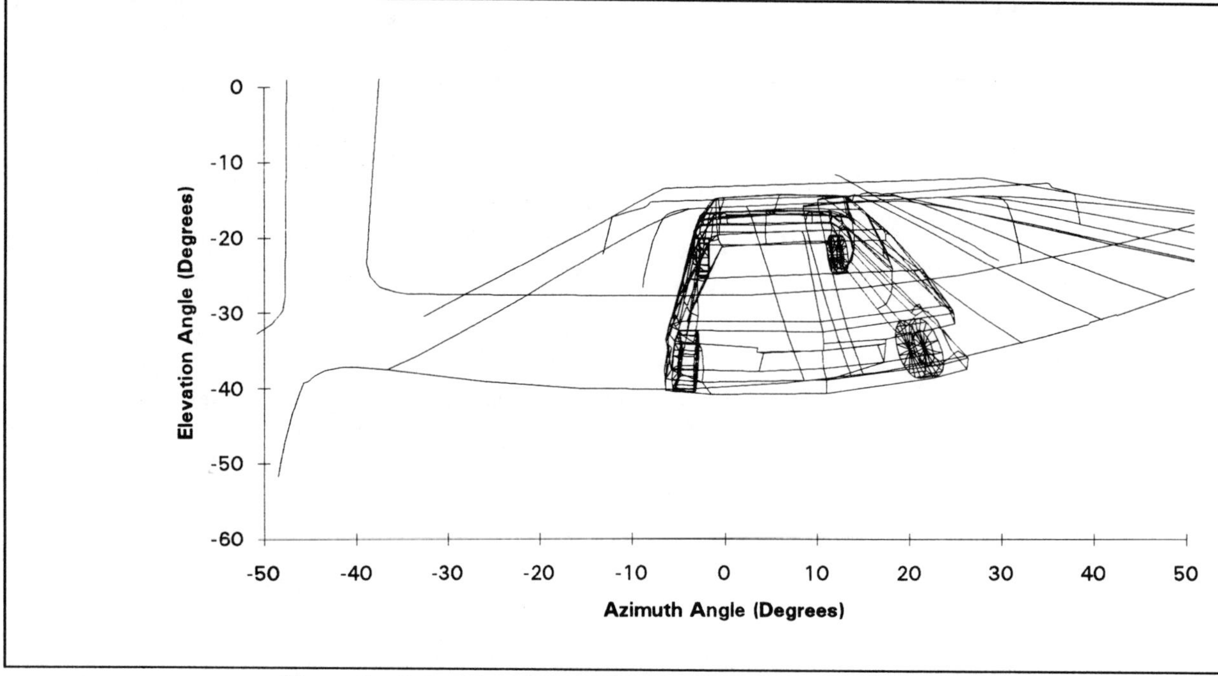

Figure 7 - Polar Plot of Truck Hood with Subcompact Hidden to Truck Driver

Figure 7 illustrates a polar plot of the same scenario. The car is seen to be hidden by the hood lines. It is possible to show the portion of the target that is not obscured, or to show the entire target including all hidden lines. More detailed hoodlines would give a more realistic representation of the hood. If the hood is foreshortened or if the hoodline were lowered or angled down, the target vehicle may not be concealed as much by the hood. The plots are eye position specific. A hood obscuration experienced by a short driver may not necessarily cause a similar obscuration to a tall driver.

MIRROR STUDIES USING REALISTIC MOVING TARGETS

Figure 8 presents a bird's-eye view of two subcompact automobiles approaching a third vehicle from behind in both adjacent lanes. The vehicles move forward, passing the vehicle in question until they disappear into the forward horizon. Recreating the same scenario with polar plots shows the view from the driver's eye point. For these plots, a single cyclopian eye point was used.

This type of scenario can be used to analyze mirror location, size and shape. The subcompacts can be moved forward in steps, perhaps 5 meters at a time, and successive polar plots can be created to show locations of the reflected images of the adjacent subcompacts while looking into a side view mirror. When the image of the target vehicle in a particular mirror is no longer visible, moving the target in smaller increments gives the exact location at which the image is no longer visible in the mirror being viewed. The polar plot can be used to determine if the target vehicles can be seen in another mirror or if the vehicle has come into the direct view of the driver. If not, the mirror can be relocated and the series of polar plots can be generated again. Additional mirrors can be provided to enable unobstructed tracking of the adjacent vehicle. This test can be repeated using drivers with different eye extremes (e.g. tall, short, leaning to the left or right, etc.) to determine an optimum mirror placement, shape, and size.

This process for one driver eye point is demonstrated in Figures 9 through 13. Figure 9 shows a vehicle with it's front bumper a little over 10.5 meters rearward of the driver's eye point. The images of the vehicle are present in both the planar mirror on top and the convex mirror below. The driver's side glass and vent window glass are also included. In Figure 10 the car has moved forward to the point where the image in the planar mirror is moving out of view. Figure 11 shows only what is visible to the driver. The image in the convex mirror at this point is visible. In Figure 12, the vehicle has come into the direct view of the driver, and is still visible in the convex mirror. Figure 13 shows the vehicle and convex reflection when the front bumper is perpendicular to the side of the vehicle at the same forward position as the driver's eye point. These plots are specific to this eye point and need to be repeated for all eye extreme points as well as the right passenger side mirrors.

Studies on the right hand side of heavy trucks can show where additional mirrors can be placed to track adjacent vehicles. Figure 14 shows the subcompact that is in the passenger side lane and the reflections in a planar mirror and two convex mirrors, one on the traditional side mount and one on the forward hood.

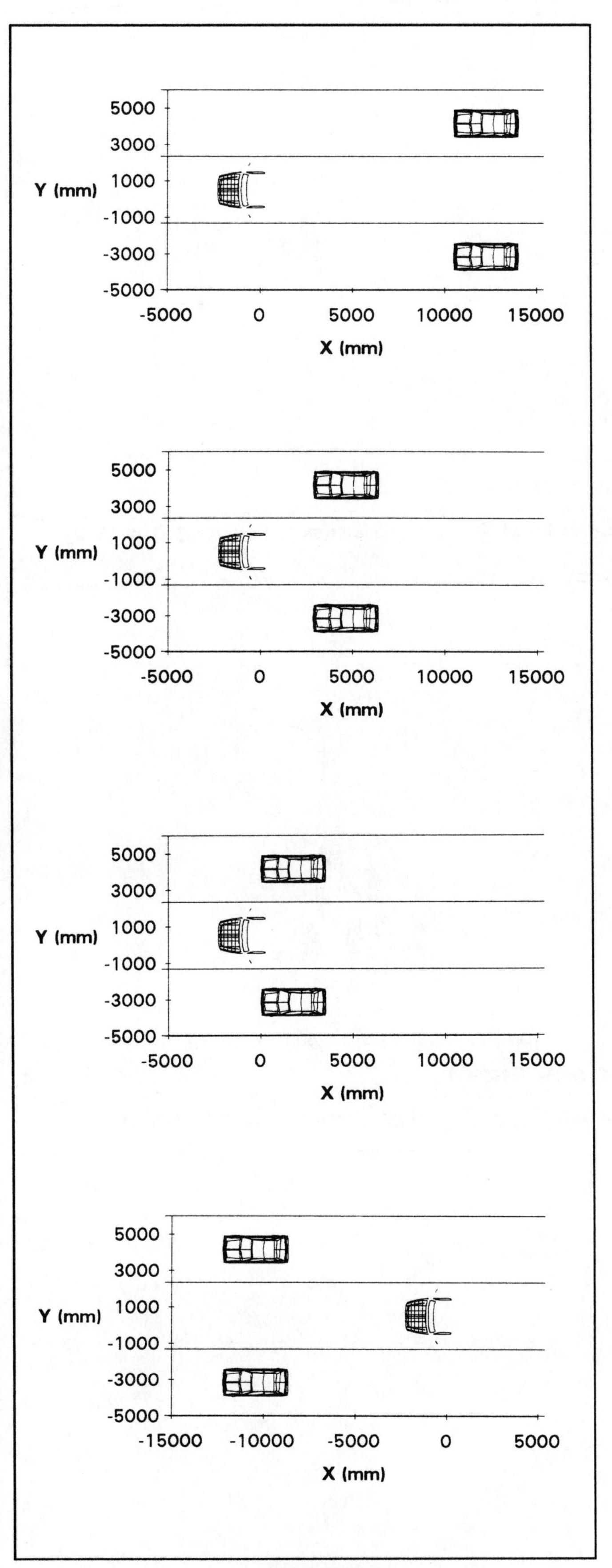

Figure 8
Bird's Eye View: Two Subcompacts in Adjacent Lanes Passing a Third Vehicle

APPLICATIONS AND TOPICS USING POLAR PLOTTING

DISTORTIONS AND QUALITY - The accuracy of polar plots in predicting driver's visual problems will depend upon the ability of the analyst to accurately represent the outlines of the objects as well as the data itself, especially the eye points.

The resulting shapes in polar plots are distorted, much like Mercator Projections in mapping distort the earth when the planet is mapped by the reverse process of converting angular measure to rectilinear. In the case of polar plots, the items that are further away are smaller, and the items closer to the driver subtend relatively larger angular size. Some distortion can also occur when straight lines as well as curves on an object are represented by too few points. As more points are used, the plot created takes on a more lifelike quality. Changing the scales also decreases or increases distortion of the figures in the plot.

HEADREST AND REARWARD VISION USING POLAR PLOTS - Polar plots can be used to create studies to determine if vision blockage occurs because of the headrests or other objects which may potentially obscure vision when backing up.

LOCATING CONTROLS AND DISPLAYS -Controls and displays should be located so that the driver can view them without any obscuration and with minimum eye and head movement. While looking directly at a detail, the driver can also acquire information with his peripheral vision without moving his/her head. That peripheral visibility region can be modeled using polar plots.

BINOCULAR POLAR PLOTS-The view created for a binocular polar plot contains images of the items of interest as seen from two different eye points. For example, these points may be the left and right eye positions of the average driver of the vehicle in question. They could also be the 95th percentile driver's left and right eye positions. The eye points used may be defined from the SAE Eyellipse Standard J941 or whatever position the designer is interested in. Perhaps there is a desire to show the difference between the vision capabilities between tall and short drivers for one vehicle design. This can be done on a single polar plot. Any combination of multiple eye points is possible.

The binocular polar plot is especially useful for looking at obscurations that might occur because of the forward or rear pillars. If a binocular plot is created, the size of the binocular angular obscuration caused by the pillar can be directly scaled from the plot to ascertain if it exceeds certain maximum limits. Models of the targets that are obscured can also be plotted to emphasize what is not visible to the driver. See Figures 15 and 16.

AMBINOCULAR POLAR PLOTS-The binocular plot also provides information for the ambinocular field. The ambinocular field is defined by determining what the driver can see with either eye. It is represented by only one outline image. It delineates that which is the periphery of the item of interest as seen by either eye, using whichever outline is the more exterior.

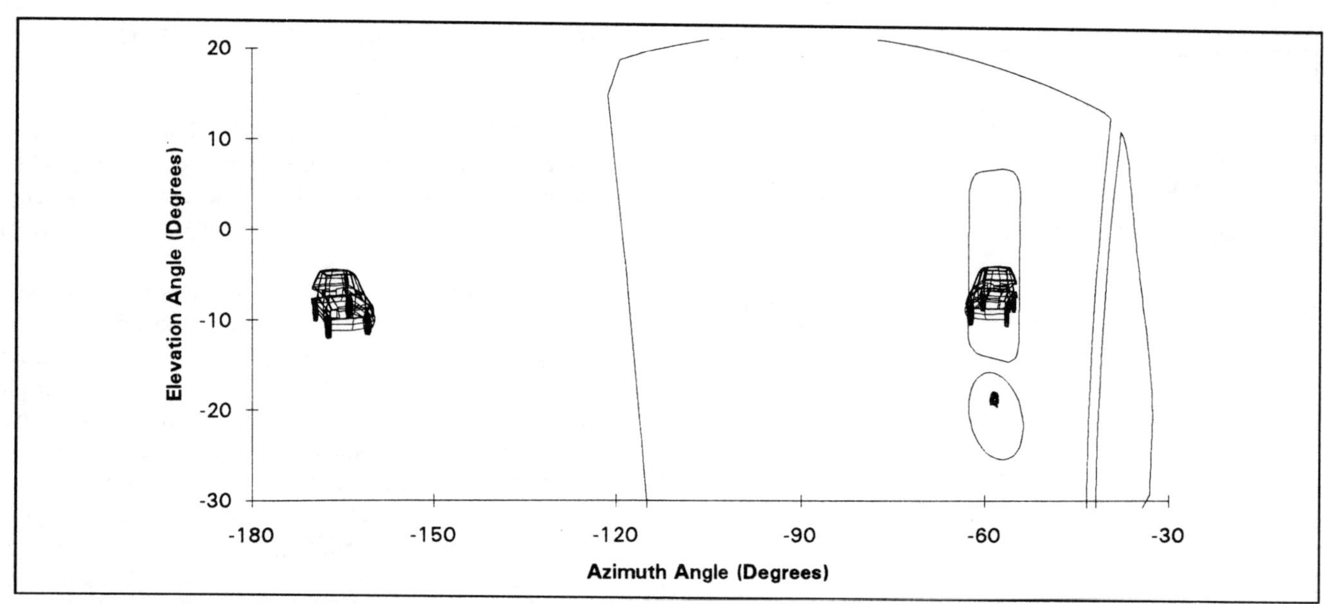

Figure 9 - Polar Plot of Subcompact in Driver's Side Lane: Front Bumper 10.5 meters Rearward of Driver's Eye

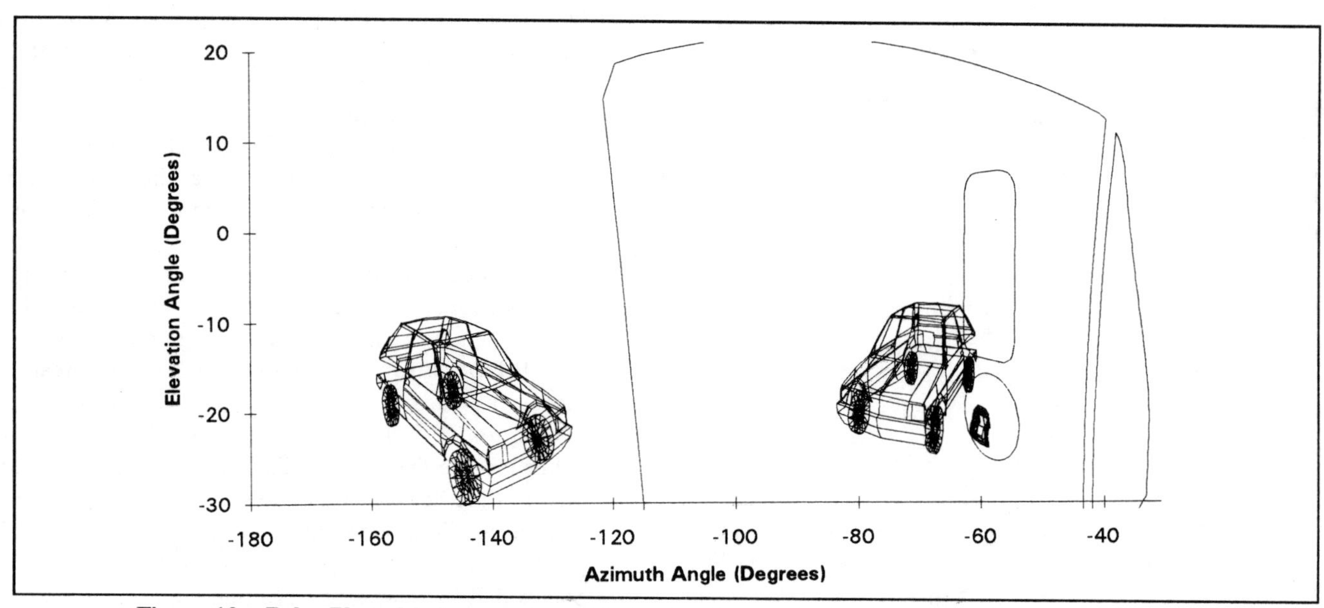

Figure 10 - Polar Plot of Subcompact and Reflection Images, Including Those Not Seen by Driver

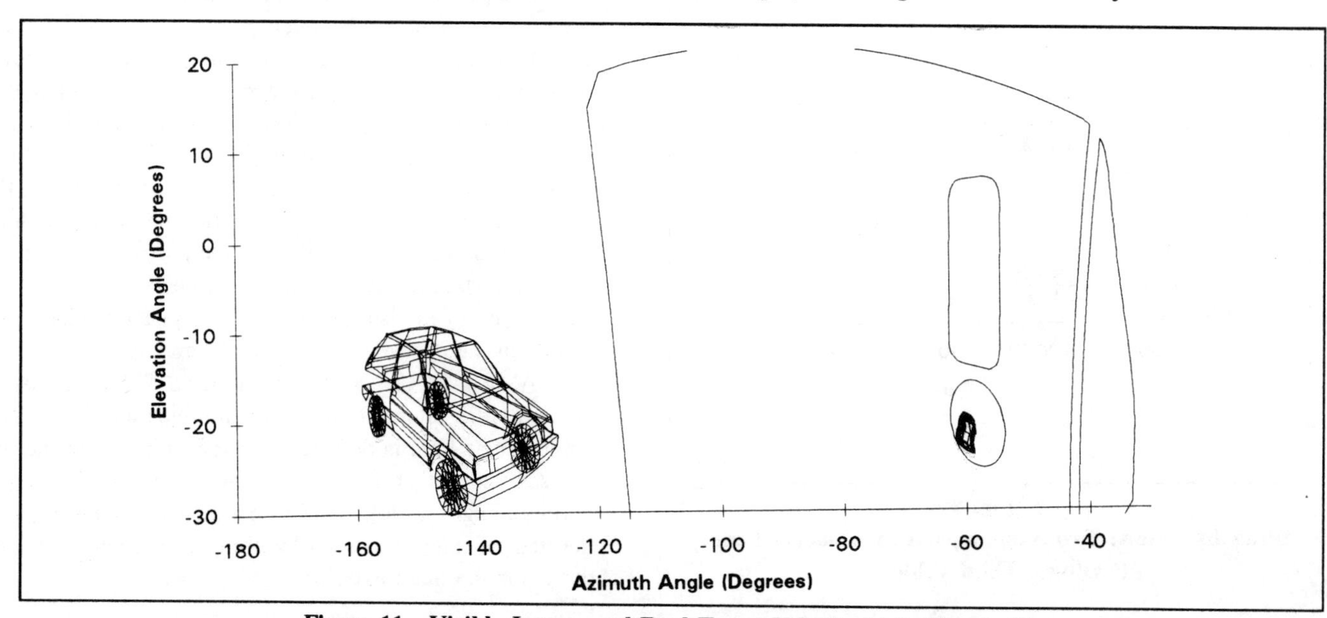

Figure 11 - Visible Images and Real Target Vehicle As In Figure 10.

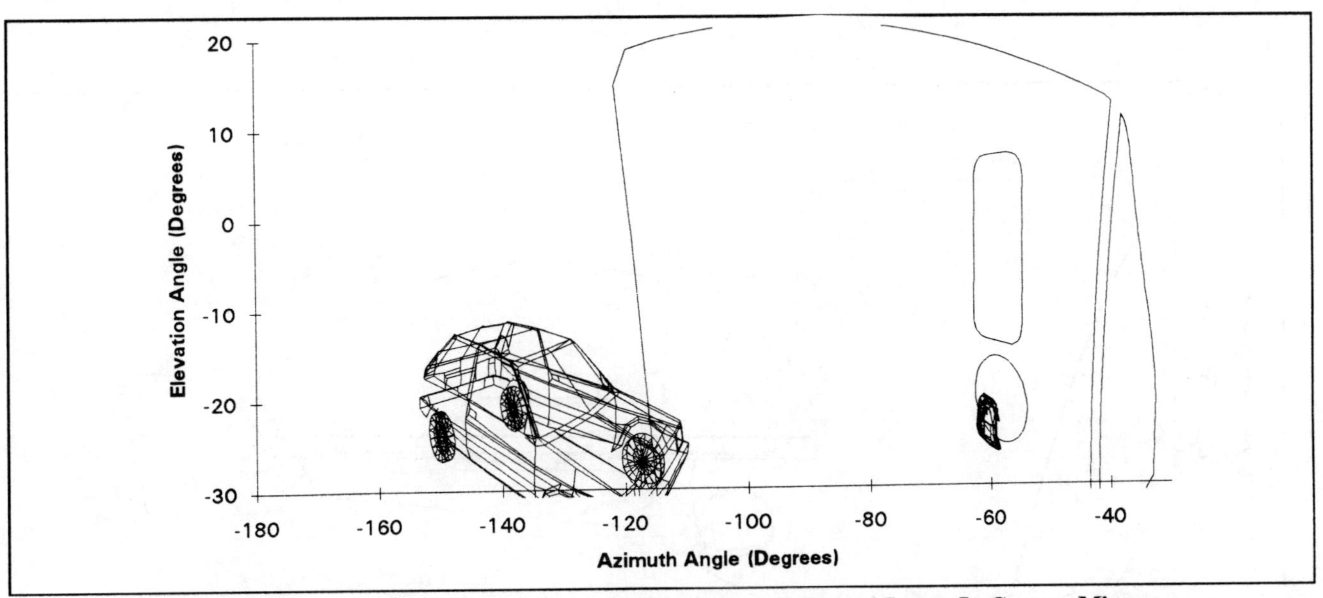

Figure 12 - Subcompact Entering Direct Field of View, Reflected Image In Convex Mirror.

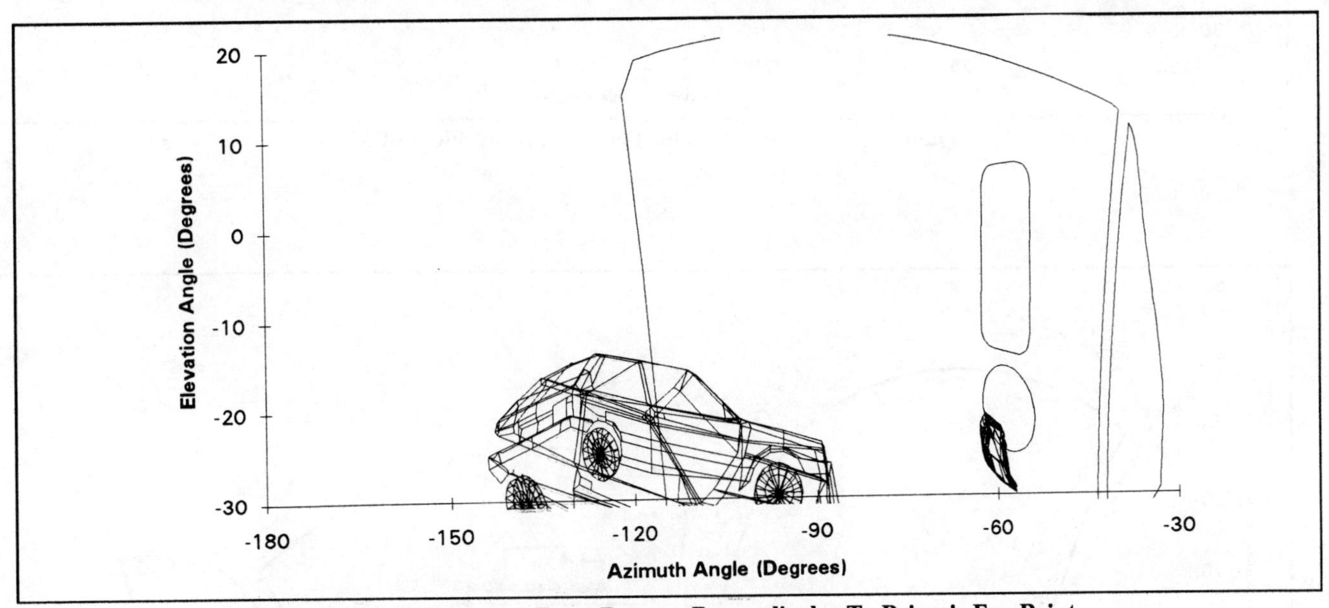

Figure 13 - Subcompact Front Bumper Perpendicular To Driver's Eye Point

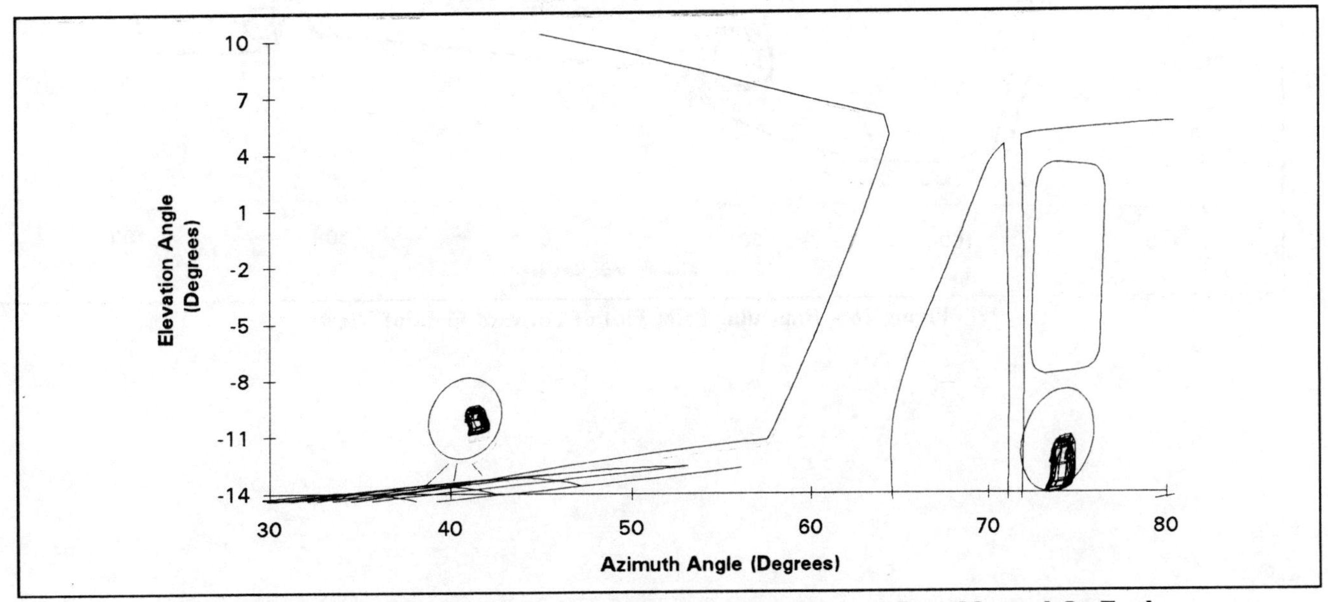

Figure 14 - Polar Plot of Passenger Side With Additional Convex Mirror Mounted On Fender

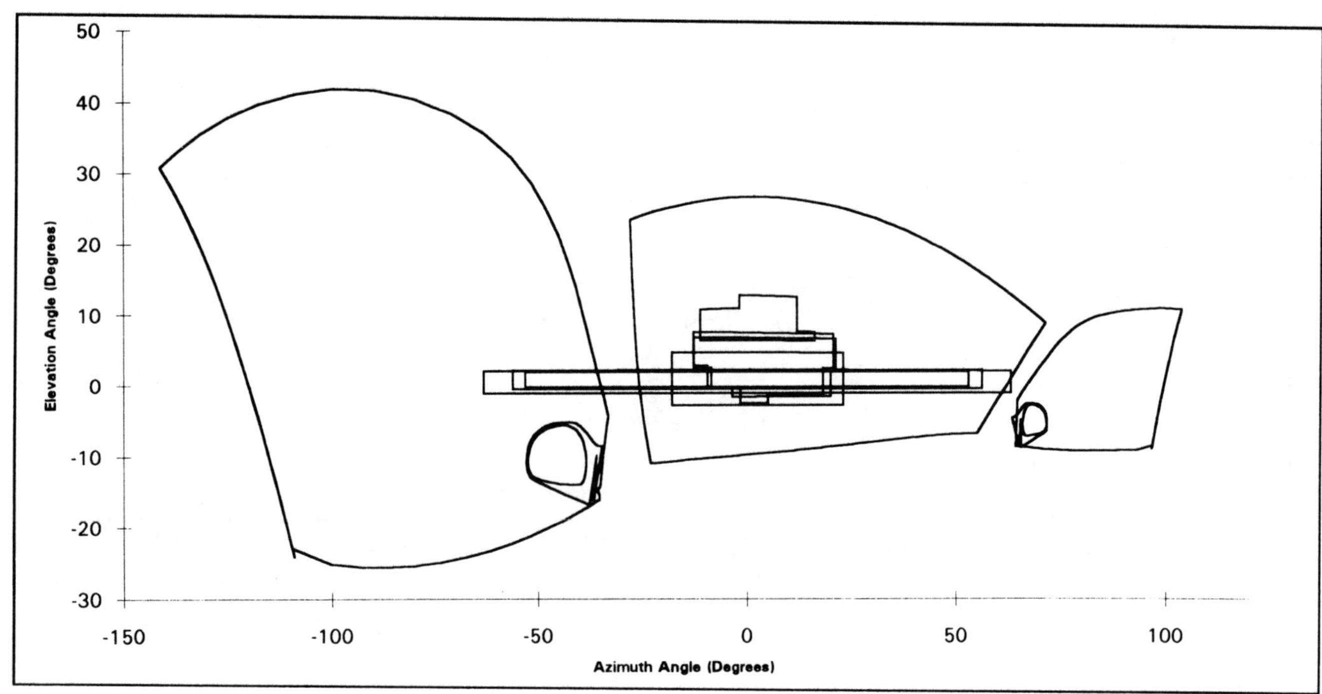

Figure 15 - Monocular Polar Plot of Forward Field of View

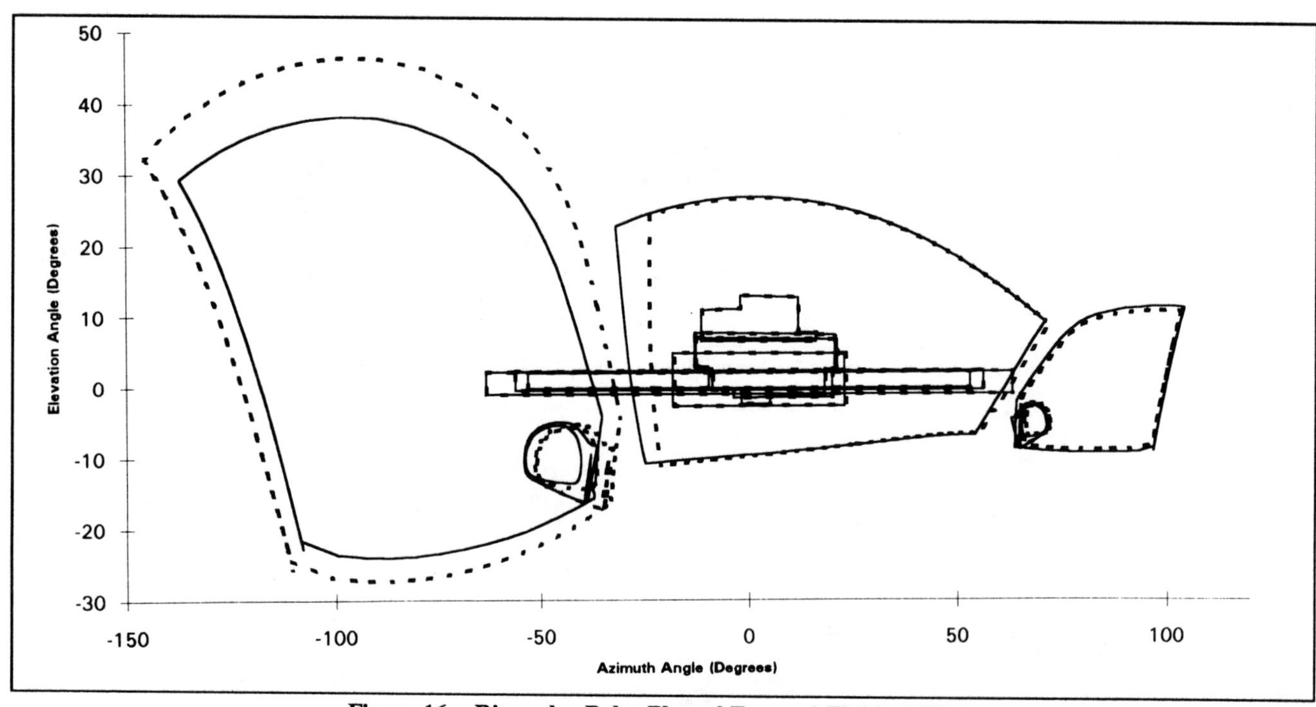

Figure 16 - Binocular Polar Plot of Forward Field of View

WIPER PATTERNS - Polar plots can be used to evaluate windshield wiper effectiveness. Plots can be made of the cleared areas from different eye points to determine the coverage area of the wiper patterns being analyzed.

VISION ISSUES RESULTING FROM DIFFERENT DRIVER SEATING POSITIONS - Evaluating the field of view of drivers with various seating locations defined by seat back angles and longitudinal (fore/aft) seat track location can be done using polar plots. Viewing the daylight openings and targets from the various eye positions can help to avoid obstructed field of view.

CONCLUSION

The paper illustrated some of the many possible applications of polar plots. We have found that polar plots are very effective in communicating complex three dimensional driver vision related issues to individuals with very different backgrounds involved in the vehicle design teams. We expect that in the future the use of polar plots will increase because of more wide spread use of computer graphics and need for faster design cycles.

ACKNOWLEDGEMENTS

The authors wish to thank Mr. Jerry Hubbell, Heavy Truck Engineering and Mr. Lyman Forbes for their support and valuable suggestions in developing the polar plotting software and in generating versions of applications illustrated in this paper.

REFERENCES

1. Society of Automotive Engineers, Inc., Describing and Measuring the Driver's Field of View - SAE J1050a Recommended Practice, July, 1977.

2. Society of Automotive Engineers, Inc., Motor Vehicle Driver's Eye Locations, SAE J941 Recommended Practice, June, 1992.

3. Society of Automotive Engineers, Inc., Vision Factors Considerations in Rear View Mirror Design - SAE J985 Recommended Practice, October, 1988.

4. Ford Motor Company, Field of View from Automotive Vehicles, Report No. SP-381, published by Society of Automotive Engineers, Inc., May, 1973.

950967

Systematic Development of a Complex MMI-Interface Shown by the Example of an Integrated Display and Information System

Ebner Roland and Spreitzer Wilhelm
Siemens AG

ABSTRACT

For the development of complex Man Machine Interfaces (MMI) we created an innovative development tool. With this system we gain significantly higher certainty in concept by earlier integration of the customer. Additional advantages are the reduction of development time and costs and an improvement of the MMI quality with regard to self-explaining operation for complex information systems.

INTRODUCTION

The importance of information, audio and communication devices in vehicles increases steadily, and people increasingly want to be entertained during the trips in their cars.

The reasons for these increased demands are low noise and more comfortable cars, which are being driven at lower speeds due to increasing traffic density. Recently developed infrastructures have become the basis for a broad spectrum of functions offered, such as car telephones, data transmission from and to the car (GSM networks) and automatic toll collection.

The current traffic situation calls for in-vehicle solutions that help people reach their destination on the safest and quickest way possible (navigation, traffic management systems).

This requires a multitude of service and information functions in the vehicle which no longer can be integrated as independent single solutions. On the contrary, the tendency is moving toward integrated and linked solutions, ergonomic control concepts and new intelligent displays.

The cross-linking of audio, communication and information devices makes it necessary to have the individual components developed in parallel processes, in order to bring them together in integrated display and control units; always considering the customers' demands, such as utmost product quality and excellent design with a reliable and simple operation (Man Machine Interface: MMI)

The typical development process of iteratively implementing and testing during the realization phase is not recommendable for complex systems. Here, a new kind of development procedure has become necessary, that complies with the demands of the automotive industry for continuous cost reduction and shorter development times.

REALIZATION

DEVELOPMENT PROCESS - The development process is basically composed of three phases: design phase, realization phase and introduction phase (figure 1). For complex systems the design phase is of considerable importance. Only an excellent design with a low risk potential leads to a successful reduction in development costs and realization times.

A convincing design, and really "experiencing" the new product, can only be reached by a simulation of the MMI. The product quality is considerably improved by the dynamically supported design phase, since fewer functional and MMI related design changes become necessary in the software, and, especially in the hardware development.

The development process within the design phase (figure 2) is basically composed of the elaboration and simulation of design, display and control alternatives, in collaboration with professional designers/stylists and ergonomic experts. After the successful decision on design, the accomplishment of an acceptance test, and, following any customer requested MMI redesign in simulation, the specification and software structure, which are the basis for the serial development, are then established automatically.

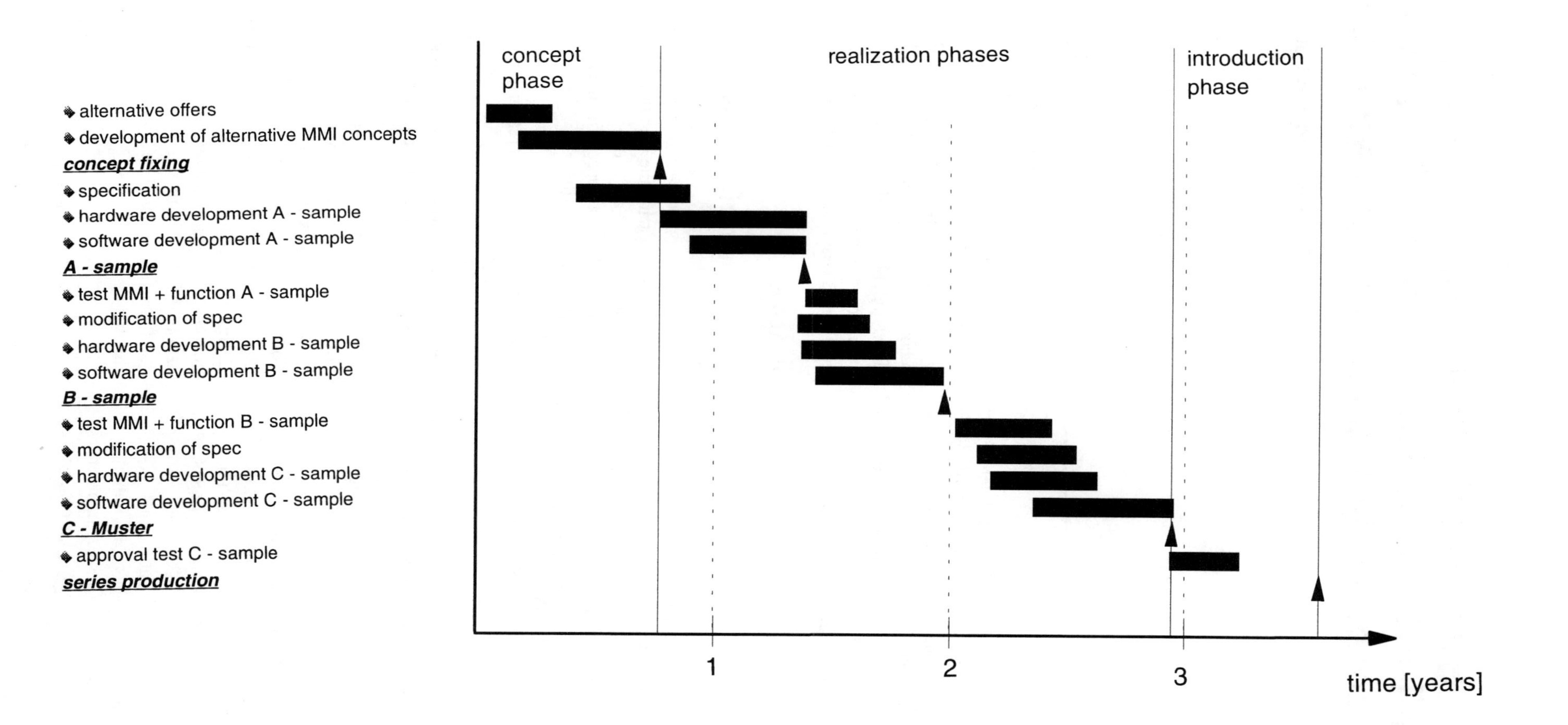

Figure 1: *Development of display systems - timing chart*

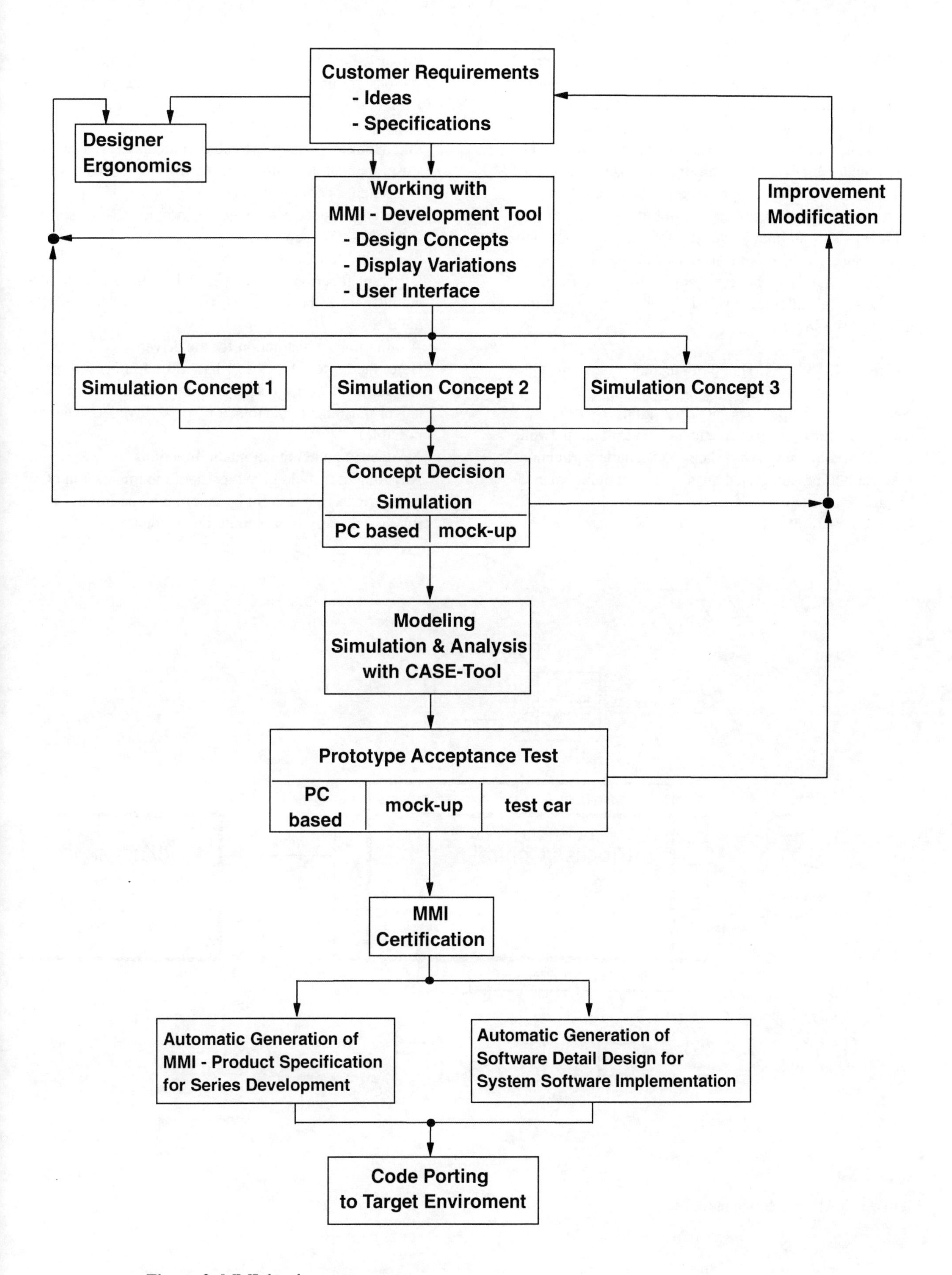

Figure 2: MMI development process

DEVELOPMENT TOOL (FIGURE 3) - The appropriate tool for the design phase is a program for visualisation and animation, which offers a photorealistic, 3-dimensional illustration of the display and the operation elements, as well as a very exact simulation of the functions using a touch-screen. Through the use of a CASE Tool, the decision for design, specifications, and software structures which are necessary for volume production, are facilitated. The automatic transfer of this data not only permits shorter development periods, but also avoids transcription errors (interpretation of specification). Furthermore, an integrated CASE-Tool offers a reliability analysis of the MMI.

SPECIFICATION - The following conditions are to be accomplished by an MMI in a vehicle:

- Self-explaining functions and illustrations
  Each function and each menu is self-explaining while the illustration may vary, though. Secondary functions that cannot be integrated in a self-explaining manner should be omitted.
- Touch operation
  Primary functions, such as volume, selection of radio stations or answering a phone call must be operable without eye contact.
- Suitability for rental cars
  The operation must be possible immediately and without a user manual; a minimum of hardkeys is necessary.
- Consistent operation concept
  Only one structure level to be provided, e.g. softkey, joystick, shaft encoder
- No double functions
  Different functions to be operated by the same hardkey would neither be understood nor accepted by most of the users.
- A minimum of distraction for the driver
  Here, the correct letter size has to be found for ergonomic reasons. Too much illustration and useless additional or flashing indicators are to be avoided wherever possible.
- No decision making automatic functions
  Anything that could give the driver the impression of being treated like a child by automatic operations or functional restrictions should be avoided.

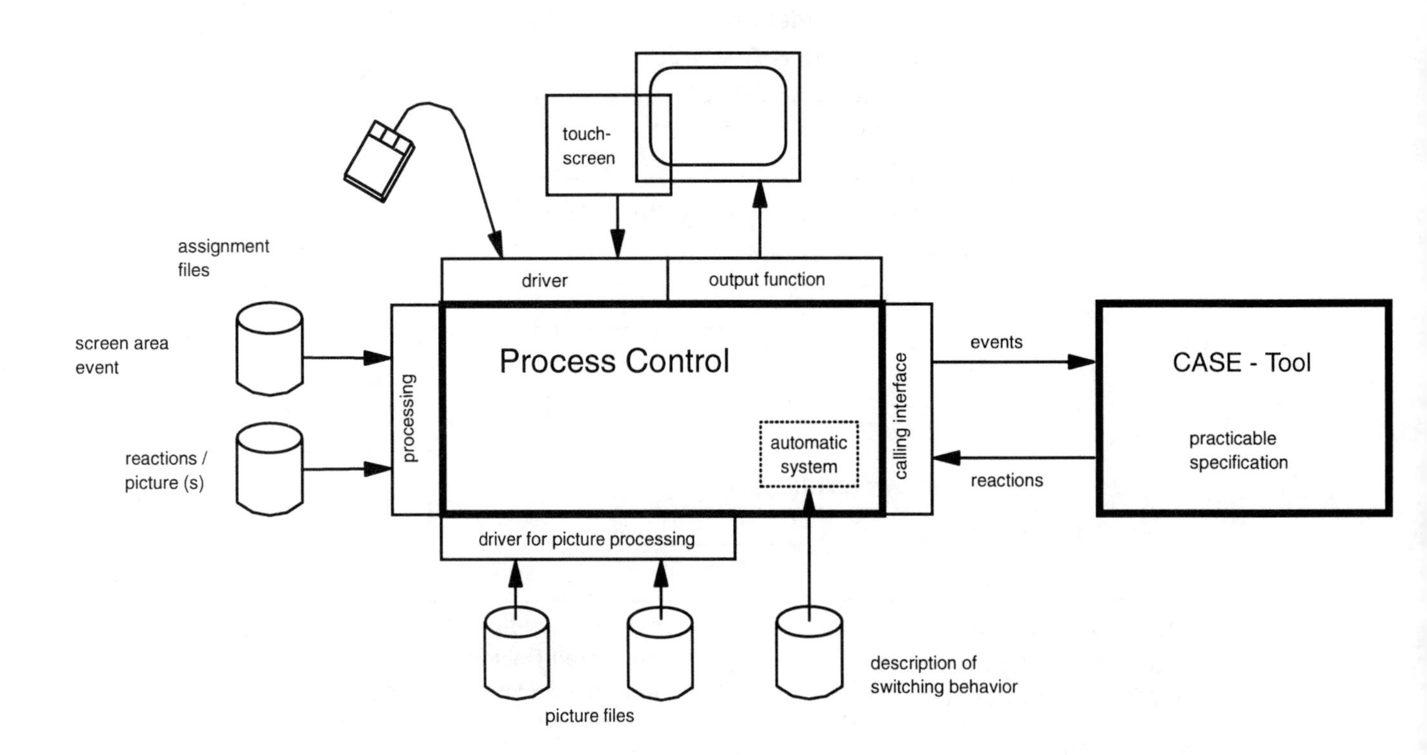

Figure 3: MMI development tool

POSITION IN THE VEHICLE

Three different installation alternatives have been examined, considering aspects as ergonomy, operationability and component integration.

SCREEN WITHIN DASHBOARD -

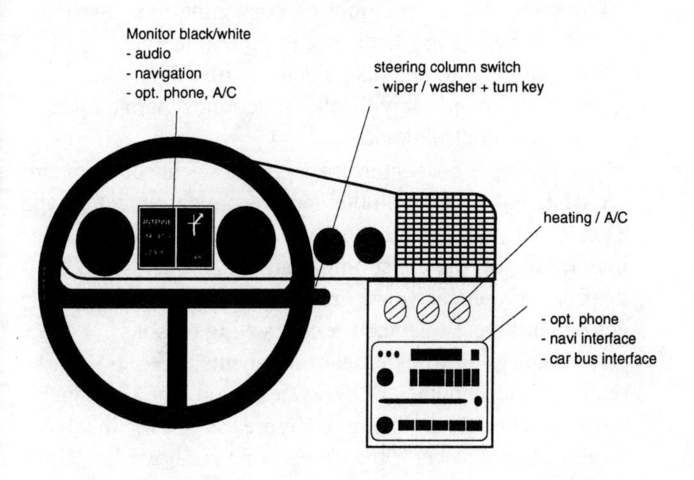

Figure 4: Installation position - screen within dashboard

The screen functions are operated by a switch that could, for example, be integrated in the lever operating the wind screen wiper, by keys on external components (audio, phone, etc) or, alternatively, by means of a remote control located at the steering wheel.

This concept offers several advantages, such as good view for the driver, ideal location for Active-Matrix-LCD regarding angle of view and anti reflection as well as the unproblematic integration of drives (audio + CD-Rom) in the external components.

There is, however, one disadvantage; the display is only partially useable for the front-seat passenger. Additional mounting space and display components become necessary in order to make an operation by the front seat passenger possible, too.

The necessary spatial connection between display and operational function is not possible in this example. The turning or pushing direction at the steering column cannot unambiguously be coordinated with the function selection from the point of view of ergonomy.

Depending on the selected wiping-mode, the operational element is situated in different positions. When turning the steering wheel, the display may be partially blocked. The multitude of variations of the dashboard will continue to increase with this type of solution.

SCREEN NEXT TO DASHBOARD -

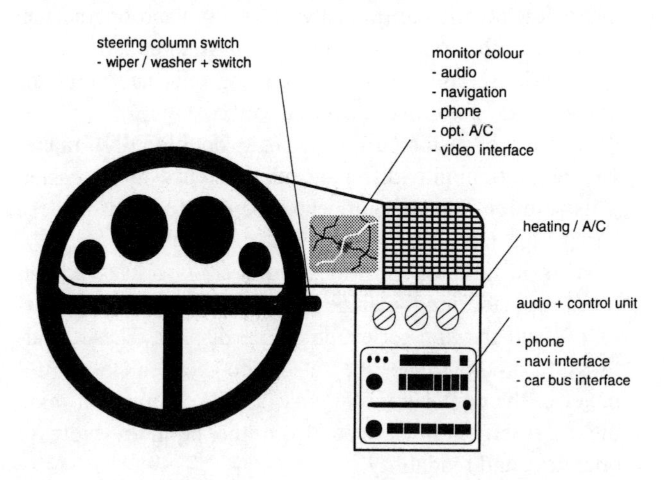

Figure 5: Installation position - screen next to dashboard

The operational devices are preferably located in the center console area. This installation position offers a perfect view for the driver and the front seat passenger as well. Both can operate the functions.

The installation position of the display is convenient for the Active-Matrix-LCD, as far as the angle of view and anti reflection is concerned. The integration of drives into external components as well as the access to necessary data carriers is possible without any problems. This solution, however, does not enlarge the variety spectrum of the dashboard.

One disadvantage for the user, though is the greater distance between display and its operation controls. The integration of the operational devices and the drive components for audio and navigation in an existing DIN cutout can only be realized with difficulties due to ergonomical reasons.

CENTER CONSOLE -

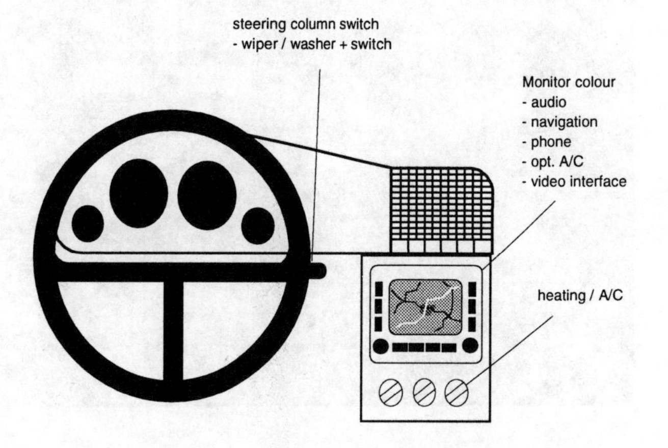

Figure 6: Installation position - center console

All operational functions are integrated. This concept offers a local connection between display and operational devices. Both, driver and front seat passenger can operate all functions. The components can be integrated into the display and operational device at low cost.

The use of a standard device (e.g. double DIN), makes the integration into different vehicle types much easier. This solution does not influence the development and the variety spectrum of the dashboard.

This installation position, however, is not as advantegous considering the angle of view of the Active-Matrix-LCD's and the anti-reflection of the display. These kinds of systems should be integrated into the vehicle control panel in the center console area; always keeping in mind the most ergonomical spot from the point of view of operation and readablity.

EXAMPLE

The systematic MMI development was initially used when developing an integrated driver information system called ZAB.

From the basically equal alternatives Fig. 5 and Fig. 6, the customer selected the middle console as the installation position for the realization. Figure 7 shows a photo-realistic simulation of this system.

The following functions have been realized:
- Audio (RDS Radio with DSP and CD changer control)
- GSM telephone (incl. digital hand free telephoning)
- Trip computer (calculations on consumption and speed, clock, timer, range distance and time of arrival)
- Navigation (autonomous guidance with GSP)
- Air conditioning (only display function; separate operational and control device)

Assuming a BUS concept as a basis - all components are designed as autonomous modules and connected with each other by a BUS suitable for vehicles - only the integration of the most important components, such as display, graphical unit, radio and telephone (basic equipment) were selected for cost saving reasons.

A connection to external components (options) can be realized via vehicle BUS systems on the customer's request. The block diagram (figure 8) shows the basic design of this system; the design of the above described ZAB is shown in figure 9.

Figure 7: MMI simulation - ZAB

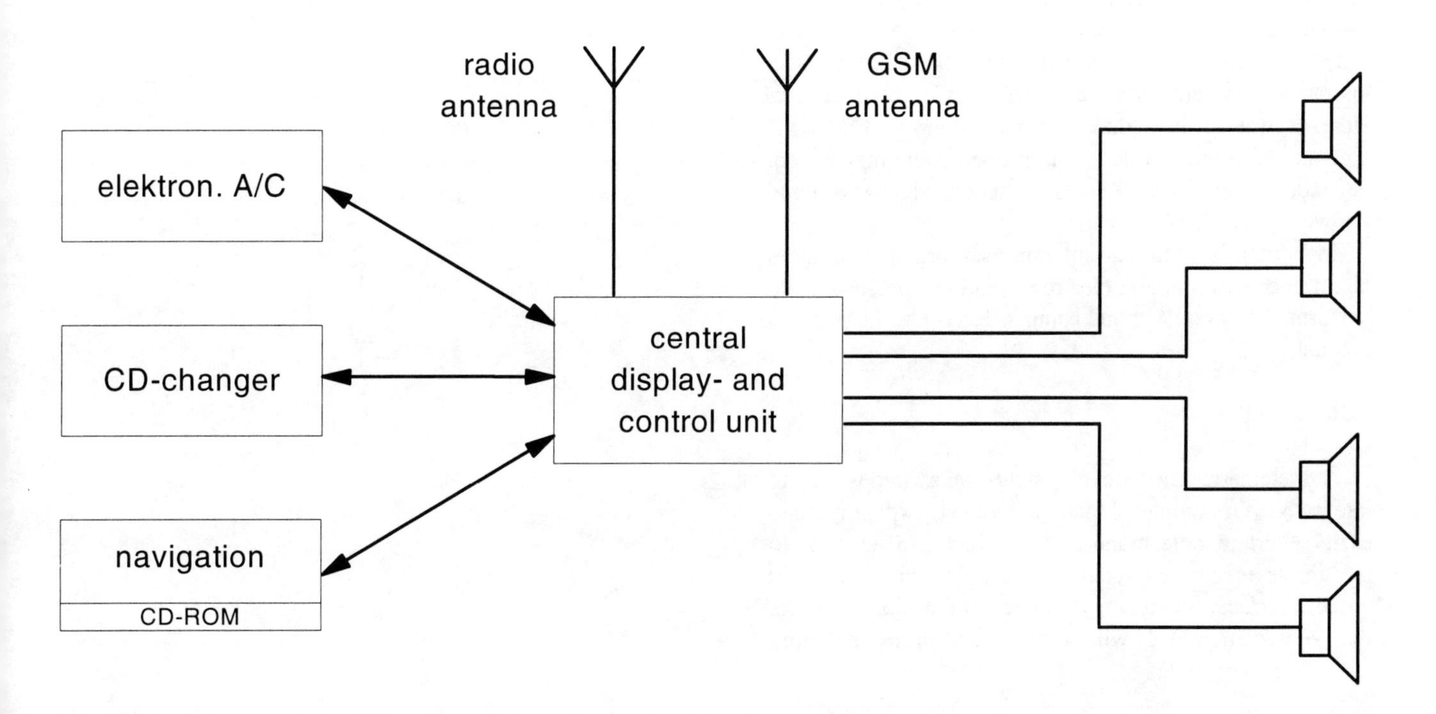

Figure 8: Block diagram ZAB - system

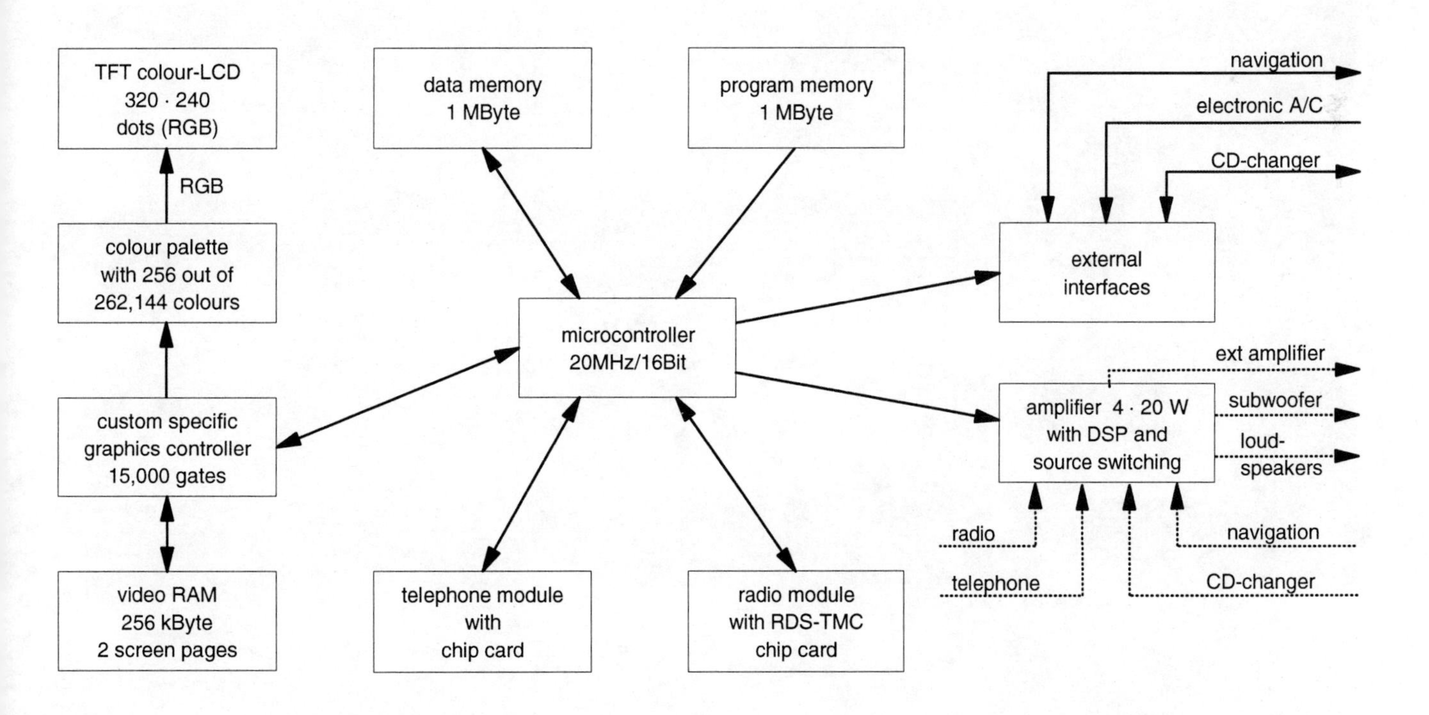

Figure 9: Block diagram ZAB - central unit

The man-machine interface was tested by 60 people, who had not known the system before, by a Prototype Automobile Cockpit (mock-up). The people were to solve tasks from all the above mentioned functional groups; in order to determine the time needed for the successful operation, the persons were recorded by a video camera. Here, it was proven, that with an optimized MMI, the amount of time needed for proper handling is not dependent on age, sex and technical background knowledge.

In comparison to current non-user oriented surfaces e.g. in the entertainment electronics, all test persons could perform the tasks without being offered any help nor a manual.

FORECAST

A systematic MMI development makes it possible to integrate even complex functions in a self-explaining and easily operationable manner in a vehicle. In addition to the above mentioned functions, further options such as video interface, TV tuner, distance warning, parking aid, rear view camera, etc. will be integrated in the operating and display concept.

950968

Using the Safety State Model to Measure Driver Performance

Edward J. Lanzilotta
Massachusetts Institute of Technology

ABSTRACT

Measurement of driver performance, with regard to safety, has traditionally posed great difficulty. While safety is often discussed in terms of risk probabilities, measurement of risk probability is hampered by several factors. First, accidents are relatively rare occurrences. Furthermore, the identifying event (the accident) occurs at the end of the time period which contains the precipitating events. As a result, it is quite difficult to monitor for potential accident scenarios. In addition, accidents are most typically the result of several compounding factors, which makes determination of causality very difficult.

The safety state network is a probabilistic model which captures the behavior of a system. Based on a finite Markov network, the safety state network models the human/machine and human/human interactions in a transportation system and forms a framework for capturing and comparing the probabilistic decision patterns of control elements in a transportation system. This paper discusses the theoretical basis of the safety state model, and applications to measuring driver performance.

INTRODUCTION

A remarkable trend in the current automobile industry is the overwhelming emphasis on the area of safety. A key component of safety in the human-machine system is the human driver. As the primary controller of the vehicle, the driver must perceive the state of the surrounding environment (which includes other vehicles, as well as roadside elements), decide the proper control strategy, and actuate the controls to carry out that strategy. In short, the driver is solely responsible for the state of the vehicle. Furthermore, the actions of the driver play a significant role in the effectiveness of any designed-in safety device or system. By altering the applied control strategy, a driver can mitigate or override the effects of a designed-in safety system. This sometimes occurs because the driver is not well-trained in operation of the safety system, while at other times it is the result of greater risk-taking because of a lower perceived risk on the part of the driver.

Let us pause for a moment to reflect on the use of the term "control strategy." In this work, the driver-vehicle system is modeled as a closed-loop control system [1][2][3], as shown in figure 1. In traditional control system engineering, the controller senses the state of the system (commonly known as the "plant"), and applies input to the system, based on a prescribed set of rules (the "control law"). In this case, the "plant" is the vehicle, which is a machine and will behave in response to its control inputs, and the "controller" is the human driver. The driver senses the state of the vehicle using sight, sound, and touch. He/she then provides input to the vehicle via the accelerator, brake, and steering controls. The "rules" used by the driver are the driving skills obtained through training and experience. These skills represent the "control strategy."

Note that any of the "components" of the control system can be replaced with computer-based automated systems. The advantages of automation, versus human control, is the precision and consistency of performance. Computers are well-suited to tasks which must be performed repetitively and with great accuracy. However, humans have a distinct advantage in two areas: flexibility and creativeness. Human controllers are able to respond and adapt quickly to widely varying circumstances. In addition, humans are able to synthesize a response to a totally new situation, based on past experience with different but similar situations.

Based on this system model, it is logical that extensive skills training is essential for the task of driving, to take best advantage of the strengths of human-based controllers. Unfortunately, the status of driver education and evaluation has not improved much in the past fifty years or so. Part of the problem lies in the difficulty of measuring driver performance, especially with respect to safety-related decision behavior. Current driver evaluation methods focus on the perception and actuation sub-tasks, simply because these task skills are observable and testable (e.g., through eye exams and simplistic road tests). However, it is the driver's decision-making skills that play a significant role in governing the interaction between a vehicle and its environment (which includes other vehicles in the system). These interactions ultimately determine the safety of the driver's actions.

This research is focused on modeling system behavior in ground transportation systems. In particular, we are interested in evaluating the decision behavior of vehicle drivers. This behavior is evaluated in terms of risk probability, as a function of the state of both the vehicle and its environment during operation.

In this research, we apply a probabilistic model onto the structural paradigm of fault tree analysis. The resulting model, known as the *safety state model*, provides a framework for tracking system behavior and driver decision behavior as a function of time. The resultant safety state trajectory allows direct determination of the time relation between events which precipitate an accident event, aiding in the identification of causality. In addition, the safety state model provides a mechanism for directly comparing the decision behavior of an individual driver to that of other drivers.

A DISCUSSION OF SAFETY AND RISK ASSESSMENT

While "safety" is broadly defined as "freedom from risk" [4], this represents an ideal (and realistically unachievable) goal. While complete elimination of risk is not achievable, reduction of risk is possible, although at a cost. Therefore, it is more appropriate to discuss the "pursuit of safety," in which the level of risk is traded off against the costs of reducing risk.

Lowrance [5] defines safety as the "judgment of acceptability of risk," with risk defined as the probability of some undesirable event. This definition separates safety into a subjective component and an objective component. The subjective component, which is the judgment of acceptability of a risk, evaluates whether a given level of risk is acceptable to the society which is affected. Based on that judgment, policies are set, which determine the trade-off between a level of risk and the effort (i.e., money) spent to reduce that level of risk. Risk judgment is typically performed by legislators, lawyers, and economists. The objective component, known as risk assessment, expresses the probability of an undesirable outcome. This expression must be made in a form that is useful for comparison among different scenarios. (For example, in transportation systems, the risk of death might be expressed in terms of fatalities per million vehicle miles.) Risk assessment is typically performed by scientists and systems analysts. It is recognized here that the quality of the subjective component of safety can be no better than the quality of risk assessment, as the former is based on the latter. This research concentrates on the objective component, risk assessment.

In the case of transportation safety, risk assessment is most often expressed in terms of fatalities, personal injury, and property damage. However, these are almost always the result of an intermediate event, known as an accident. Therefore, we can further divide risk assessment into two sub-components, which are the probability and outcome of the intermediate event (i.e., the accident). The probability of intermediate event occurrence estimates the relative likelihood that such an event will occur. The event outcome estimates the ultimate outcome of the intermediate event, in the terms of interest (i.e., injuries and deaths), for a given set of conditions regarding the intermediate event. (Because an accident represents the inability for a transportation system to meet its goals, it can be considered a system failure. The terms "accident" and "failure" will be used interchangeably throughout this paper.) These components of risk assessment parallel the concepts of "active safety" and "passive safety" devices, respectively. "Active safety" is a term typically applied to those devices or systems which assist in preventing accidents (such as ABS and traction control), while "passive safety" devices are those which reduce the severity of an accident when it does occur (such as airbags and door guard beams).

Risk probability, especially in transportation systems, is not a static quantity. Instead, risk probability varies as a function of the state of the system. The system includes the human operators, the machines being controlled, and the environment in which those machines operate. Failure scenarios are typically the result of compounding several individual failures. Some of these may be driver errors [6], while others may be machinery failures in vehicle or wayside equipment. The collected set of potential failure conditions can be considered as a system state, with regard to safety.

Time is an integral component of risk probability. Probability theory models the relative likelihood of the occurrence of an event. In this discussion, we consider the event of interest to be an accident. However, the risk probability of this event only makes sense if we compare its occurrence to the alternative event, which is the non-occurrence of an accident. Since nothing "happens" during the non-occurrence, it can only be considered with respect to time. Therefore, the probability of an accident must be expressed with reference to some time frame. A fixed time period is selected as a basis for measurement, and is referred to as a *time slice*. The risk probability is expressed in terms of the percentage of time slices that result in an accident. An alternate form of expression is in terms of the mean time between occurrence of accidents. This form is commonly known as the mean time between failures (MTBF), and is used extensively in the field of reliability engineering.

Taking a fatalistic viewpoint, a truly probabilistic event (i.e., an event that is well-modeled by probability theory) will eventually occur, given enough time [7]. The only way to avoid the occurrence of such a probabilistic event is to "get out of the game" before that event occurs. (In fact, this is what happens to most of us with regard to rare catastrophic events.) This is the basis of risk exposure—given a constant risk probability, the expected number of failures over a prescribed period rises with the size of the period. The colloquial notion of risk exposure is that the "laws of probability catch up with you." While this presents a convenient scapegoat, in fact the relationship is reversed—the occurrence of the accident provides statistical data to validate the probabilistic model.

THE SAFETY STATE MODEL

The safety state model is an extension of the more familiar event tree and fault tree models. An event tree is a representation of possible scenarios that can occur from a fault-precipitating event [8]. A fault tree, by contrast, works backward from a system failure to identify the logical combination of all of the potential causes of that failure [9]. The safety state model has been inspired by these methods of system safety analysis, and represents a step forward in generality.

EVENT TREE ANALYSIS — Event tree analysis is a method which is commonly used for human reliability analysis (HRA). The goal of the method is to identify the probability of system failure from the occurrence of a precipitating event. Along the way, it is possible to identify points in the failure process where human reliability is problematic, and use that knowledge to suggest improvements to the procedures and/or equipment.

The analysis starts at the precipitating event. From the occurrence of that event, tree limbs are constructed to all of the possible next events, along with the probabilities of occurrence of those next events. Then, from each of the next events, tree limbs are constructed for subsequent events, with associated probabilities. Once the tree has been completed, the overall probability of each possible event path can be calculated. A brief example of an event tree is shown in figure 2.

In the process, a complex failure process has been transformed into a collection of probabilities of simpler failures. In general, it is much easier to determine these simpler probabilities. Swain and Guttman [8] have developed a database of human failure probabilities for these simple actions. In the case of machine failures, standard reliability data may be used. The event tree is then used to evaluate the more complex combination of probabilities to arrive at the overall probability of failure.

Event tree analysis is especially well suited for systems which are procedural by nature, as it effectively measures deviation from an ordered sequence of events. In this regard, it has proven very useful in the nuclear power industry. However, in the case of a system which has many potential independent contributing events, event tree analysis becomes unwieldy, as the tree must include every alternative events from each node. The requirement for event independence is the greatest weakness of event tree analysis when applied to transportation systems.

FAULT TREE ANALYSIS — Fault tree analysis is another method commonly used in human reliability analysis. Where event tree analysis is considered a "forward-looking" approach (i.e., the analysis begins at the precipitating event and moves forward in time), fault tree analysis is considered "backward-looking." The analysis begins with the occurrence of the failure, and works backward to identify the combinations of contributing factors.

An important feature of fault tree analysis is the logical combinations of preceding conditions. Through the use of the logical "and" and "or" operators (which correspond to intersection and union in set theory), the fault tree describes the combinations of precursor faults which lead to a system failure. In some cases, probabilities are assigned to the various failure events, which allow determination of the probability of a failure path. A simple example of a fault tree is shown in figure 3.

As with event tree analysis, the greatest weakness of the fault tree is coverage — in the case where contributing events are independent, the analysis becomes quite cumbersome. Fault tree analysis does, however, provide a very useful role in safety state analysis by providing a qualitative method for identifying the most significant contributing conditions.

STRUCTURE OF THE SAFETY STATE MODEL — The safety state model represents an extension of event tree and fault tree analysis. By assuming that the conditions contributing to system failure are truly independent, the safety state model generalizes these techniques.

Consider a collection of n conditions which could possibly contribute to an accident scenario. These conditions include actions taken by the driver (such as acceleration or braking), the state of another vehicle in the system (such as vehicle ahead braking), and the state of the system environment (such as a red traffic light ahead). These conditions are constrained to be binary conditions. That is, the condition is defined such that it has only two possible values. The complete set of possible combinations of such a set of conditions can be represented by a binary word which is n bits long. The total number of possible combinations is 2^n.

Let us now consider each of the possible combinations (i.e., each number in the set of possibilities) to represent a state in a Markov network. These states are identified by the associated number, which is within a range of 0 to $(2^n - 1)$, inclusive. The accident scenario is identified as an additional state, labeled 2^n. The state number is termed the *safety state* of the system, and the resultant Markov network is known as the *safety state network*. An example of a three condition safety state network is shown in figure 4.

The state transition of the Markov network is defined to occur at regular time intervals, with the time period of the interval fixed at h. The value of h is set such that only one condition may change its state (within reasonable probabilistic bounds). At each state transition "instant" (i.e., at the end of each state transition interval), the model will transition from the current state $S(i)$ to the next state $S(j)$ with probability p_{i_j}. The probability p_{i_i} represents the holding probability for the state $S(i)$, which is the probability that the state will not change at the next transition. The collection of transition probabilities for a given state $(p_{i\_j}, j = 0,1,2,\cdots,2^n)$ represents a probability distribution, and the sum of these probabilities must be one (equation 1).

$$\sum_{j=0}^{2^n} p_{i\_j} = 1 \qquad (1)$$

The safety state corresponding to the accident scenario (state $S(2^n)$) is a trapping state, which means that, once entered, the process can never exit that state. This notion correlates with the reality that the occurrence of an accident is permanent and cannot be "undone." As a trapping state, the accident state has a holding probability of one, and the probability of transition to any other state is zero.

The collection of probability distributions for the entire set of possible safety states can be represented in matrix form (equation 2). The row of this state transition matrix (P) represents the current state, while the column number represents the next state.

$$P = \begin{bmatrix} p_{0\_0} & p_{0\_1} & \cdots & p_{0\_2^n} \\ p_{1\_0} & p_{1\_1} & \cdots & p_{1\_2^n} \\ \vdots & \vdots & \ddots & \vdots \\ p_{2^n\_0} & p_{2^n\_1} & \cdots & p_{2^n\_2^n} \end{bmatrix} = \begin{bmatrix} P_O & P_F \\ 0 & I \end{bmatrix} \qquad (2)$$

Note that, because of the structure imposed earlier in the development, we can partition the state transition matrix in a convenient manner. The rightmost column ($p_{i\_j}, i = 0,1,2,\cdots,2^{n-1}$) represents the failure probabilities from any of the non-failure states, and can be represented by the vector P_F. The lowest row is the probability distribution for the failure state, which is all zeros with the exception of the holding probability. The remaining sub-matrix represents the transitions among the non-failure states. These can be considered the operational states, and the sub-matrix of operational state transition probabilities is labeled P_O.

When considering this network topology, it is important to keep in mind the issues of scale. The number of states in the safety state network grows as a power of two with an increasing number of conditions, and the number of elements in the state transition matrix grows as the square of the number of safety states. So, for example, a ten condition network has roughly 1000 states and 1,000,000 elements in the state transition matrix. When applying this method to actual systems, the analyst must keep in mind the effects of scale, and choose the conditions carefully to avoid having an unnecessarily large and unwieldy safety state network.

ESTIMATING RISK — While the state transition matrix itself is interesting, the ultimate power of this model lies in the ability to estimate the probabilities of future states. Consider the current state $S(i)$ to be represented as a vector $\overline{\theta}(k)$, of dimension 2^{n+1} by 1,in which the ith element is one and the remaining elements are zero. (In this notation, k represents the transition number, as the process progresses in time.) We can calculate the probability distribution of the next state, shown as $\overline{\theta}(k+1)$, using equation 3.

$$\overline{\theta}^T(k+1) = \overline{\theta}^T(k)P \qquad (3)$$

Using this strategy, we can look beyond the next transition to determine the probability of reaching a given state in any number of transitions in the future (equations 4). This quite a powerful concept—using it, we can evaluate the probabilistic behavior as far into the future as we like. We can summarize future probabilistic behavior in the $\Phi(\tau)$ matrix, which expresses the ability to transition from one state to another in τ transitions (equation 5). Note that the $\Phi(\tau)$ matrix can be partitioned in exactly the same manner as the state transition matrix (equation 2).

$$\begin{aligned} \overline{\theta}^T(k+2) &= \overline{\theta}^T(k+1)P = \overline{\theta}^T(k)P^2 = \overline{\theta}^T(k)\Phi(2) \\ \overline{\theta}^T(k+\tau) &= \overline{\theta}^T(k)P^\tau = \overline{\theta}^T(k)\Phi(\tau) \\ \overline{\theta}^T(k+\infty) &= \overline{\theta}^T(k)P^\infty = \overline{\theta}^T(k)\Phi(\infty) \end{aligned} \qquad (4)$$

$$\Phi(\tau) = \begin{bmatrix} \phi_{0\_0}(\tau) & \phi_{0\_1}(\tau) & \cdots & \phi_{0\_2^n}(\tau) \\ \phi_{1\_0}(\tau) & \phi_{1\_1}(\tau) & & \vdots \\ \vdots & & \ddots & \vdots \\ \phi_{2^n\_0}(\tau) & \cdots & \cdots & \phi_{2^n\_2^n}(\tau) \end{bmatrix} \qquad (5a)$$

$$\Phi(\tau) = \begin{bmatrix} \Phi_O(\tau) & \Phi_F(\tau) \\ 0 & I \end{bmatrix} = P^\tau \qquad (5b)$$

Our real goal is to determine the probability of reaching the failure state from any given current state. Because our Markov network is considered a finite single chain network, the ultimate state will be the failure state. (This corresponds to our fatalistic notion that, given enough time, a probabilistic failure will eventually happen.) However, the theory of Markov processes provides a mechanism for calculating the mean time to another state, from a known state. In the case of the safety state network, we are interested in identifying the mean time to the failure state.

To calculate the mean time to failure, we first need to express the probability that the failure will on a specific state transition in the future. As shown in equation 6, $\overline{\Psi}(\tau)$ is a vector quantity that provides the probability that the failure will occur on the τth state transition in the future, as a function of the safety state. The mean time between failures, as a function of the safety state, is the expected value of the number of transitions until the failure (equation 7).

$$\Psi(\tau) = \Phi_F(\tau) - \Phi_F(\tau - 1) \qquad (6)$$

$$\text{MTBF} = \overline{M} = \sum_{\tau=1}^{\infty} \tau\Psi(\tau) \qquad (7)$$

Knowing the mean time between failures, we can estimate the risk probability, as a function of safety state, by taking the inverse of the MTBF (equation 8).

$$\bar{F} = 1/\bar{M} = \begin{bmatrix} 1/M_1 \\ \vdots \\ 1/M_n \end{bmatrix} \tag{8}$$

In summary, this section details the methodology used to derive the risk probability and mean time between failures, given a static state transition matrix. Both the risk probability and mean time between failure are expressed as a function of the system safety state. Provided there is an obtainable state transition matrix that characterizes average driver behavior, these results may be used to compare the performance of drivers in several ways. Some of these are discussed below.

MODEL CALIBRATION — In the previous section, we assumed that we had access to a state transition matrix which is characteristic of the average driver behavior. In this section, a method is derived for obtaining that matrix.

A fundamental assumption is that the state transition matrix will be derived from experimental data only. To do this, a representative system is observed, and the various conditions used in the safety state network are observed. A record is made of each change in state of the individual conditions. This record includes a time stamp, using a calibrated time base.

From this data, the safety state trajectory can be constructed as a function of time. From the safety state trajectory, important data can be derived regarding each point at which the state changes. The necessary information derived includes the current state, the next state, and the amount of time spent in the state.

The next phase of analysis involves reorganizing and processing that data so that we have the transition statistics and mean holding time for each of the operational safety states. These quantities are then the basis for calculating the state transition probabilities. Based on the average holding time for each state, the holding probability is computed, as shown in equation 9. Given the holding probability p_{i_i}, the sum of the probabilities for leaving the state must be $(1 - p_{i\_i})$. This is divided up among the paths from the state, according to the relative likelihood of the target state (equation 10).

$$p_{i\_i} = \left(1 - \frac{1}{\tau_{av}}\right) = \left(\frac{\tau_{av} - 1}{\tau_{av}}\right) \tag{9}$$

$$p_{i\_j} = (1 - p_{i\_i}) \frac{\#\ \text{transitions to}\ S(j)}{\#\ \text{transitions from}\ S(i)} \tag{10}$$

Clearly, given the statistical nature of this calibration method, increasing the amount of input data will improve the quality of the derived state transition matrix. One of the potential future outgrowths of this research is determination of the tradeoff between the amount of calibration data provided and the quality of the derived state transition matrix.

DRIVER PERFORMANCE MEASUREMENT — Assuming that we have obtained a state transition matrix of sufficient quality, it is now possible to compare the performance of subject drivers. Data is collected on the subject driver in the same manner as used for calibration data—by monitoring the safety state conditions and recording the changes of state. This data provides enough information to derive the current safety state, either in real-time or in post-processing.

Using the safety state as an index, a simple look-up table access provides us with the risk probability associated with that state. This is possible because the risk probabilities have been pre-computed, based on the system average state transition matrix.

The risk probability, as a function of safety state, is immediately useful for comparing the individual driver performance against the norm. The average risk probability for the system average can be pre-computed by taking the weighted average of the risk probability vector elements, where the weights are the relative amount of time spent in that state. By taking a running average of the risk probability for the individual driver, it is possible to compare that behavior to the average, as a function of time.

AN EXAMPLE

In order to best illustrate the method of safety state analysis, we will consider an example. One example appropriate to both highway and rail applications is the frontal impact scenario. The details as presented are from a rail setting, but can be easily transformed to highway application.

In this example, the goal is to identify the risk probability of striking another vehicle (or obstruction) with the front of the vehicle. The first step is to select the set of conditions that are considered potentially contributory to this failure event. The following set of conditions are used:

<u>condition 0: throttle actuated</u> — this condition is true if the throttle is applied, and false if the throttle is not actuated

<u>condition 1: brake applied</u> — this condition is true if the brake is applied, and false if the brake is not actuated

<u>condition 2: brake failure</u> — this condition is true if the braking system of the train has failed in any way (even if the driver is not aware of this condition), and false if the braking system is functioning properly

<u>condition 3: overspeed</u> — this condition is true if the train is traveling at a speed greater than that allowed by either the static speed limit or the signaling system, and false if the train is being operated within the allowable speed bounds; the overspeed condition includes the case of passing a stop signal (entering an occupied block)

condition 4: obstacle — this condition is true if there is an obstruction on the track (possibly another vehicle) within the stopping distance of the train, and false if the track is clear; for the purposes of this example, the stopping distance is defined as the distance required to bring the train to a full stop from the current speed using full service braking, and is a function of speed

This set of conditions includes measurement of human control actions (throttle and brake actuation), vehicle state as a result of human control actions (speed condition), on-board machinery failures (brake failure), and external conditions (obstacle). Since each condition represents a binary condition, it is possible to combine this set of conditions into a single number by concatenating the bits into a single binary "word." The resulting set of safety states are shown in table 1. The states are numbered from 0 to 31, and the equivalent conditions are shown in the table.

The application of throttle and brake are included as separate conditions because they represent two different types of control actions which may be taken by the driver. There is a tacit assumption here that the throttle and brake are operated by separate control interfaces. This is true for highway vehicles, as for many types of rail vehicles.

An additional state is added in the table, state 32. This state represents the occurrence of the failure event, which is a frontal collision in this example. Also identified in the table is the (qualitative) possibility of having the failure occurrence directly from each state. For example, it is not possible to directly have a collision from states 0 through 15 because there is no obstacle present.

The next step in the procedure is to collect operational data, so that the model can be calibrated. This is accomplished by monitoring several different operators in a human-machine system. The system used for monitoring can be either an actual operational system, or a human-in-the-loop simulation. The quality of the calibration is a function of the number of different operators and total amount of time monitored — in general, more operators and time will result in a more representative calibration. The data recorded includes the changes in state (i.e., true or false) of the individual conditions, and the points in time of these changes.

Once the data has been recorded, the mean time in each state and counts of the state transitions are compiled, and are then used to calculate the state transition probabilities (equations 9 and 10). After the state transition matrix has been formulated, it is then possible to calculate the MTBF from each state (equation 7), which leads directly to the risk probability as a function of the safety state (equation 8).

New subjects can be evaluated by recording the safety state information and converting the resultant safety state trajectory into a risk probability trajectory. In addition, since we have recorded the safety state trajectory for the human subjects that contributed to the calibration, each can be converted to a risk probability trajectory. The risk probability trajectories associated with the human subjects can be compared against one another to determine relative performance with regard to risk sensitivity.

An alternate use of the model is to provide a synthesis of probabilistic human behavior. Once the model has been sufficiently calibrated, the model can be operated as a Monte Carlo simulation. In this scenario, the safety state model is used to replace the human as a vehicle controller. The safety state model, operating in this manner, captures the stochastic behavior commonly exhibited by human drivers.

CONCLUSIONS

This research is a response to a need for methods in which the safety-related decision performance of vehicle operators can be evaluated. Some of the difficulties in this area include event rarity, compound and interacting errors, and related difficulty in determining causality. This work identifies the human operator as a key component in the safety of transportation systems. The driver-vehicle system is modeled as a closed-loop control system, and a probabilistic model of system behavior is presented.

Based on the work of Lowrance, an organization for the efforts involved in the pursuit of safety is identified. Safety-related work is divided into subjective and objective components. Risk assessment, the objective component, is further divided into two components. One component, risk probability, measures the probability of occurrence of a necessary intermediate event, while risk outcome measures the outcome of these events in terms of the ultimate risks. In the case of transportation systems, the ultimate undesirable outcome is injury and death, and the necessary intermediate event is termed an accident. Based on this organizational description, the focus of this research is in the area of risk probability assessment.

A probabilistic model for system behavior (the safety state model) is developed. This model is based on finite Markov processes, using event tree and fault tree techniques as an inspiration. From the safety state model, a method for determining the mean time between failure (MTBF) and risk probability is developed. Both of these quantities are expressed as a function of system state. A method for calibrating the safety state model is presented, based on experimental data. Finally, a method is presented for measuring individual driver performance and comparing it to average driver behavior. Suggestions are made regarding possible alternate applications.

REFERENCES

1. Sheridan, Thomas *Telerobotics, Automation, and Supervisory Control*, MIT Press (1992).

2. Forbes, T. W. (ed.) *Human Factors in Highway Traffic Safety Research*, Wiley-Interscience (1972).

3. Salvendy, Gavriel (ed.) *Handbook of Human Factors*, Wiley-Interscience (1987).

4. *Webster's New World Dictionary, Second College Edition* (1979).

5. Lowrance, William W. *Of Acceptable Risk*, William Kaufmann, Inc., Los Altos, CA (1976).

6. Reason, James *Human Error*, Cambridge (1990).

7. Drake, Alvin *Fundamentals of Applied Probability Theory*, McGraw-Hill (1967).

8. Swain, A. D. and H. E. Guttman *Handbook of Human Reliability Analysis with Emphasis on nuclear Power Plant Applications*, Sandia Natl. Labs., NUREG CR-1278, Washington, D, U.S. Nuclear Regulatory Commission (1983).

9. Marshall, Gilbert *Safety Engineering*, Brooks/Cole Engineering Division (1982).

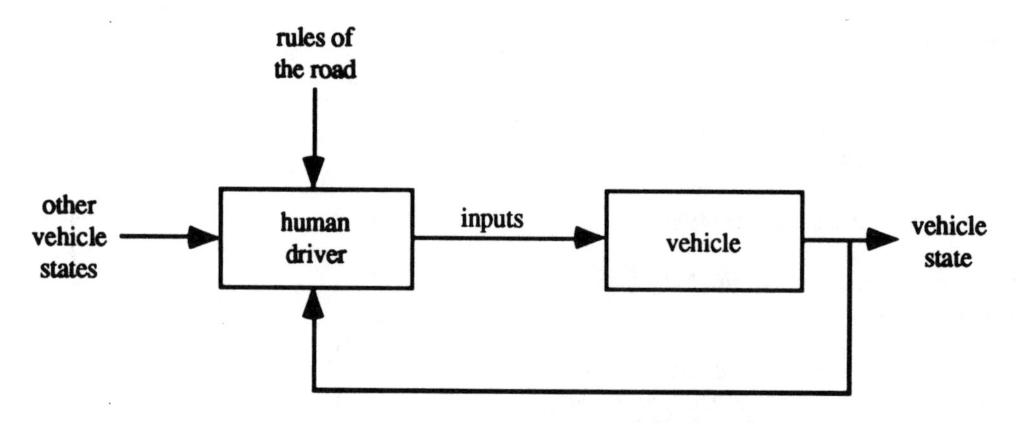

Figure 1 — Closed-Loop Driver-Vehicle System

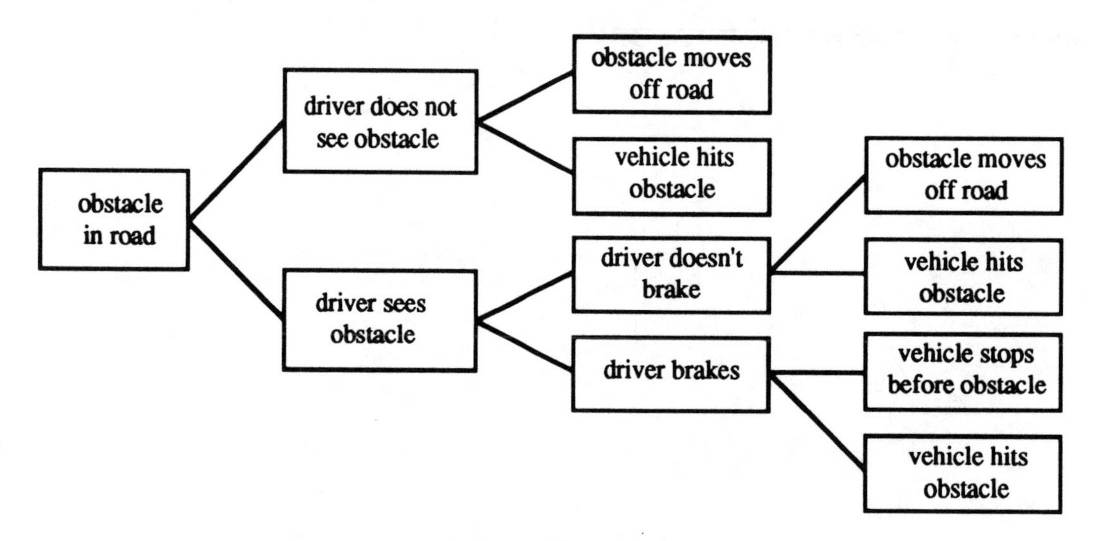

Figure 2 — Sample of Event Tree Analysis

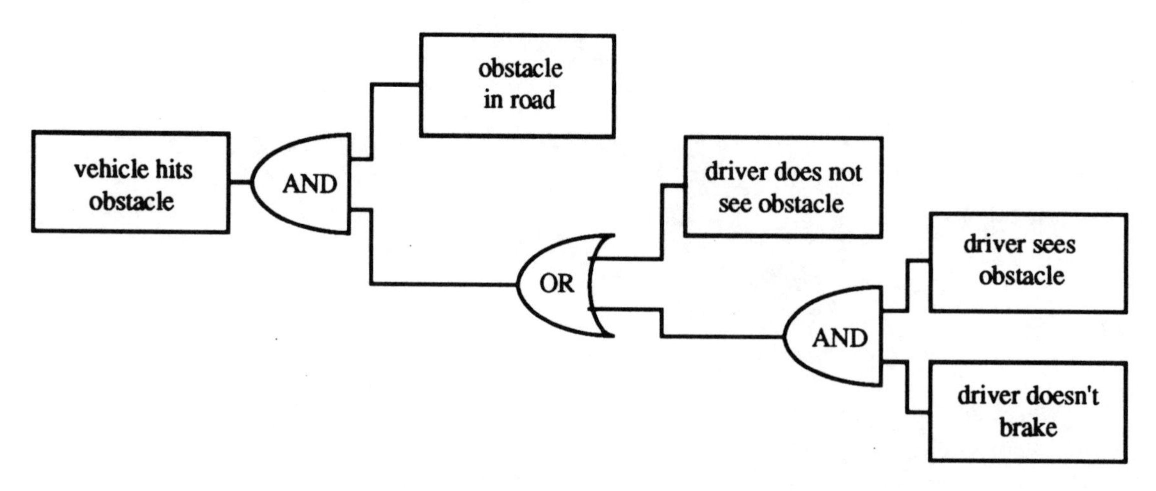

Figure 3 — Sample of Fault Tree Analysis

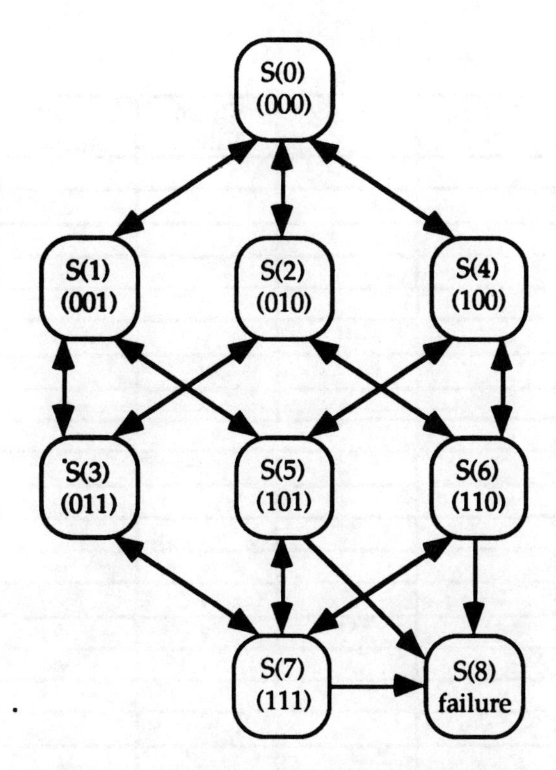

Figure 4 — Example of a Safety State Network

Safety State	Binary Form	Obstacle	Overspeed	Brake Failure	Brake Applied	Throttle Applied	Failure Possible?
0	00000						
1	00001					√	
2	00010				√		
3	00011				√	√	
4	00100			√			
5	00101			√		√	
6	00110			√	√		
7	00111			√	√	√	
8	01000		√				
9	01001		√			√	
10	01010		√		√		
11	01011		√		√	√	
12	01100		√	√			
13	01101		√	√		√	
14	01110		√	√	√		
15	01111		√	√	√	√	
16	10000	√					√
17	10001	√				√	√
18	10010	√			√		√
19	10011	√			√	√	√
20	10100	√		√			√
21	10101	√		√		√	√
22	10110	√		√	√		√
23	10111	√		√	√	√	√
24	11000	√	√				√
25	11001	√	√			√	√
26	11010	√	√		√		√
27	11011	√	√		√	√	√
28	11100	√	√	√			√
29	11101	√	√	√		√	√
30	11110	√	√	√	√		√
31	11111	√	√	√	√	√	√
32	100000						

Table 1 — Safety State Description

950970

Automatic Target Acquisition Autonomous Intelligent Cruise Control (AICC): Driver Comfort, Acceptance, and Performance in Highway Traffic

James R. Sayer, Paul S. Fancher, Zevi Bareket, and Greg E. Johnson
The University of Michigan Transportation Research Institute

ABSTRACT

This study investigated levels of driver comfort and acceptance for an autonomous intelligent cruise control (AICC) system driven in an actual highway environment. Objective measures of driving performance and behavior are compared with participants' subjective assessments when operating under manual control, conventional cruise, and AICC. Included in the comparison are measures of driver velocity and braking behavior.

Participants drove at slightly higher mean velocities under the manual condition as compared with AICC. Participants applied the brakes least frequently when driving manually. Participants rated the AICC system favorably for comfort, ease of use, and convenience. However, participants did express limited concerns associated with the use of AICC.

INTRODUCTION

This study is the first of several to be performed by The University of Michigan Transportation Research Institute (UMTRI). These studies will investigate drivers' comfort, acceptance, and driving behavior with Autonomous Intelligent Cruise Control (AICC). AICC is an extension of conventional cruise control technology that is intended to enhance driving comfort by controlling vehicle speed relative to traffic conditions (i.e., the presence of other vehicles).

An electronic sensor is mounted on the front of a vehicle. When another vehicle is detected in the path of the test vehicle, an electronic controller evaluates the situation and determines whether the cruising speed set by the driver can be maintained without encountering the forward vehicle. If the speed set by the driver can not be maintained, the AICC system will slow the vehicle.

PROCEDURE

TEST VEHICLE - The test vehicle in which the AICC system was evaluated is a 1993 Saab 9000 Turbo. An ODIN 3 sensor system, provided by Leica AG, with a controller algorithm designed by UMTRI was used. The headway maintenance algorithm employed a coast-down deceleration associated with throttle-off to slow the test vehicle, when necessary, to maintain a headway of 1.4 seconds behind preceding vehicles. The rate of deceleration associated with the throttle-off condition was approximately 0.04 g. When this level of deceleration on the part of the system was not sufficient to maintain a "comfortable" headway, or prevent a collision, the participants were required to intervene by using the brake pedal or maneuvering the car out of the path that presented an obstacle. The UMTRI-modified Leica-based system employed a fixed monobeam infrared sensor with a beam width of 3.6 m at a distance of 70 m. A complete description of the vehicle and the AICC system employed in this study can be found in Fancher et al. (1994).

Participants were always able to over-ride both the conventional cruise control (CC) and the AICC system through use of the accelerator pedal, brake pedal, and controls located on the turn signal stalk. The same control mechanism was used for both CC and AICC operation. When participants used the accelerator to over-ride CC or AICC, once released, the system always reverted to the set speed before the intervention. Manual braking caused both the CC and AICC systems to disengage, requiring the driver to re-engaged the system using the controls located on the turn signal stalk. Both the CC and AICC systems were designed such that increases in the set speed after the system was engaged were accomplished in increments of 2 mph.

The term "adaptive cruise control" was substituted for the term Autonomous Intelligent Cruise Control during the study to prevent participants from over estimating the capabilities of the AICC system. The term "adaptive" was found to be more descriptive of the AICC system's functioning to lay participants.

ROUTE - Each participant drove a predetermined route on local highways. The length of the route was 55 miles, and took approximately 50 minutes per trial to complete. This time was believed to be sufficient to allow participants time to experience, and become accustom to, controlling the research vehicle under the three cruise control states. The route was traversed three times by each participant, once each under manual control, conventional cruise control, and AICC. A map of the route is provided in Figure 1. Annual average 24-hour traffic volumes for the selected route are provided in Table 1. Participants drove only when weather and road conditions permitted. Data collection only took place between the hours of 9 a.m. - Noon and 1:30 p.m. - 4:30 p.m. to avoid large fluctuations in traffic density associated with "rush hours." Similarly, no data was collected on Friday afternoons.

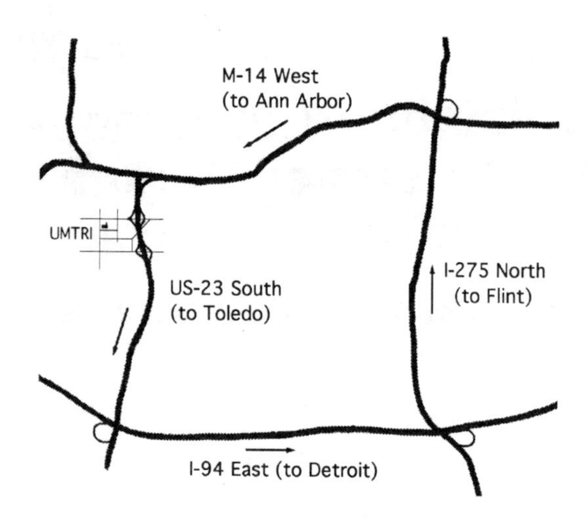

Figure 1. Map of the selected route through Ann Arbor and the Metropolitan Detroit area.

Table 1. Annual average 24-hour traffic volumes for the selected route.

Segment	Average Volume	Lanes
US 23 (South)	44,000 - 56,000	2
I-94 (East)	60,000 - 91,000	2-3
I-275 (North)	45,000 - 112,000	3-4
M14 (West)	43,000 - 70,000	2-3

Source: Michigan Department of Transportation (1993)

PARTICIPANTS - Thirty-six licensed drivers were recruited from a local Secretary of State's office (Department of Motor Vehicles), as well as through newspaper advertisements, to participate in the study. Prospective individuals were required to meet the following criteria:

a. possess a valid, unrestricted, driver's license,
b. have a minimum of two years driving experience,
c. and appeared not be under the influence of alcohol, drugs, or any other substances that could impair their ability to drive.

The participant population was balanced for gender, age, and experience in the use of conventional cruise control. The three age groups examined were 20 - 30, 40 - 50, and 60 - 70 years of age. The average yearly mileage driven by participants was 13,500 miles. Experience with conventional cruise control was divided into two groups; those who frequently used cruise control and those who never, or very rarely, used cruise control.

Among those who never, or rarely, used cruise control, having a car that was not equipped with cruise was cited most often as the reason it was not used (57.1%). Among users of cruise control, reduced work load was cited most often as the reason for its use (64%). When the participants were asked to describe their cruising speed on the open freeway, 57.1% reported that they drove 5 mph above the speed limit, 22.9% reported driving at the speed limit, 2.9% reported driving 5 mph below the speed limit, and 17.1% reported driving at some other speed. In addition, 44.4% of the participants reported regularly driving at speeds consistent with the flow of traffic, while 55.6% drove at a speed with which they felt comfortable.

In the event participants encountered a slower moving vehicle, and the adjacent lane was free, 75% of the participants responded that they would pass the vehicle and return to the lane even if momentary acceleration was necessary. Another 16.7% would maintain their speed even if it meant moving to another lane and remaining. While 8.3% reported that they would adjust their speed and remain in the lane if the other vehicle were only slightly slower.

STUDY DESCRIPTION - Individuals were briefed as to the nature of the study. Prospective participants were asked to read an information letter describing the study and the associated benefits and risks. Individuals who agreed to participate, and meet the previously mentioned criteria, provided informed consent.

Participants were shown the research vehicle, and instructed on its operation. Specific attention was paid to locating and identifying controls and displays. Instruction on the use of the two cruise control devices was also provided. Participants were asked to adjust the driver's seat and vehicle mirrors. All participants were required to wear the seat belt.

EXPERIMENTAL TRIALS - Each of three experimental trials began and ended at the UMTRI facility. In each trial, participants were instructed to drive a pre-determined route along local highways (an experimenter was present at all times to aid participants in route guidance). On each trial a different form of speed assistance (cruise control) was evaluated: no cruise control, conventional cruise control, and AICC. The orders in which participants experienced cruise control types were counter balanced to eliminate order effects. The same route was followed for each trial. At the end of each experimental trial, participants returned to the UMTRI research facility to complete a questionnaire. A ten minute break was provided to participants at the end of each trial.

OBJECTIVE MEASURES - Measures of velocity and braking behavior were made for all participants on each of the three trials. Only velocities at, or above, highway speeds (55 mph) were included in the analyses. Additional data, which included time spent driving to and from the UMTRI facility to the highway, were not included. Measures of braking behavior including braking frequency and duration were also determined for each participant. Additional measures, but which are not addressed in this publication, included headway (range and range rate), yaw rate, steering wheel position, accelerator pedal position, cruise control set speed, and throttle position.

SUBJECTIVE MEASURES - Participants completed questionnaires at the end of each trip. The questionnaires required participants to assign levels of comfort and ease-of-use to the cruise control device they had most recently experienced. Six questions were posed to participants for each of the three driving conditions. On numerical rating scales from 1 to 7, with adjectival anchors, participants were asked to rate their driving experience for each of the three cruise control states.

RESULTS

OBJECTIVE MEASURES - Measures of driver performance and behavior related to vehicle velocity and braking were made for each of the three cruise control states. Histograms that describe driving velocities, collapsed across all participants, and instances of braking, are provided below for each of the three cruise control states.

Figures 2 through 4 provide velocity histograms for the no cruise, conventional cruise, and AICC conditions, respectively. The mean velocities, collapsed across all participants, are lowest for the AICC condition

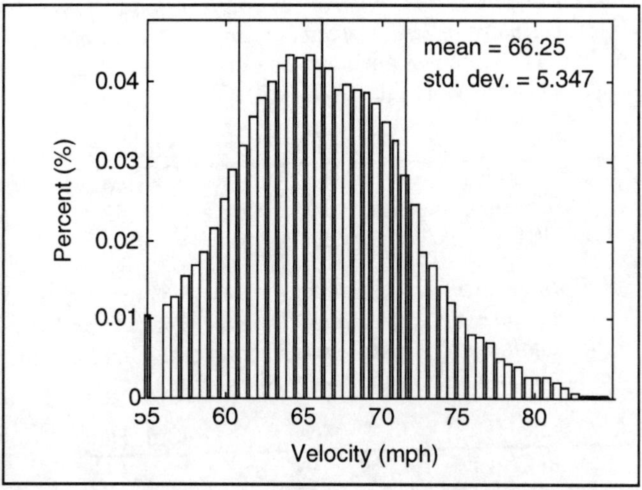

Figure 2. No cruise control velocity histogram, collapsed across all participants.

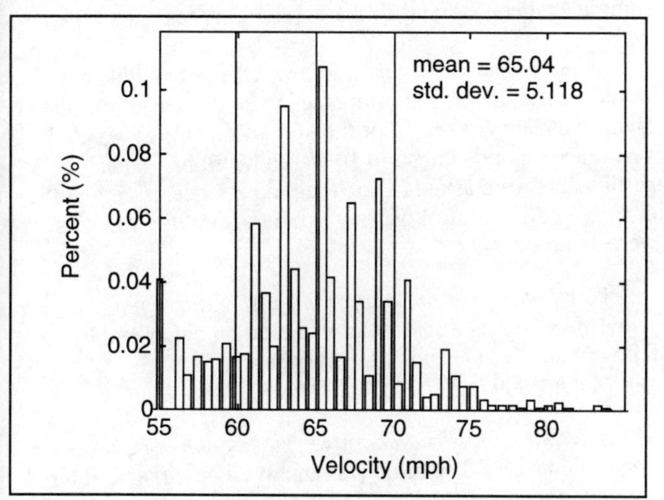

Figure 3. Conventional cruise control velocity histogram, collapsed across all participants.

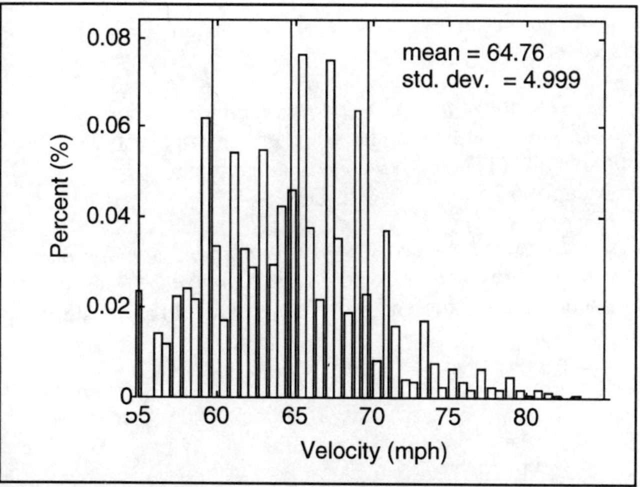

Figure 4. AICC velocity histogram, collapsed across all participants.

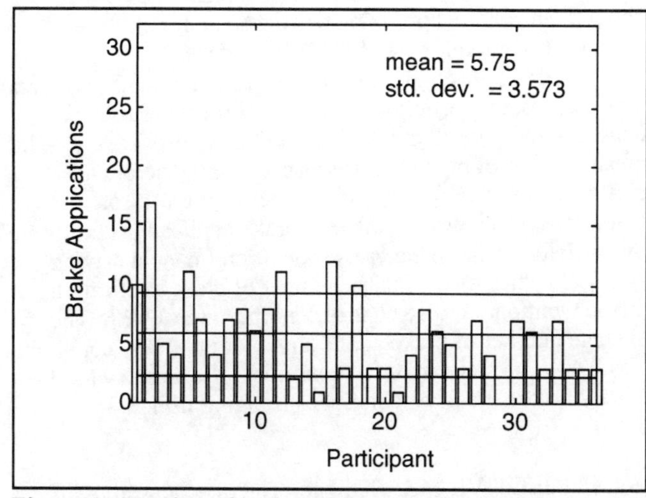

Figure 5. Number of brake applications, by participant, in the no cruise control state.

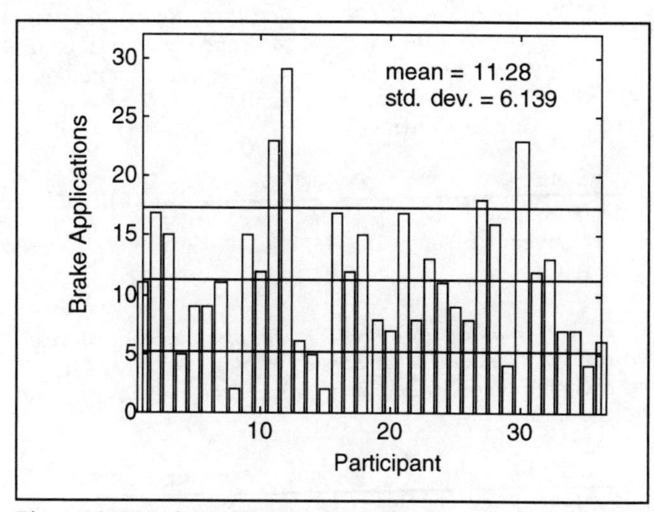

Figure 6. Number of brake applications, by participant, in the conventional cruise control state.

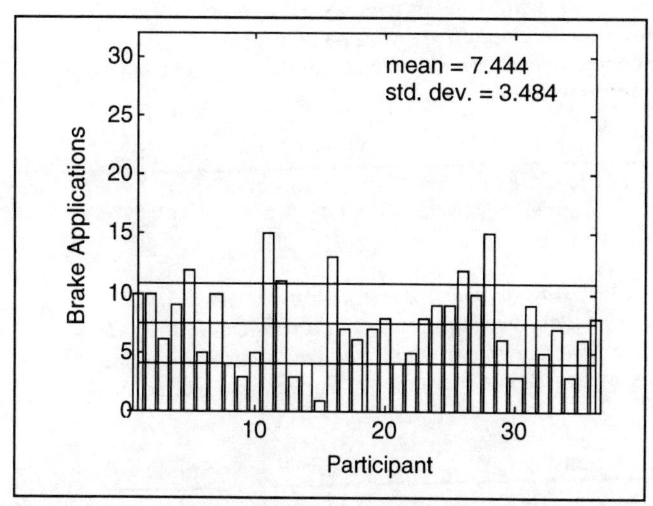

Figure 7. Number of brake applications, by participant, in the AICC state.

(mean = 64.76 mph) and highest for the no cruise control condition (mean = 66.25 mph). The standard deviations about the mean are approximately the same for all three conditions (ranging from 4.99 to 5.35 mph).

Figures 5 through 7 display the absolute number of brake applications, by participant, for each of the no cruise, conventional cruise, and AICC conditions, respectively. The mean number of brake applications, collapsed across all participants varied widely for the three cruise control conditions. The mean number of brake applications performed by participants in the no cruise condition (manual driving) was 5.75, whereas the means number of brake applications in the conventional cruise control condition was 11.28. The average number of brake applications for the AICC condition was 7.44. The standard deviations about the means varied widely across the three conditions (ranging from 3.48 to 6.14).

SUBJECTIVE MEASURES - The questions asked of participants at the end of each trial, as well as means and standard deviations of the responses (collapsed across all participants), are presented below.

1. *How comfortable, from a safety standpoint, did you feel driving the car with no cruise control/conventional cruise control/adaptive cruise control?* The scale was anchored on either end by "Not Comfortable" (1) and "Very Comfortable" (7) respectively, as shown below.

State	Mean	Std. Dev.
No Cruise	6.17	1.28
Conven. Cruise	5.75	1.05
AICC	6.00	1.22

2. *How easy did you find it to maintain a safe distance between your car and other cars in front of you?* The scale was anchored on either end by "Not Easy" (1) and "Very Easy" (7).

State	Mean	Std. Dev.
No Cruise	5.86	1.50
Conven. Cruise	5.14	1.62
AICC	6.33	1.17

3. *How comfortable did you feel with the ability to pass other cars while driving with no cruise control/ conventional cruise control/adaptive cruise control?* The scale was anchored on either end by "Not Comfortable" (1) and "Very Comfortable" (7).

State	Mean	Std. Dev.
No Cruise	6.36	0.96
Conven. Cruise	5.67	1.22
AICC	5.72	1.56

4. *Using no cruise control/ conventional cruise control/adaptive cruise control, do you feel that you drove either faster or slower than you would normally?* The scale was anchored on either end by "Slower than Normal" (1) and "Faster than Normal" (7).

State	Mean	Std. Dev.
No Cruise	5.17	1.13
Conven. Cruise	3.86	1.15
AICC	3.69	1.43

5. *Using no cruise control/ conventional cruise control/adaptive cruise control, do you feel that you applied the brakes more or less frequently than usual for comparable traffic?* The scale was anchored on either end by "Less than Usual" (1) and "More than Usual" (7).

State	Mean	Std. Dev.
No Cruise	4.42	1.46
Conven. Cruise	4.39	1.52
AICC	2.47	1.52

6. *In general, how similar was your driving to the way you would normally drive under the same types of road and traffic conditions?* The scale was anchored on either end by "Not at all Similar" (1) and "Very Similar" (7).

State	Mean	Std. Dev.
No Cruise	5.97	1.36
Conven. Cruise	5.33	1.43
AICC	4.72	1.95

DETAILED AICC EVALUATION - Following the completion of all three trials, each participant was asked to complete a detailed questionnaire regarding the use of the AICC system.

When participants were asked whether an audible tone would be useful when a difference in vehicle speeds required them to use the brakes 17 of the 36 participants responded yes; nine responded no, and 10 were uncertain. When participants were asked to rate their experience driving with adaptive cruise control on 1 to 7 point scales, their responses were as follows:

1. *What impact did adaptive cruise control have on your sense of safety?* The scale was anchored on either end by "I felt very unsafe" (1) and "I felt very safe" (7).
Mean = 5.97 Std. Dev. = 1.08

2. *What impact did adaptive cruise control have on you sense of comfort?* The scale was anchored on either end by "I felt very uncomfortable" (1) and "I felt very comfortable" (7).
Mean = 6.25 Std. Dev. = 1.10

3. *Did the system ever make you feel <u>too</u> comfortable, as if someone else had taken control of the car for you?*
Yes = 11
I am not certain = 3
No = 22

4. *How convenient did you find using adaptive cruise control?* The scale was anchored on either end by "It was very inconvenient" (1) and "It was very convenient" (7).
Mean = 6.25 Std. Dev. = 1.23

5. *How similar to your own driving behavior do you think the adaptive cruise control system operated?* The scale was anchored on either end by "Not similar" (1) and "Very similar" (7).
Mean = 4.91 Std. Dev. = 1.68

When participants were asked what aspects of the AICC system were bothersome, the reasons cited by participants were:

• Loss of target on curves	25.00%
• Slow rate of acceleration; lane change	12.50%
• Inability to track merging traffic	12.50%
• Location of cruise controls	12.50%
• Wrong target on curves	6.25%
• Headway too short	6.25%
• Headway too long	6.25%
• Lack of brake lights during deceleration	6.25%
• Questions about reliability of the system	6.25%
• Inability to use on entrance/exit ramps	6.25%

DISCUSSION AND CONCLUSION

The subjective impressions of participants in response to questions concerning the velocities they maintained while driving under the three cruise control states are consistent with objective measures. Specifically, participants reported, and in fact did, drive at a slightly higher mean velocity when controlling the research vehicle manually as compared to the conventional cruise control and AICC conditions. Participants frequently reported the "urge" to keep up with other vehicles when traffic volume was heavy. This behavior may have contributed to the slightly higher mean velocities observed when participants drove the research vehicle under manual control. However, additional evaluation of the data collected under these conditions is necessary.

Participants' subjective responses were not consistent with objective measures regarding the frequency of brake applications. Participants reported applying the brakes least frequently when driving under the AICC condition, when in fact braking was required least in the manual (no cruise) control condition. The reduction in brake applications under manual control is likely the result of accelerator pedal modulating, performed unconsciously, by participants in response to changes in headway between the research vehicle and preceding vehicles. Additional evaluation of the data is required to confirm, or deny, this hypothesis.

In general, participants rated the AICC system favorably for their sense of comfort, ease of use, and convenience. However, participants did express concerns associated with the use of the AICC system on curves and the inability to track merging traffic. These concerns are consistent with previous investigations under similar conditions and circumstances (Fancher and Ervin, 1994).

ACKNOWLEDGMENTS

This work was sponsored under contract with the National Highway Transportation Safety Administration (DTNH22-94-Y-47016), with supporting instrumentation provided under contract by the Michigan Department of Transportation "Smart Cruise Platform" (93-2165). The authors would like to thank Jack Ferrence and Art Carter, Office of Crash Avoidance, NHTSA; Mary Lynn Mefford, Bob Ervin, and Michael Sivak, UMTRI; James Haugan, Haugan and Associates; and Leica Heerburgg AG.

REFERENCES

Fancher, P. S., Ervin, R. D., Bareket, Z., Johnson, G. E., Trefalt, M., Tiedecke, J., and Hagleitner, W. (1994). Intelligent cruise control: Performance studies based upon an operating prototype. In Moving Toward Deployment, Proceedings of the IVHS AMERICA 1994 Annual Meeting (pp. 391-399). Atlanta, GA: IVHS AMERICA.

Fancher, P. S. and Ervin, R. D. (1994). Implications of intelligent cruise control (ICC) systems for the driver's supervisory role. In Towards an Intelligent Transportation System: Proceedings of the First World Congress on Applications of Transport Telematics and Intelligent Vehicle-Highway Systems (pp. 2071-2078). Paris: ERTICO.

Michigan Department of Transportation (1994). Michigan 1993 annual average 24-hour traffic volumes, Lansing, MI.

950971

SUSI Methodology: Evaluating Driver Error and System Hazard

Richard Stobart and Jeremy Clare
Cambridge Consultants Ltd.

ABSTRACT

The introduction of more intelligence into vehicle control systems increases functionality but at the same time threatens to overload the driver. A second and potentially more serious effect is that the driver's understanding of how the vehicle is behaving may be incorrect. The user interface may have the capacity to misrepresent important information.

The SUSI™ methodology devised to assess hazard driving system design is directed towards this problem. SUSI™ exploits modem software design methods to represent human and machine behaviour in a uniform context. A form of HAZOP is then used to draw out potential hazards from which risk assessment and risk mitigation actions can be developed.

SUSI™ has been applied in the automotive environment and has shown its utility at various stages of the design process.

INTRODUCTION

The increasing use of intelligent systems in vehicles raises new facets of the safety issue. The analysis of safety for automotive systems has traditionally made a distinction between those failures where failure can lead to an incident and those failures where the driver can reasonably be expected to control the vehicle in a safe manner in the presence of the failure condition. However, the desire to provide an easier driving environment may lead to situations where the driver may misunderstand the mode in which the vehicle system is operating.

A failure in such a circumstance may lead the driver into inappropriate behaviour, in turn leading to a hazardous condition. For example, failure of the anti-lock brake system will not cause a hazard in itself. However, if the driver believes it to be functioning effectively then the he/she may make assumptions about braking capability. While such an example would appear to be easy to interpret, the failure of the display system to adequately inform the driver may be regarded as a contributory cause in any incident.

In the face of complex interactions between the driver and the vehicle systems there needs to be an effective approach to safety analysis. Such an analysis needs to take account of driver behaviour and the interactions with the vehicle system.

In other industries such as petro-chemical, rail transport and air transport the consequences of a failure could be catastrophic and they have realised that evaluating and controlling the risk to humans arising from failures of such systems is vital.

There has been an increasing realisation in these industries that human behaviour is an important consideration in system design. The critical issue is that as systems become more complex the human operators are increasingly prone to 'error'. In this case 'error' is used to describe behaviours that were not the designer's intent. In part, such behaviours occur because the operators are not able to comprehend their role with the system. However, a key part is that operators choose to do something different to the designer's intent because of new working procedures, or because the system did not perform as intended. However, once an incident occurs then we normally identify human error as the root cause, Westrum (1991).

When we consider the design and implementation of systems in general we find that there are two viewpoints: that of the system and, separately, that of the user.

In order to build effective, safe systems it is clear that all viewpoints must relate to a common design concept. However, the reality is that for most systems the two areas have their own specialist vocabulary and models of the system, and they tend to work independently of each other. In addition, industries which have an established record of building safety critical systems are likely to have specialists in safety. These personnel, again, have their own particular jargon and often may not be familiar with the techniques of developing computer based systems. This lack of co-ordinated coverage during system development has the potential to lead to hazardous situations.

We have developed a methodology that we believe shows promise in dealing with the above problem. The methodology is called SUSI™, standing for Safety analysis of User System Interaction. The SUSI™ methodology comprises two parts:

- A common representation of all entities in a systems so that communication between specialists is enabled; coupled with
- a structured, hazard identification procedure which addresses features that are particular to human-machine interactions.

The remainder of this paper describes the approach and gives examples of its use applied to a collision warning system.

SUSI™: THE NEED FOR A METHODOLOGY

A COMMON REPRESENTATION IS NECESSARY AND POSSIBLE - In developing such systems, the specialists all tend to build their own models of the system which are not easy to correlate with the other models. In addition, the people who want the system, those who will use it and those who might have to judge its safety, will need to have an understanding of the system. However, they all need to communicate with each other to ensure they share a common understanding of the system. There also needs to be a stable system which can be viewed from each of the perspectives. Gaining that understanding is not easy with a plurality of system models and it is difficult to imagine the consequences to the rest of the system of changes in one representation.

An analysis of the features that need to be described in understanding human activities, Stammers et al (1990), and those used to describe a software system, Hatley and Pirbhai (1988), show a commonality of entities:

Human	Software
task	process
information flow	dataflow
interactions	control
documentation	database

Commonality of Human Activities and Software Description

With the existing widespread use of software data and control flow analysis and description tools there would thus appear to be a basis for a common representation. In adopting a dataflow and process model for the human components of a system we can now generate an integrated representation of the overall system. Using this we can explore the consequences of failure in a consistent manner across the whole system. In Figure 1, part of a collision warning system description is shown to illustrate the convention used in this type of representation. The key components are circles to represent a process (human or machine), a solid line is a dataflow, a dashed line is a control flow (start, stop etc), and two parallel lines represent a data store (we show the driver display as a data store because data may be displayed, but there has to be an explicit human process to read the data).

We have found that a dataflow and process model for both the human and the computer parts of a system allows us to generate an integrated representation of the overall system. This model can then be used to explore various properties of the system. A major advantage we have found is that such models are easy to understand by non-computer specialists and that users of systems are able to comment on and critique designs at a very detailed level.

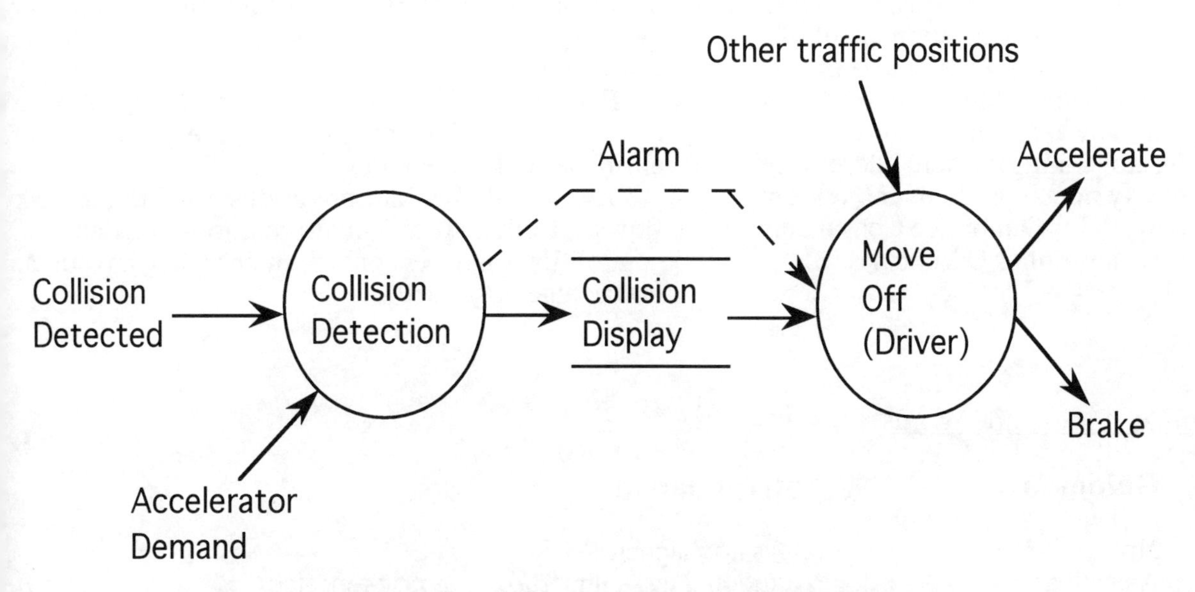

Figure 1 : Collision Warning Process

ADAPTION OF HAZOP PROCEDURE

The process of assuring safety of systems has been well established over many years. However, only in more recent years has the potential for computer systems to fail in ways that might impact safety been recognised formally in standards work, IEC65A (1991) and DEF STAN 00-56. Emerging standards and existing practice in established industries use the same basic lifecycle approach to addressing safety which may be summarised as follows:

- System definition: generating a concise and complete description of the system under review;
- Hazard identification: identifying the potential for hazardous events;
- Risk analysis: judging the safety risk of the system as defined;
- Risk acceptability: determining if the risks are acceptable;
- Risk mitigation: activities, as necessary, to modify the system definition or to include additional measures in order to reduce the risk to an acceptable level.

Hazard identification is thus a key step in the process and a number of structured procedures have been developed to provide confidence that the safety analysis is complete and thorough. In this section we introduce one of the main hazard identification methods - HAZOP. We explain how we have adapted the HAZOP procedure to address user system interaction and we describe some advantages and limitations of the approach.

WHAT IS A HAZOP? - The full name of HAZOP, HAZard and OPerability study, says a great deal about its purpose. It is a critical review of the hazards that may occur and the method of operation of a system. The technique was developed by ICI in the late 1960's and has grown to be well established in the petrochemical industries. An excellent introduction to the technique is given in the CISHEC Safety Committee publication (1990).

In the process industries it is usual to describe plant designs in the form of Piping and Instrumentation Diagrams (P&IDs). The HAZOP is carried out by a small team with the following members: a study leader; a recorder; personnel who have detailed knowledge of operation of similar systems; personnel who have detailed knowledge of the design intent of the system; and specific technical specialists as necessary. The team work logically through the P&IDs, examining deviations from normal operation asking "can the deviation happen?" and if so, "would it cause a hazard?" (a hazard could be things such as a fire or release of toxic material). To guide the process a series of attributes and guidewords are used. Thus for liquid in a pipe, a relevant attribute is "flow" and potential deviations are "high, low, no, reverse". For fluids another attribute is "pressure" with guidewords "high, low". In theory, each attribute/guideword combination should be applied to each process line and vessel. In practice, an experienced team leader will judge the correct detail of questioning for each area.

MODIFYING HAZOP FOR USER SYSTEM INTERACTION - A critical element of the HAZOP process is the interpretation of attribute and guidewords. We believe the guidewords for petrochemical plant are applicable, with a few additions relating to timing, when addressing computer based systems and user system interaction. Other work by Cambridge Consultants and their parent company Arthur D Little has led to modifications of the HAZOP approach for computer based systems and this is reported in Chudleigh and Catmur (1992).

For our work with user system interaction we have developed the following interpretations of attributes and guidewords.

Table 1 shows interpretations for human task flows. Table 2 shows interpretations for data and control flows independent of whether they are human or machine originated.

Table 1: Human Task Guide Words

Attribute	Guideword	Interpretation
Process/ Task	No	Does not happen
	As well as	Additional tasks undertaken, or part repeated
	Part of	Only part of task executed
	Other than	Wrong task executed
Control Process	No	Control action does not take place
	Part of	Action incomplete
	Other than	Action incorrect
	As well as	Additional unwanted action takes place
Sequencing	Before/after	Action out of expected order
	Early	Action takes place before expected
	Late	Action takes place after expected
	No	No action

Table 2: Data Flow Guide Words

Attribute	Guideword	Interpretation
Flow (of data or control)	No	No information flow
	Part of	Information passed is incomplete
	Other than	Information complete but incorrect
	More	More data passed than expected
	As well as	There is some corruption of the information
Flow of data to/from data store	Other than	Data changed in store
	No	Data not stored or not found
	As well as	Data store contains more items than expected
	Part of	Data store is incomplete
Data rate	More	Data rate is too high
	Less	Data rate is too low
Data value	More	Data value is too high
	Less	Data value is too low

We have found that the traditional HAZOP team structure and general approach can be used without change. However, it has been found essential to have independent technical personnel who are experienced in system design and human factors as part of the HAZOP team.

ADVANTAGES AND LIMITATIONS OF THE MODIFIED HAZOP - The main advantages of the approach are:

- it is done by a team,
- it gives a top-down approach to the system,
- it can be used both on new system designs and on existing systems.

The team approach brings a variety of expertise and viewpoints onto a common problem and concentrating on hazard identification leads to productive sessions. Also the team, by providing a variety of viewpoints, helps to avoid excessive investigation of non-credible hazards.

The top-down approach, examining the whole system first, allows homing-in on key issues based on the potential hazards and is very good at assessing system-wide implications. The approach of looking at deviations from design intent, then their causes and consequences, encourages exploration of non-obvious interactions both of the user/operator with the automated system and of the automated system with its hardware environment. The use of a HAZOP provides guidance towards the most critical areas to concentrate upon in any subsequent low level investigation.

The HAZOP fits naturally at all stages of the life of a system, from concept through to operation.

There are, however, limitations to the HAZOP approach. We have found that straightforward application of the deviation guidewords is not sufficient: the process relies on the experience and intuition of the team members (especially of the independent technical experts). Further, the choice of an experienced HAZOP leader is key. It is the leader who controls the pace of the study and it takes significant experience to guide the team discussion to the most critical areas while still ensuring full coverage within usually tight time constraints.

THE USE OF SUSI™ APPLIED TO A COLLISION WARNING SYSTEM

The SUSI™ methodology has been applied to an outline design for a collision warning system (CWS). The role of the CWS is to alert the driver if a low speed shunt collision is detected. This would typically occur at a rotary or other junction where a queue of traffic has formed. The driver of a second or third car would see the car(s) in front start to move off and start to follow. At the same time the driver would be looking to see if there is a sufficient gap to allow entry into the traffic flow. During this time the driver in front decides not to join the traffic, resulting in a collision.

The CWS detects the potential for such a condition and alerts the driver through an audible or tactile signal. Visual signals cannot be used as the driver's gaze is not determined. Figure 2 shows a top level view of the system.

The collision warning system was segmented into three top level processes; a data fusion module which collated data from various sources, including a sensor system detecting cars in front of the car; a situation assessment module, which interpreted the data available to identify if a collision was likely; and finally, a collision warning module which alerted the driver of a potential collision condition.

In order to understand the process more fully, state transition diagrams were also used to identify specific conditions relating to the onset of warnings. An example is shown in Figure 3.

During the HAZOP procedure, a number of hazard conditions were identified. Examples of these are:

Alarm: None - driver fails to recognise the alarm signal.
Recommendation - use tactile feedback via accelerator pedal.
Vehicle detection: Other than - multiple vehicles in sensor beam causes confusion.
Recommendation - improve sensor to detect nearest of many targets.
Collision detected: Other than - false target identified.
Recommendation - draw drivers attention to area covered by sensor.

In carrying out the HAZOP procedure on this type of representation we have found that the operators/users have no difficulty in understanding the process. In many cases they find the notation is natural for thinking about both system functions and operator tasks.

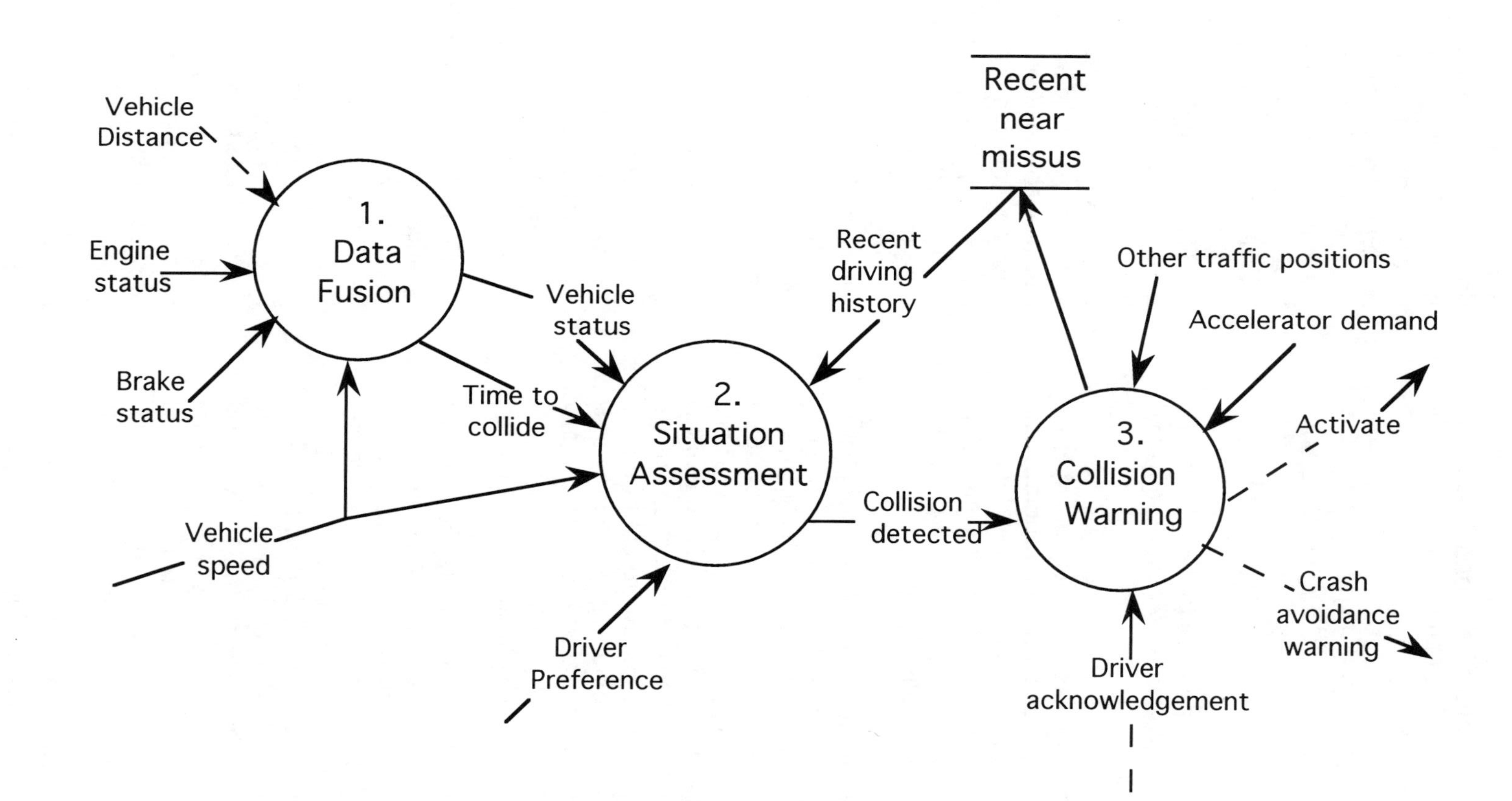

Figure 2 : Top Level View of Collision Warning System

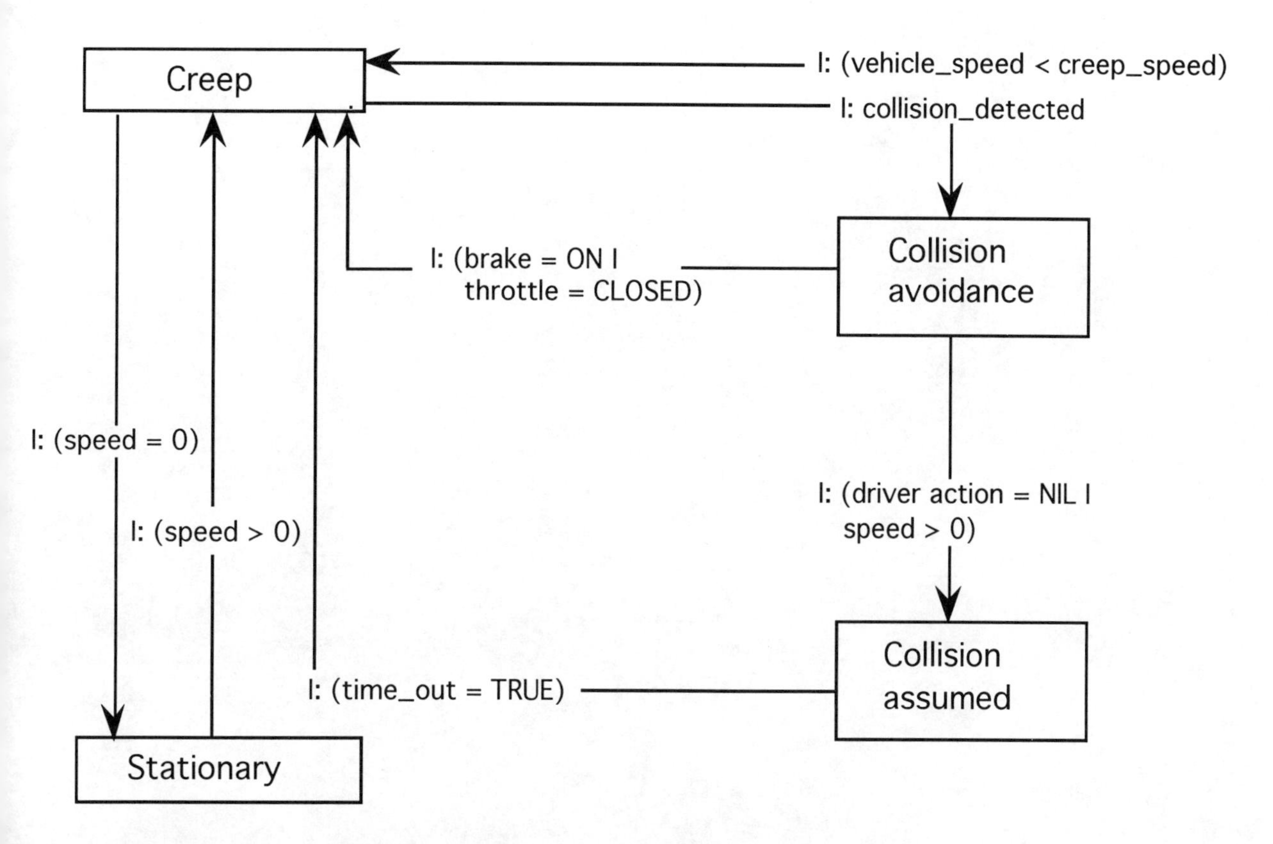

Figure 3 : State Transition For Collision Warning

CONCLUSIONS

We have found that the SUSI™ methodology provides a very good basis for reviewing human centred systems. It provides a notation which is easily understood by system designers, human factors specialists and operators. It provides an important basis for a common view of system functionality. This in itself leads to easier to understand designs and hence improved confidence in safe system functionality.

By using standard computer based tool sets for developing the representation we gain the benefit of checking for internal consistency of the representation. Using the HAZOP procedure on this type of representation has proved a simple adaptation. This provides confidence that the overall process gives a good basis for critically reviewing the safety of the system.

REFERENCES

CISHEC Safety Committee, "A Guide to Hazard and Operability Studies", Chemical Industries Association Limited, London, 1990.

Chudleigh M, "Hazard Analysis Using HAZOP: A Case Study"

Chudleigh M, Catmur J, "Safety Assessment of Computer Systems using HAZOP and Audit Techniques", *Safety of Computer Control Systems*, (Safecomp '92) pp 285-292, Frey (ed) *S*

Hatley D, Pirbhai I, *Strategies for real-time system specification.* Dorset House, 1988.

Stammers R, Carey M, Astley J, "Task Analysis" *Evaluation of Human Work*, , pp 134-160, Wilson J, Corlett E (eds), Taylor & Francis, London, 1990.

Westrum R, *Technologies and Society*, Wadsworth Inc., California, 1991.

Functional Safety of Electrical/Electronic/Programmable Systen Generic Aspects, IEC 65A (Secretariat) 123, 1991.

Interim Def Stan 00-56, *Hazard Analysis and Safety Classification of the Computer and Programmable Electronic System Elements of Defence*, Directorate of Standardisation, Ministry of Defence.

950972

Correlation of Driver Confidence and Dynamic Measurements and the Effect of 4WD

Midori Kubota and Takayuki Ushijima
Fuji Heavy Industries

Jac Brown
Subaru Research & Design

ABSTRACT

Engineers understand the advantages of four wheel drive on low coefficient of friction surfaces. The advantages of four wheel drive for other surfaces and for handling and stability are not well documented. When surveyed, customers tell us that four wheel drive feels safer for every road condition including dry road. The object of this paper is to determine what factors contribute to the driver's feeling of safety and how different drive systems affect driver confidence.

We tested three different drive systems; front wheel drive, front wheel drive with traction control, and four wheel drive. Acceleration, steering angle, and yaw velocity were measured for these vehicles on a wide variety of road surfaces. Subjective ratings of stability were made by professional test drivers and average drivers. Acceleration and handling tests on various surfaces were also measured.

A correlation was found between driver ratings and measured parameters. Improved damping in yaw velocity had the strongest effect in improving driver rating in handling and stability tests.

We also found differences in drive systems. Four wheel drive was found to improve performance in every road condition. Front wheel drive with traction control was found to improve performance in some conditions but has a small effect on dry roads. We found that traction control improves the performance of average drivers but not trained test drivers. Four wheel drive improved performance of all drivers.

INTRODUCTION

Today's market demands safe automobiles. This includes both passive safety and active safety. Passive safety involves systems to protect the occupants in a crash including air bags, seat belts, and side impact beams. Active safety involves systems for avoiding a crash including Anti-Lock Brakes (ABS), Four Wheel Steering, Four Wheel Drive (4WD), and Traction Control (TCS). Active safety systems often provide another advantage by increasing driver confidence and reducing driver stress.

In order to understand the advantages of various active safety systems, we have developed tests to evaluate the performance of these systems. In particular, we are interested in understanding the market value of different traction systems. Customers tell us that they like 4WD and use comments like; "feels safe", "easy to drive", and "comfortable in bad road conditions." These comments indicate a sense of Driver Confidence that is an important advantage of these traction systems.

This paper attempts to correlate driver confidence with measured dynamic parameters. The intent is to develop a repeatable measurement method which will allow the evaluation of the customer advantages of a given traction system. A secondary objective is to define the relative performance of FWD,

FWD+TCS, and 4WD. This is especially important in the current U.S. market where some auto makers claim that Traction Control (TCS) is almost as good as 4WD.

TEST METHOD

Vehicles

FWD was used as a baseline and the performance of FWD+TCS and 4WD were compared to FWD. Two vehicles were used, a FWD+TCS and a 4WD vehicle.

The FWD vehicle uses a full range TCS which cuts individual cylinder fuel injectors to modulate engine torque and uses brake control independently on each driven wheel to control wheel speed (Figure 2). A TCS ON/OFF switch was provided for this car to enable direct comparison of FWD and FWD+TCS.

Torque transfer to the rear axle of the 4WD vehicle is controlled by a computer modulated, multi-plate clutch (Figure 1). A 4WD ON/OFF switch was provided for this vehicle which cuts hydraulic pressure to the transfer clutch making the vehicle FWD. The result is a test vehicle where the only difference is the traction mode, FWD or 4WD.

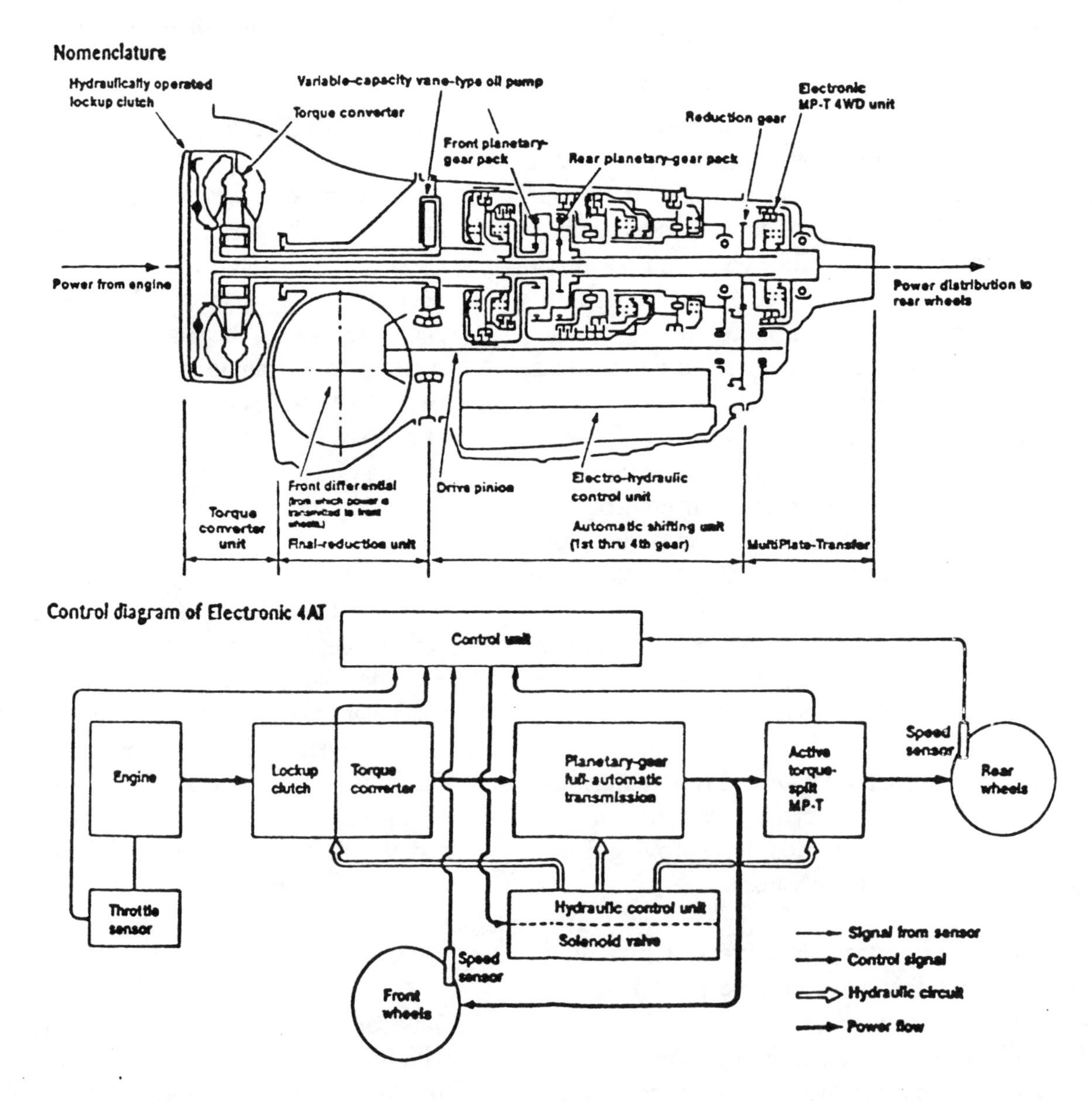

Fig. 1 4WD System

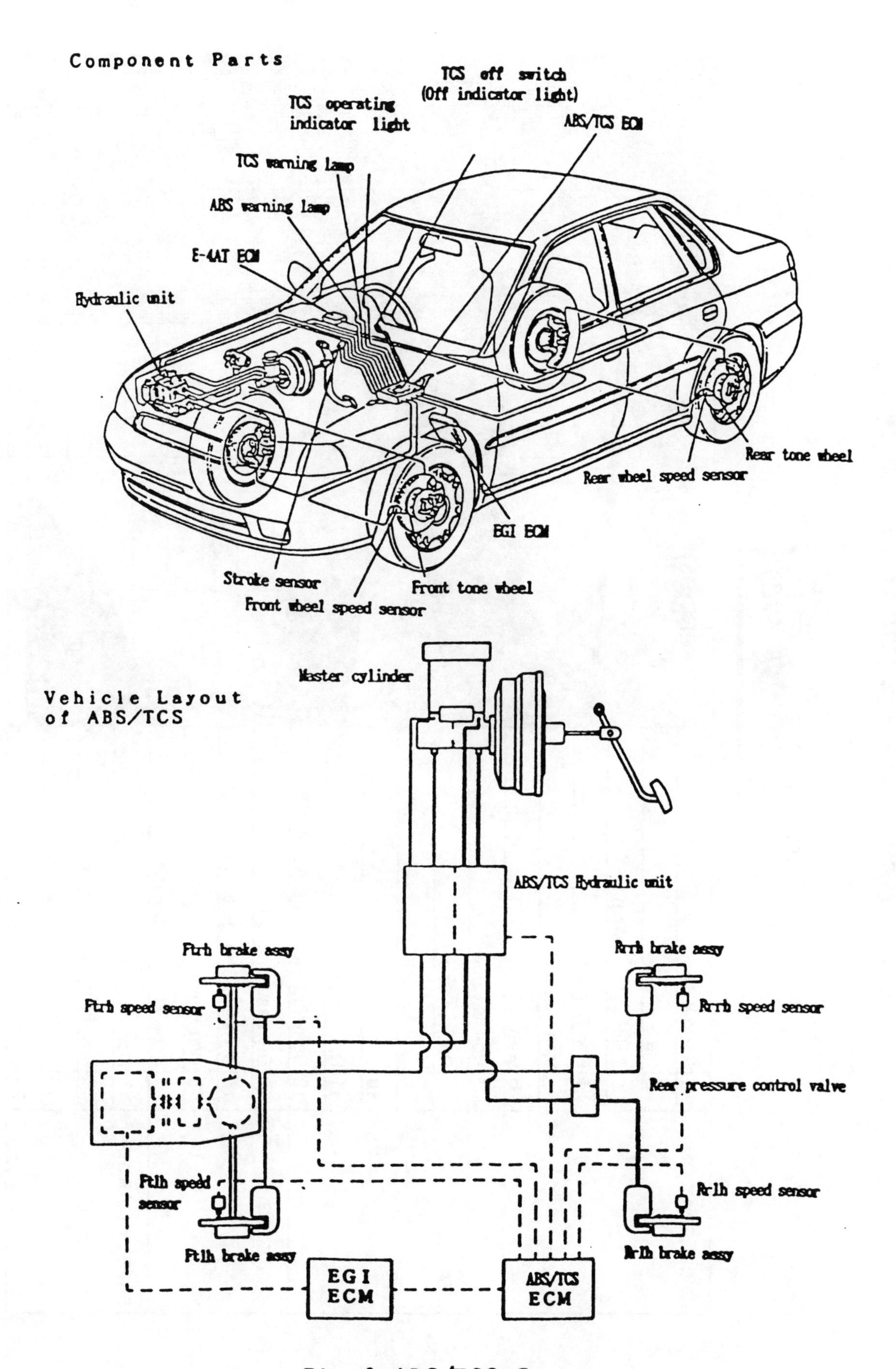

Fig. 2 ABS/TCS System

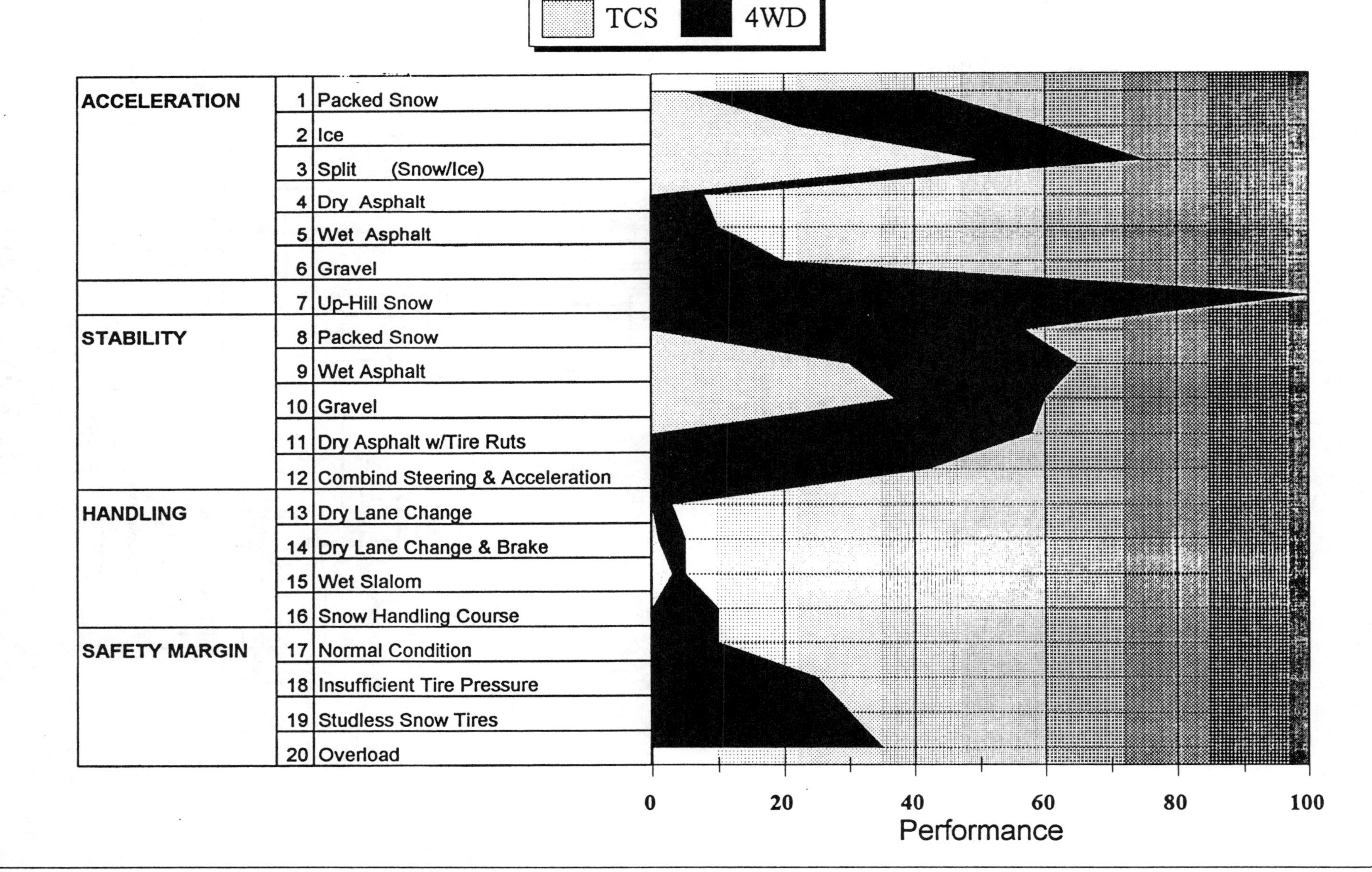

COMPARISON of FWD with TCS and 4WD
Figure 14
TCS
4WD
ACCELERATION
1 Packed Snow
2 Ice
3 Split (Snow/Ice)
4 Dry Asphalt
5 Wet Asphalt
6 Gravel
7 Up-Hill Snow
STABILITY
8 Packed Snow
9 Wet Asphalt
10 Gravel
11 Dry Asphalt w/Tire Ruts
12 Combind Steering & Acceleration
HANDLING
13 Dry Lane Change
14 Dry Lane Change & Brake
15 Wet Slalom
16 Snow Handling Course
SAFETY MARGIN
17 Normal Condition
18 Insufficient Tire Pressure
19 Studless Snow Tires
20 Overload
0
20
40
60
80
100
Performance

Drivers

Six drivers were used for subjective evaluations, three average drivers and three professional test drivers. The three average drivers were women with 3 to 6 years driving experience and no additional driver training. The professional test drivers were Subaru development engineers with 10 to 26 years experience.

TEST RESULT

20 test conditions were established which would show differences between traction systems. These tests have been divided into four categories; Acceleration, Stability, Handling, and Safety Margin. Figure 14 shows a summary of the test results comparing FWD, FWD+TCS, and 4WD. Several of these tests show significant differences in measured performance, however, we were especially interested in the tests that correlated to a subjective feeling of driver confidence. Of these tests, we selected three tests for subjective evaluation by both average and professional test drivers.

Test 1 - Stability on a Wet Asphalt Highway

The test vehicle is evaluated on a wet, straight asphalt road with puddles at a constant speed of 100 km/h. Water depth on the road varies randomly from 3 to 10 mm depth. Figure 3 shows a photo of the test road.

Test 2 - Combined Steering and Acceleration

This test simulates entering a busy highway from a side street or T intersection. The test vehicle makes a standing start at full acceleration while turning 90 degrees onto the highway (Figure 4). Road condition is dry asphalt .

Test 3 - Combined Lane Change and Brake

This test is an accident avoidance maneuver on dry asphalt. It uses a single lane change as shown in Figure 5 with an entry speed of 100 km/h. The object of the test is to brake to a stop as quickly as possible while completing the lane change.

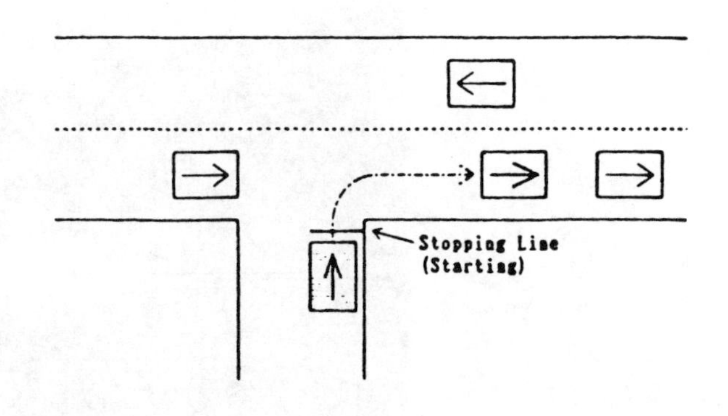

Figure 4
Combined Steering and Acceleration

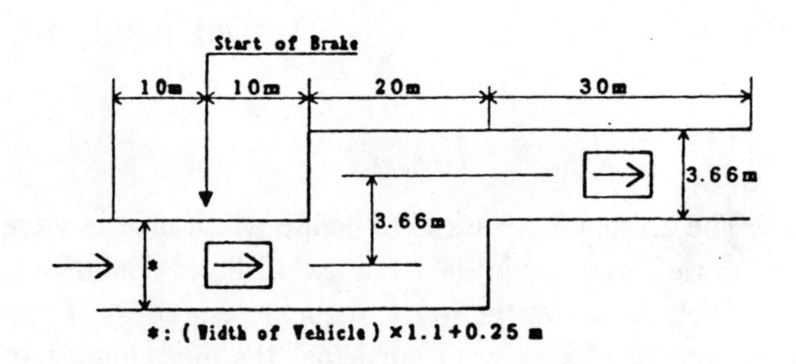

Figure 5
Combined Lane Change and Brake

Table 1 shows the subjective ratings for these three tests for our average and professional drivers. It also compares three different traction systems; FWD, FWD+TCS, and 4WD.

- The tests shows clear differences in subjective rating for each traction system.
- FWD+TCS shows improvement over FWD in the first two tests.
- 4WD shows a clear advantage in all three tests.
- The differences in traction system ratings are consistent between average and professional test drivers.

Figure 3

Straight Road, Wet Asphalt, Random Depth

The drivers were asked to define which factors were most important in their ratings of driver confidence. Table 2 shows the most frequent responses. From these results, it was found that the most important factors in subjective driver confidence are:

- Steering Pull or Kickback
- Yaw Overshoot or Reduced Yaw Damping
- Perceived Reduction in Directional Control of the Vehicle
- Longitudinal Shock Caused by Repeated Slip and Grip of the Tires

Average drivers were found to be most sensitive to ocillations in vehicle behavior or vehicle damping. Professional drivers were found to be sensitive to both vehicle damping and vehicle control gain.

VEHICLE MEASUREMENTS

During these tests, each vehicle was instrumented for vehicle dynamic measurement. These measurements show clear differences in steering angle, steering force, yaw velocity, lateral acceleration, and longitudinal acceleration between different traction systems. These measurements show good correlation to the subjective ratings of driver confidence.

Figures 6 and 7 show sample data comparing of FWD and 4WD on Test 1. It shows a significant reduction in all measured data for 4WD for both average and professional test drivers. Figure 8 shows the effect of TCS on Test 1 for a professional test driver. Figures 9 and 10 show the effect of 4WD on Tests 2 and 3. In all cases a good correlation is shown to subjective ratings and the factors mentioned in subjective comments.

Having established a good correlation between subjective ratings of driver confidence and dynamic measurements, we used the measurement method to evaluate the differences in traction systems for additional test conditions. Figures 11 to 13 show measured data for Packed Snow, Loose Gravel, and Dry Asphalt with Tire Ruts. All of these show significant advantages for 4WD and clear differences between traction systems.

Table.1 Subjective Ratings of Driver Confidence

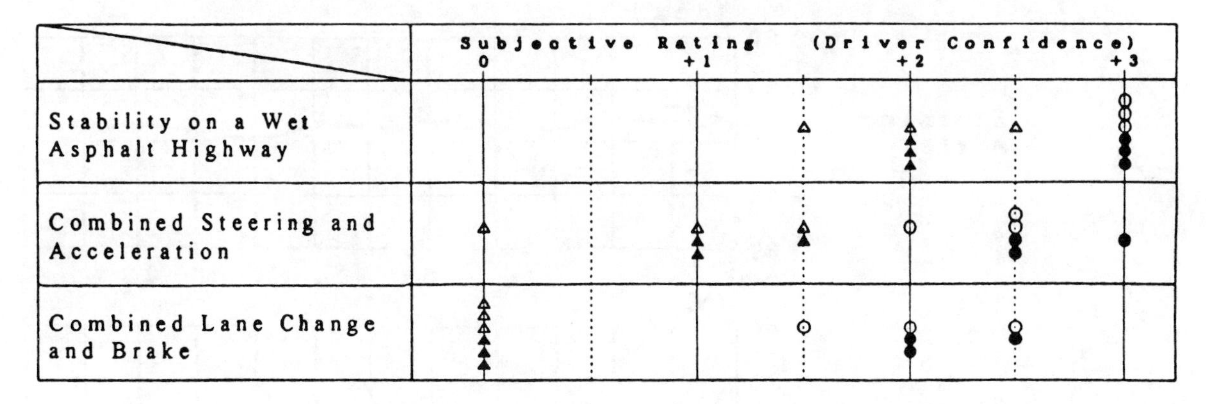

Table. 2
Factor Analysis of Negative Comments

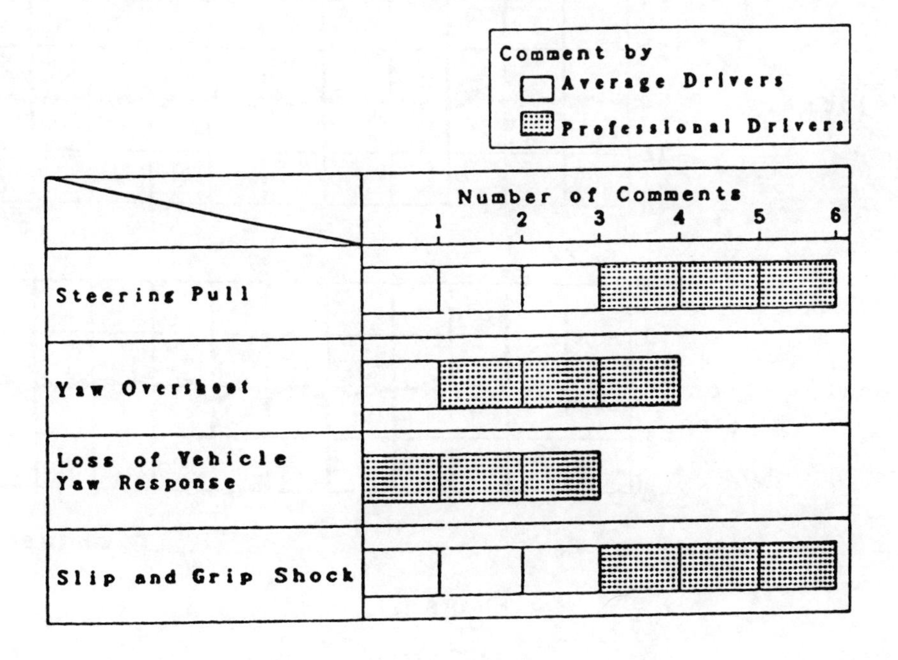

'93 Legacy Wagon
100 Km/h
Same Car Tested

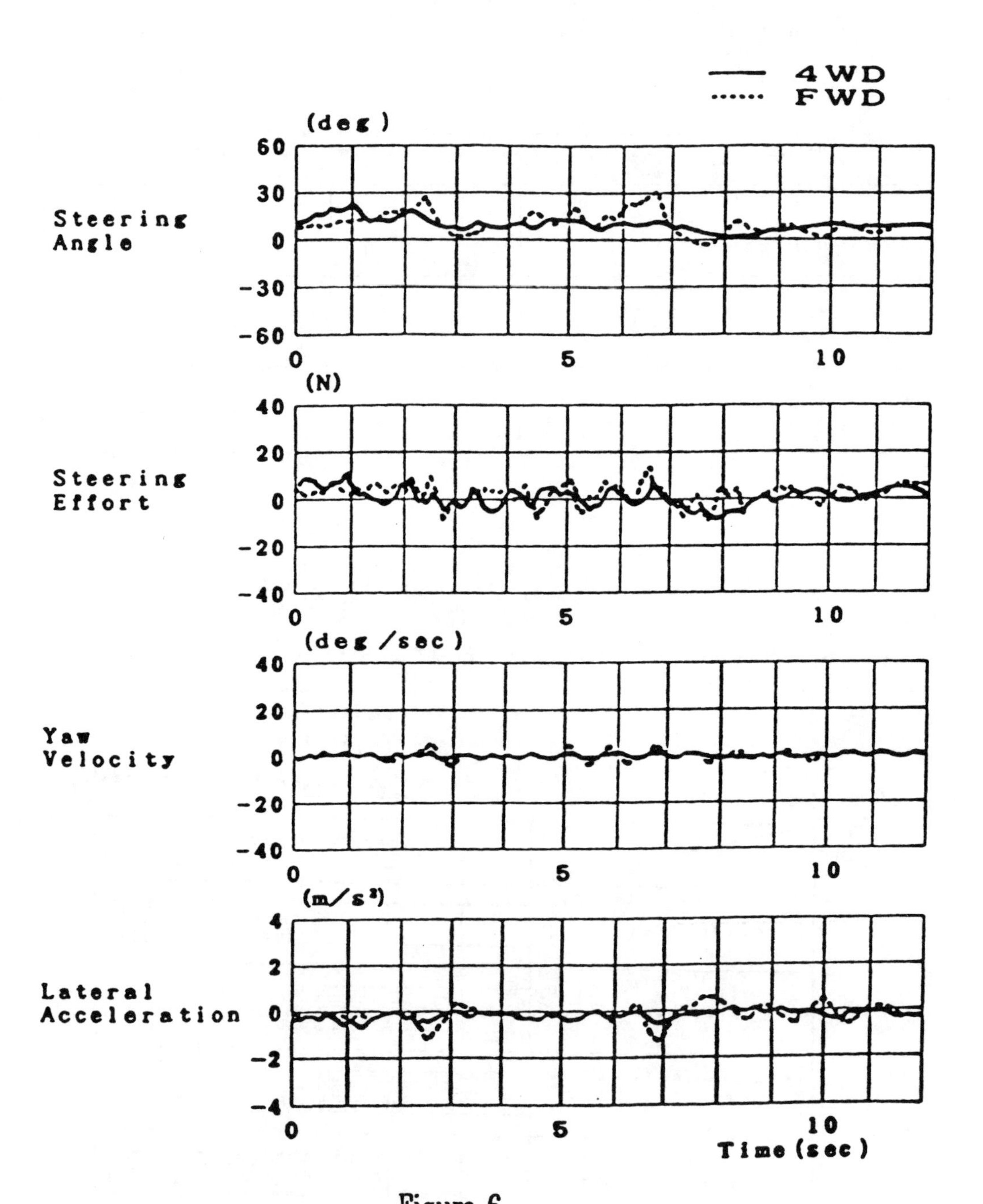

Figure 6

Comparison of FWD vs. 4WD

Straight Road Wet Asphalt (random depth, 3-10 mm)

Average Driver

'93 Legacy Wagon
100 Km/h
Same Car Tested

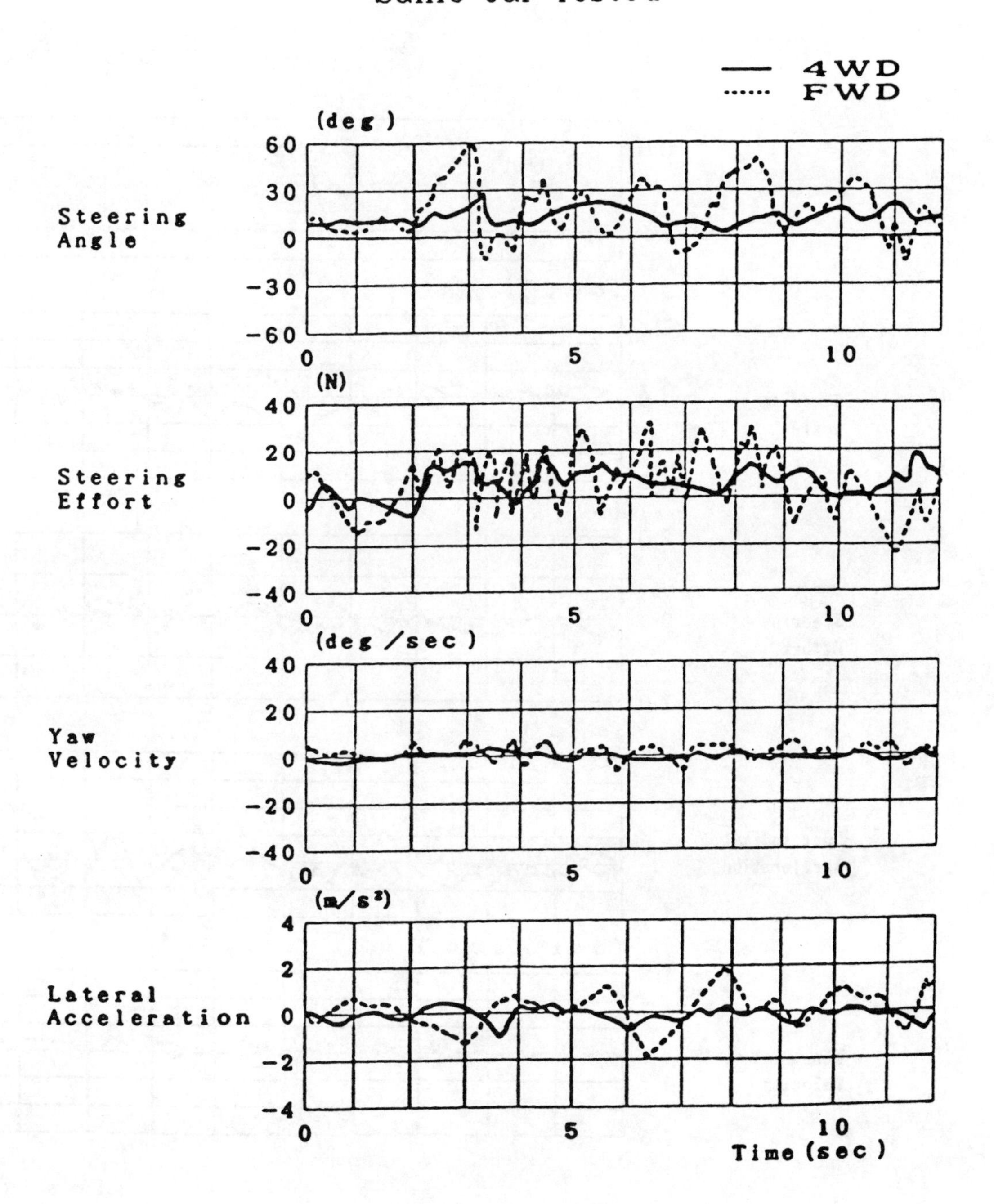

Figure 7

Comparison of FWD vs. 4WD

Straight Road Wet Asphalt

Professional Test Driver

'95 Legacy Wagon
100 Km/h
Same Car Tested

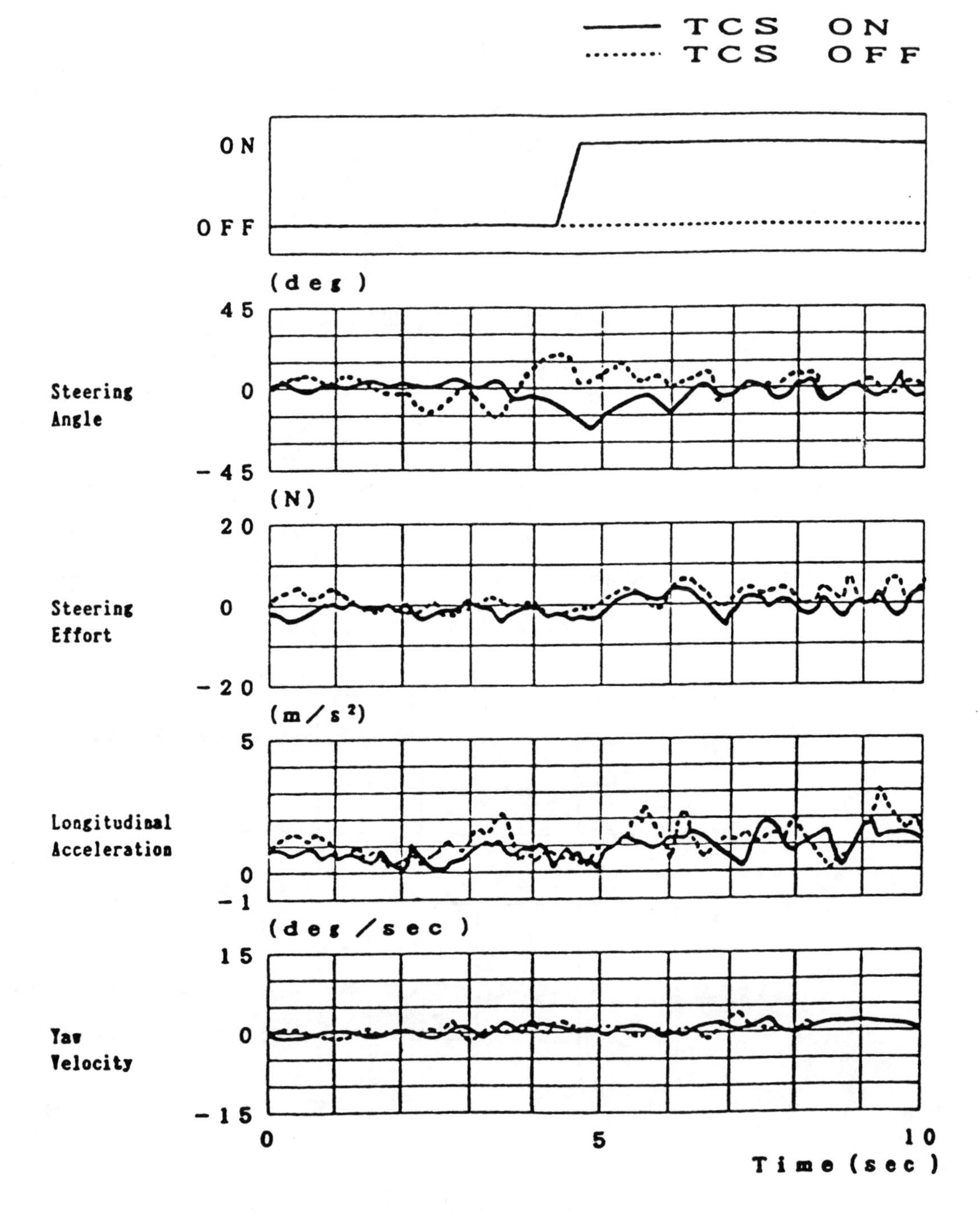

Figure 8
Comparison of TCS ON-OFF
Straight Road Wet Asphalt
Professional Test Driver

'93 Legacy Wagon
Maximum Acceleration
Same Car Tested

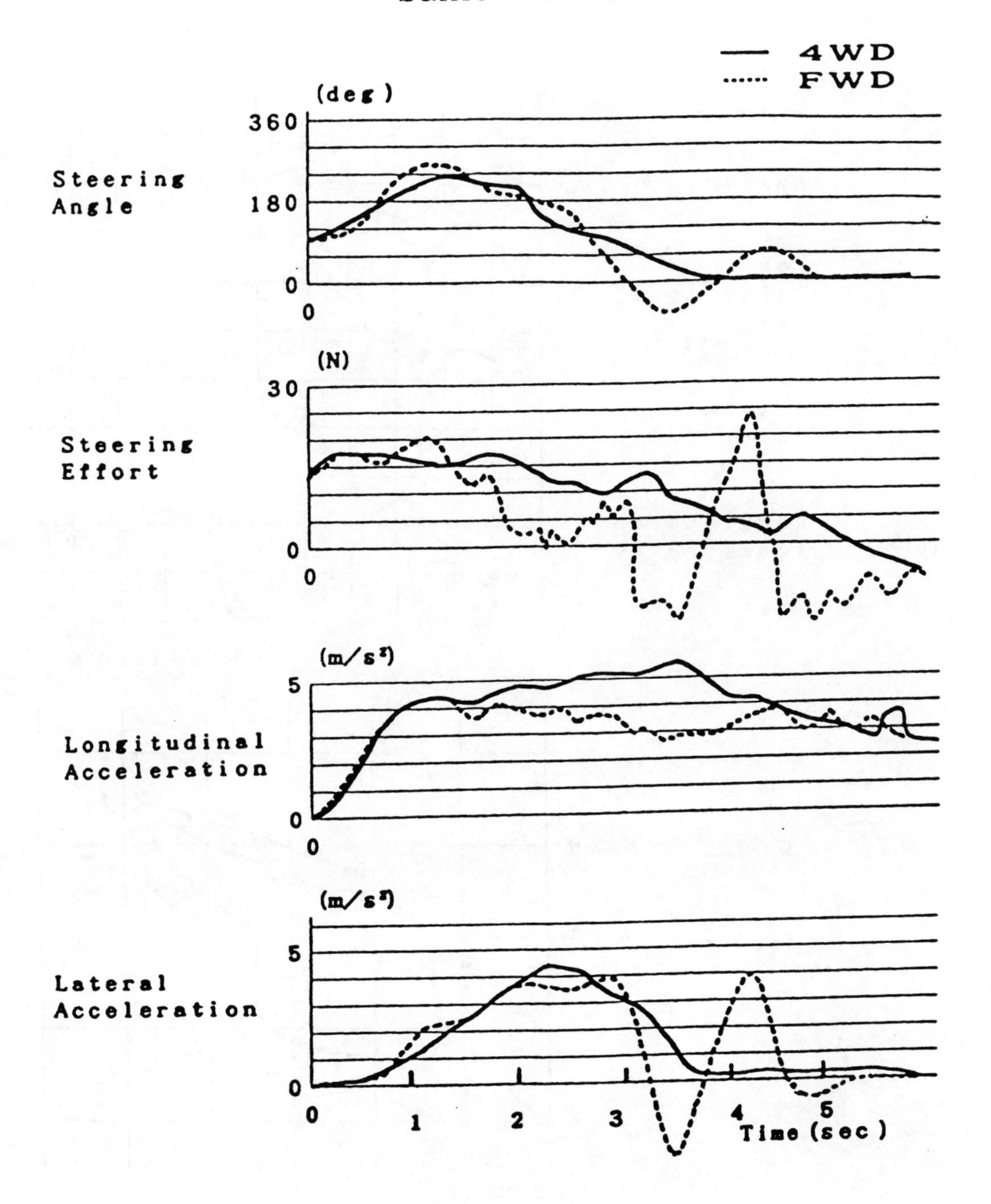

Figure 9

Comparison of FWD vs. 4WD

Combined Steering and Acceleration

Professional Test Driver

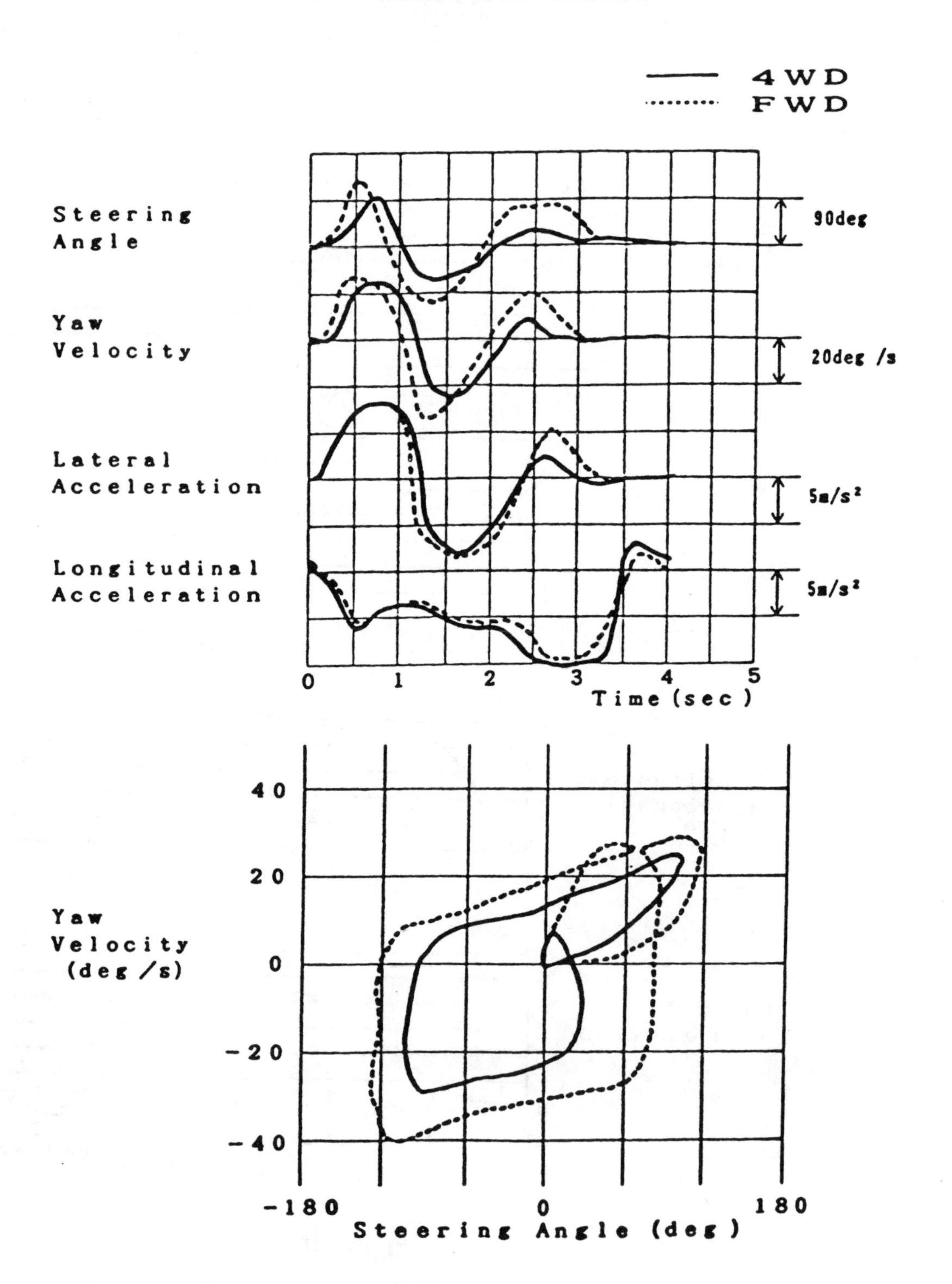

Figure 10

Comparison of FWD vs. 4WD

Combined Lane Change and Brake

Professional Test Driver

'93 Legacy Wagon
100 Km/h
Same Car Tested

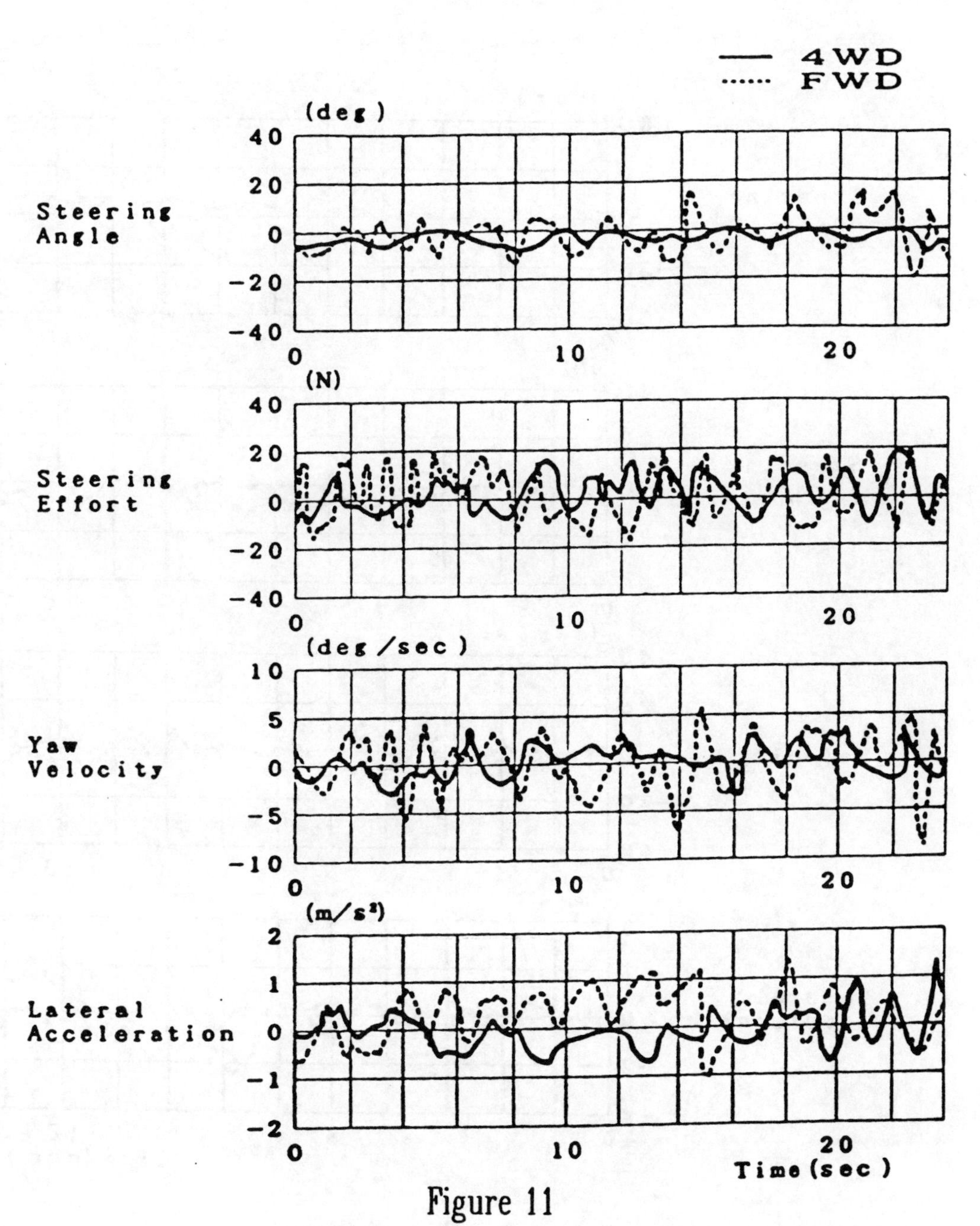

Figure 11

Comparison of FWD vs. 4WD

Straight Road on Packed Snow

Professional Test Driver

'93 Legacy Wagon
100 Km/h
Same Car Tested

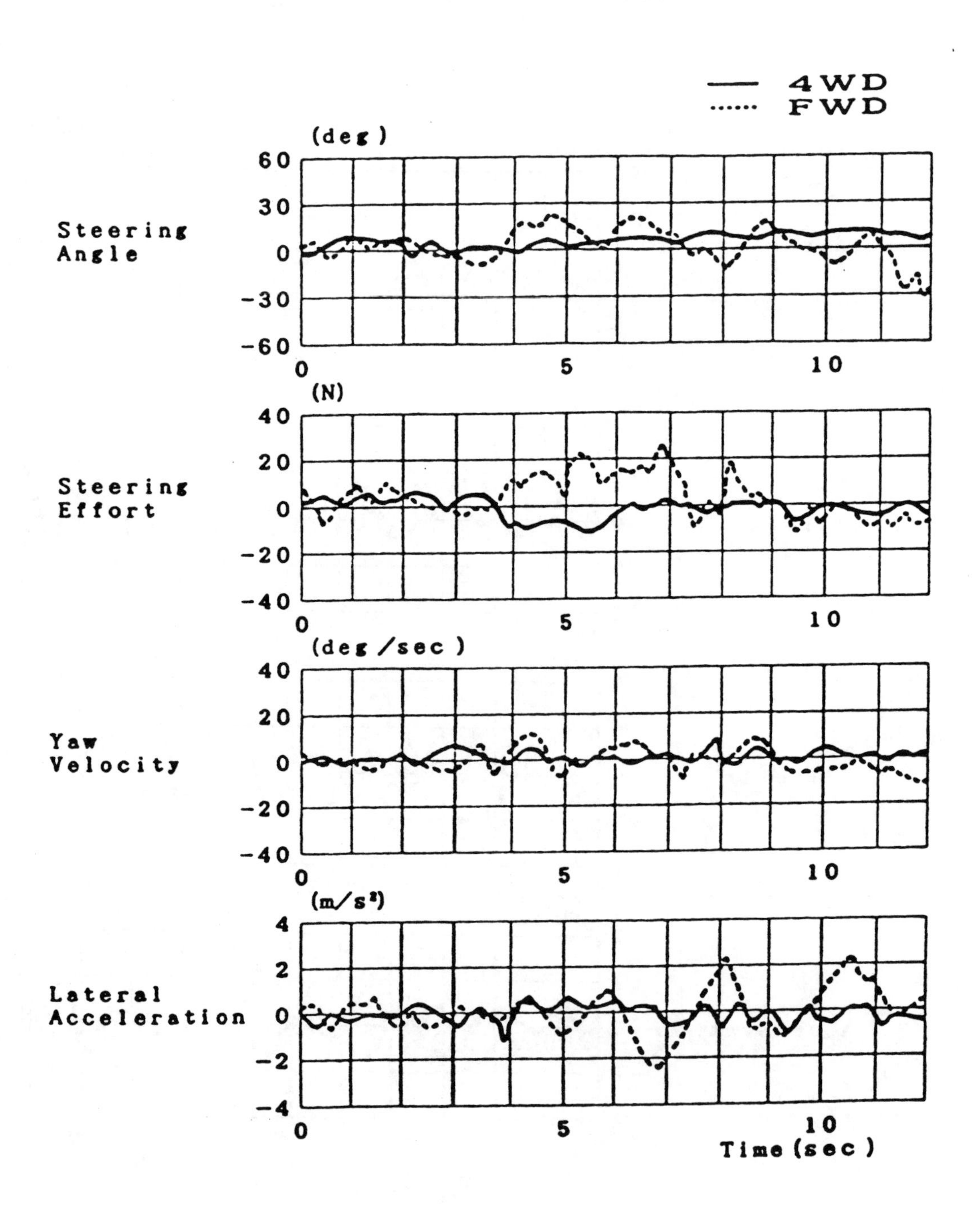

Figure 12

Comparison of FWD vs. 4WD

Straight Road on Loose Gravel

Professional Test Driver

'93 Legacy Wagon
50 Km/h
Same Car Tested

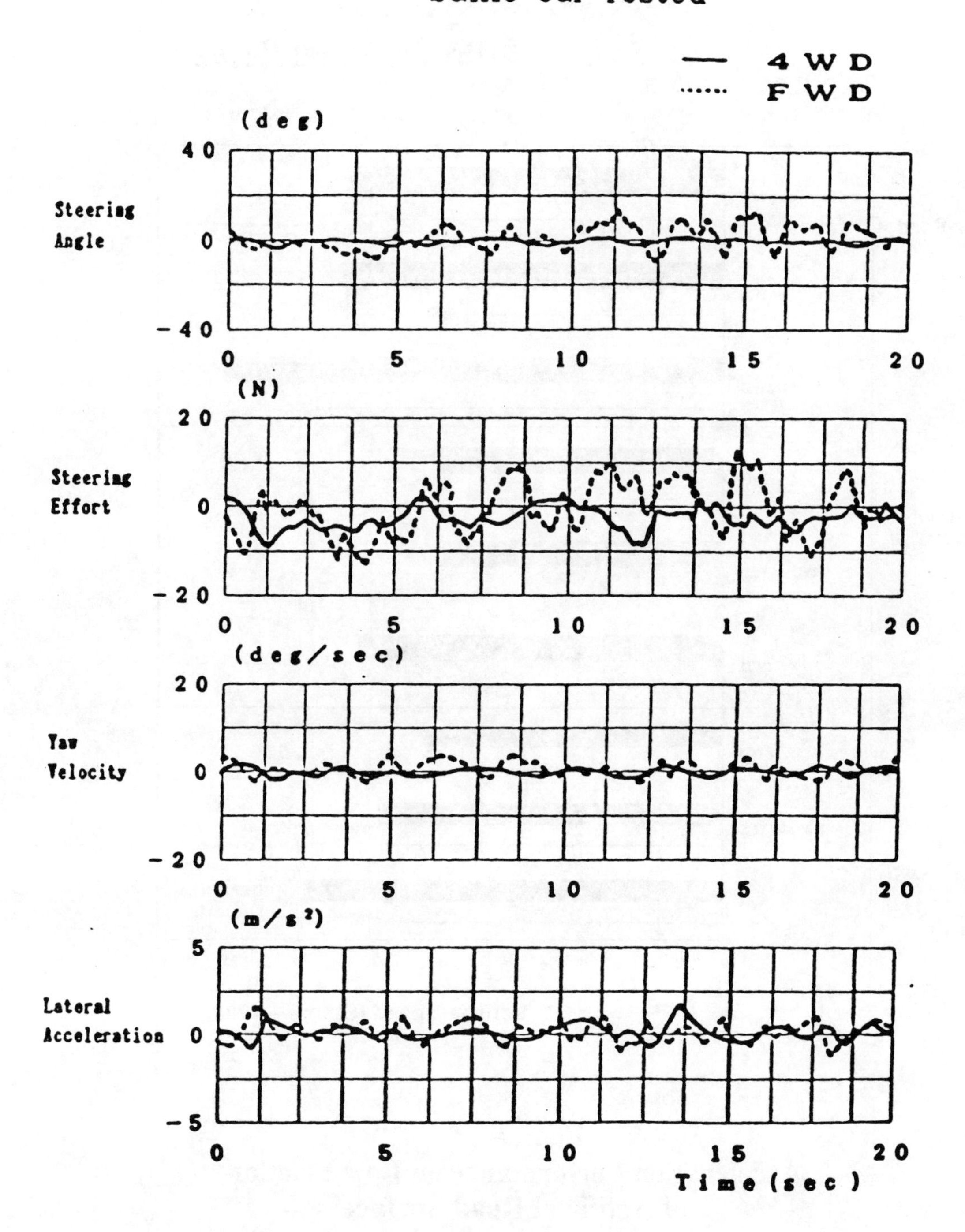

Figure 13

Comparison of FWD vs. 4WD

Sraight Road on Dry Asphalt with Tire Ruts

Professional Test Driver

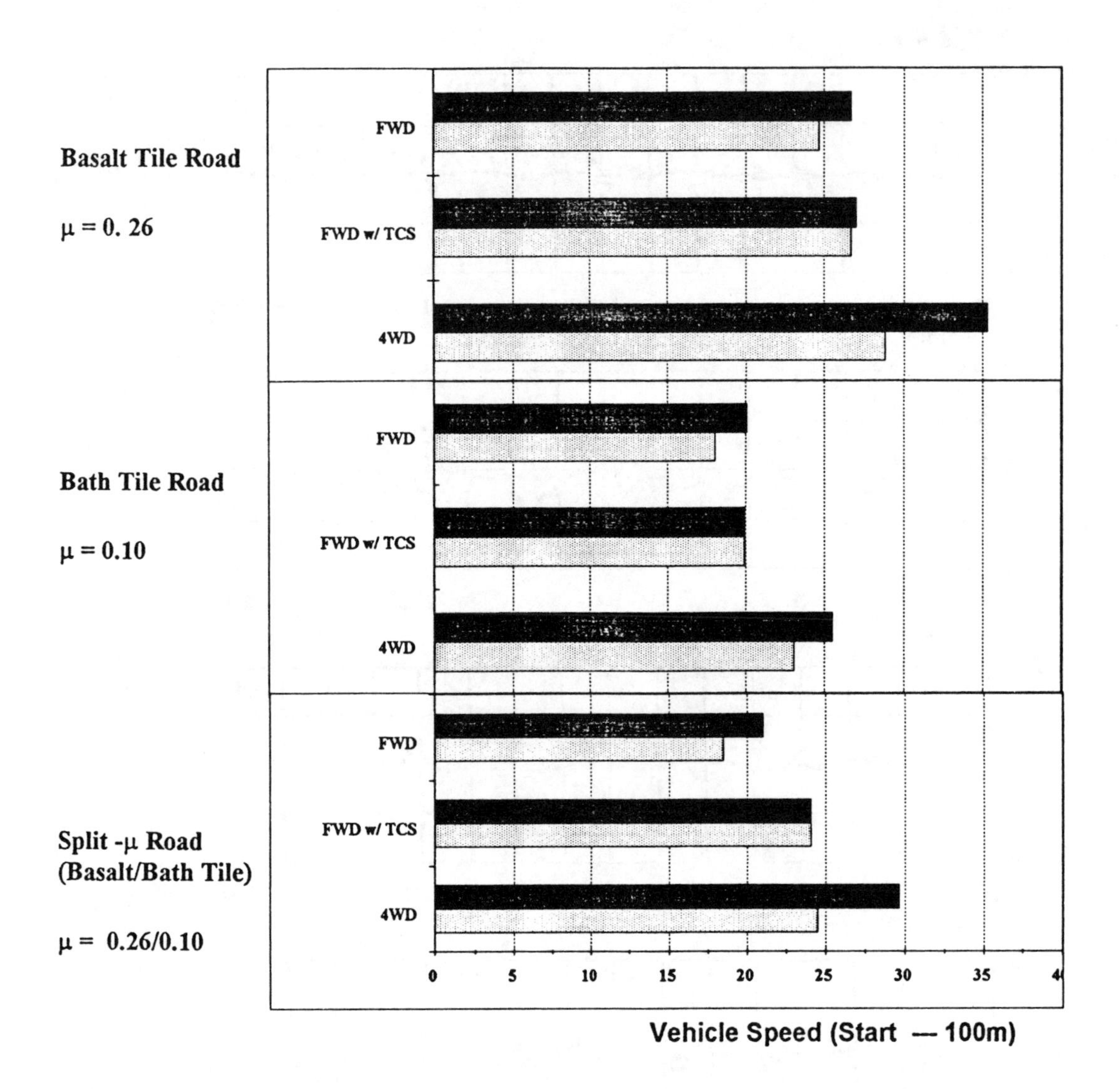

Figure .15
Acceleration Performance on Low Friction of Artificial Road Surface

Effect of Traction System and Driver Skill

It has been shown that 4WD offers significant advantages in driver confidence and feeling of safety under a wide variety of driving conditions. It has also been shown that TCS offers smaller but measurable advantages. Clearly these traction systems also affect the available vehicle performance. Do these systems offer the same advantage to average drivers as to professional test drivers? To answer this question, the acceleration performance of both driver groups was measured on a 100 meter acceleration test on three low friction surfaces. Figure 15 shows the speed at 100 meters for both driver groups and three traction systems.

It was observed that TCS is effective in improving the performance of average drivers but does not improve the performance of the professional test drivers except on split - μ conditions. 4WD shows dramatic improvements for both driver groups under all conditions.

SUMMARY

Returning to Figure 14, we can see a summary of the data presented in this paper. It measures percent improvement over FWD for TCS and 4WD for each test case. Tests 1 to 6 measure time for a 100 meter acceleration. Test 7 is a measure of the maximum grade that can be climbed on packed snow. Tests 8 to 12 measure the percent reduction in vehicle dynamic measurements from FWD as is shown in Figures 6 to 13. Tests 13, 15, and 16 measure the maximum speed for the test course and Test 14 measures stopping distance. Tests 17 to 20 measure the maximum speed at which the driver is subjectively comfortable driving on the highway under the given vehicle conditions. Normal condition indicates design conditions; Insufficient tire pressure indicates a low rear tire pressure condition; and Overload simulates excess cargo loading.

CONCLUSIONS

1. There is a strong correlation between subjective ratings of driver confidence and the parameters of steering angle, steering force, yaw velocity, and lateral and longitudinal acceleration.

2. Four Wheel Drive (4WD) and Traction Control (TCS) are effective in improving driver confidence and a feeling of driving safety.

3. Four Wheel Drive (4WD) provides significant advantages in both driver confidence and vehicle performance on a wide range of driving conditions.

4. Four Wheel Drive (4WD) offers significant improvement in both driver confidence and vehicle performance for both average and skilled drivers.

5. Traction Control (TCS) offers improvements primarily on slippery surfaces.

6 Traction Control (TCS) is reasonably effective for average drivers but offers little improvement for skilled drivers.

REFERENCES

1. S. Oobayashi, et al., " Characteristics of Maneuverability of 4WD Vehicle", Journal of JSAE, Vol. 39, 1985 260-265 (in Japanese)

2. M. Kubota, et al., "Maneuverability and New Suspension for Four-Wheel Drive Vehicles", 10th ESV Conference, England, July 1985

3. S. Ikeda, et al., " Trend of Traction Control Systems", Journal of JSAE, Vol. 42, No. 3, 1988 336-341 (in Japanese)

4. K. Ise, et al., " the 'Lexus' Traction Control (TRAC) System", SAE Paper 900212

5. M. Ito, et al., "The Present and Future Trends of Traction Control System", Journal of JSAE, Vol. 46, No.2 1992 32 - 37 (in Japanese)

VERMONT STATE COLLEGES
0 0003 0611275 6